Molecular Processes in Space

PHYSICS OF ATOMS AND MOLECULES

Recent volumes in the series:

ATOMIC INNER-SHELL PHYSICS
Edited by Bernd Crasemann

ATOMIC PHOTOEFFECT
M. Ya. Amusia

ATOMIC SPECTRA AND COLLISIONS IN EXTERNAL FIELDS
Edited by K. T. Taylor, M. H. Nayfeh, and C. W. Clark

AUTOIONIZATION: Recent Developments and Applications
Edited by Aaron Temkin

COHERENCE IN ATOMIC COLLISION PHYSICS
Edited by H. J. Beyer, K. Blum, and R. Hippler

COLLISIONS OF ELECTRONS WITH ATOMS AND MOLECULES
G. F. Drukarev

ELECTRON-ATOM AND ELECTRON-MOLECULE COLLISIONS
Edited by Juergen Hinze

ELECTRON-MOLECULE COLLISIONS
Edited by Isao Shimamura and Kazuo Takayanagi

ELECTRON-MOLECULE SCATTERING AND PHOTOIONIZATION
Edited by P. G. Burke and J. B. West

ISOTOPE SHIFTS IN ATOMIC SPECTRA
W. H. King

MOLECULAR PROCESSES IN SPACE
Edited by Tsutomu Watanabe, Isao Shimamura, Mikio Shimizu, and Yukikazu Itikawa

PROGRESS IN ATOMIC SPECTROSCOPY, Parts A, B, C, and D
Edited by W. Hanle, H. Kleinpoppen, and H. J. Beyer

QUANTUM MECHANICS VERSUS LOCAL REALISM: The Einstein–Podolsky–Rosen Paradox
Edited by Franco Selleri

RECENT STUDIES IN ATOMIC AND MOLECULAR PROCESSES
Edited by Arthur E. Kingston

THEORY OF MULTIPHOTON PROCESSES
Farhad H. M. Faisal

ZERO-RANGE POTENTIALS AND THEIR APPLICATIONS IN ATOMIC PHYSICS
Yu. N. Demkov and V. N. Ostrovskii

Molecular Processes in Space

Edited by

Tsutomu Watanabe and
Isao Shimamura

RIKEN, The Institute of Physical and Chemical Research
Wako, Japan

Mikio Shimizu and
Yukikazu Itikawa

Institute of Space and Astronautical Science
Sagamihara, Japan

Springer Science+Business Media, LLC

Library of Congress Cataloging-in-Publication Data

Molecular processes in space / edited by Tsutomu Watanabe ... [et al.].
p. cm.
Includes bibliographical references.
ISBN 978-1-4612-7881-8 ISBN 978-1-4613-0591-0 (eBook)
DOI 10.1007/978-1-4613-0591-0
1. Molecular astrophysics. 2. Takayanagi, Kazuo. I. Watanabe, T. (Tsutomu)
QB462.6.M66 1990
523.01'96--dc20 89-77539
CIP

Originally published by Plenum Press, New York in 1990
Softcover reprint of the hardcover 1st edition 1990

PREFACE

Like a river, the progress of science has a tendency to run fast or slow. Once the water meets a dam, it may stop for a while, but eventually it will flow over the top and run fast again. In scientific research, a breakthrough to overcome a similar barrier is often made by a small number of scientists, or perhaps by a single person of special creativity, extraordinary talent and unusual perseverance. Through such individuals science can proceed in great strides. No one can deny that Professor Kazuo Takayanagi is one of these special individuals who have played a leading role in the field of atomic and molecular physics, as well as space physics. This book is dedicated to Professor Takayanagi on the occasion of his retirement from the Institute of Space and Astronautical Science.

Professor Takayanagi was born in 1926 and grew up in Tomakomai in Hokkaido, the northern island of Japan. In his boyhood, he was interested in natural sciences, particularly astronomy. On 5th February, 1943, when he was attending secondary school, a solar eclipse was seen in his town. He organized a group of students from his school to observe the eclipse. He still remembers the scene: it grew so dark during the eclipse that two stars, Vega and Arcturus, could be seen. After graduation from the University of Tokyo in 1948, he entered the graduate school there. First he studied both the theory of elementary particles and nuclei and the theory of molecular structure. His first paper was on the evaporation process in nuclei. His interest then shifted gradually to atomic and molecular processes, and he wrote a thesis under the guidance of Professor Masao Kotani, who at that time was one of the pioneers of atomic physics in Japan. The title of the thesis was "The theory of collisions between two diatomic molecules".

In 1950 he was appointed Research Associate at the Department of Physics, Saitama University. There he was promoted to Associate Professor and finally to Professor. In 1955-1957, he was awarded a scholarship as an exchange scientist from the British Council and studied with Professor Harry S.W. Massey at University College, London. He also spent one year at JILA (Joint Institute for Laboratory Astrophysics) in Boulder, Colorado, in 1963-1964. In 1966, he moved to the Institute of Space and Aeronautical Science, University of Tokyo. Later the Institute was detached from the University and renamed the Institute of Space and Astronautical Science (ISAS). He retires from ISAS in March, 1990. In 1988, he was elected a member of the Science Council of Japan as one of the five physicists assigned to the Council.

In the early stage of his work (1955), Professor Takayanagi had already pointed out the problem of non-orthogonality between the wave functions of the initial and final states in the charge transfer process. To overcome this problem, he proposed a method of expanding the electronic wave function in terms of an orthogonalized set of basis functions using the Schmidt orthogonalization. His contributions to atomic and space physics, however, have been concerned mainly with low-energy collisions including electron-molecule, ion-molecule and molecule-molecule collisions. Only a few examples are mentioned here.

For the study of molecule-molecule collisions, he introduced the concept of "modified wave number". This is based on the characteristics of intermolecular potential at a large distance, particularly its slight dependence on collisional angular momentum. This method has facilitated a quantitative treatment of the complicated process of rotational and vibrational relaxation of molecules in a gas. He performed a series of applications of this approach, starting from a study of H_2 - H_2 collisions. With regard to ion-molecule collisions, he paid special attention to the nature of the long-range, orientation-dependent interaction between an ion and a polar molecule. For such collisions he devised the "perturbed rotational state (PRS)" approach, which is based on the fact that the angular component of the relative motion is rather rapid in comparison with its radial component. The PRS function is, therefore, determined by solving an eigenvalue problem with the intermolecular distance fixed. He and his colleagues applied this method to problems of rotational excitation, chemical reaction, transport phenomena, etc. and extended it to more general cases (e.g., molecules with a quadrupole moment). In the field of astrophysics, he suggested for the first time that the H_2 - H_2 collision could be a powerful cooling process in interstellar clouds when heated by cloud-cloud collisions. This idea was very fresh at that time and promoted leading Japanese astrophysicists to begin a study of star evolution, resulting in the well known classification of stars into types I and II.

After returning from the U.K., Professor Takayanagi personally organized a research group on atomic collision physics. Since then he has been working hard to promote atomic collision research in Japan. In October, 1976, the Society for Atomic Collision Research (Japan) was founded, and he was elected the first chairman of the Society. In 1979 the XIth International Conference on the Physics of Electronic and Atomic Collisions (XI ICPEAC) was held in Kyoto under the joint sponsorship of the Society and the Science Council of Japan. As Local Chairman of the Conference, Professor Takayanagi made a great contribution to the eventual success of the Conference. Professor R.F. Stebbings, the Chairman of the next ICPEAC, said in his closing remarks at the Conference, "The XI ICPEAC was so well organized that it can be a model of conferences in the future". This was largely a reflection of the way in which Professor Takayanagi ran things.

This book contains ten chapters describing photon- and electron-molecule collisions, collisions between heavy particles (ions, atoms and molecules) and molecules, from both theoretical and experimental aspects, and details of these processes as they occur in the upper atmosphere of the Earth and other planets, interstellar space, and other situations.

These are subjects on which Professor Takayanagi has been working for almost 40 years. In these fields remarkable progress has been made over the last two decades. We anticipate that this book will prove a useful guide to these recent developments, as well as serving as an introduction to these fields.

We express our sincere appreciation to the authors, who all willingly accepted our invitation to write a chapter and completed their manuscripts in a rather short time. In the course of editing and preparing this book, we received valuable assistance from our colleagues, in particular Drs. A. Ichimura, K. Onda and K. Sakimoto, who helped with proofreading and other tasks, and to whom we are indebted.

September 1st, 1989

Tsutomu Watanabe

Isao Shimamura

Mikio Shimizu

Yukikazu Itikawa

CONTENTS

MOLECULAR PROCESSES IN SPACE: INTRODUCTION

A. Dalgarno

Harvard-Smithsonian Center for Astrophysics
60 Garden Street
Cambridge, Massachusetts 02138
U.S.A.

1. INTRODUCTION

Molecules are durable species with the ability to survive in the hostile environments encountered in space. Molecules have been detected in a remarkable array of extraterrestrial bodies ranging from the planetary atmospheres and comets of the solar system to massive clouds in active galaxies and quasars. Observations of molecular absorption and emission lines have provided a powerful diagnostic probe of the physical conditions in which molecules are formed, excited and destroyed. From the spectroscopic data, physical characteristics such as density, temperature, velocity, elemental abundances, radiation field and magnetic field can often be inferred. The presence of molecules profoundly influences the nature and evolution of the objects in which they exist. Molecules play a critical role in the birth and death of stars, in the formation and evolution of galaxies and in the early Universe. They constitute a significant component of the newest astronomical entity, supernova 1987A.

The characteristics of different regions of space vary widely and the molecular processes that occur are correspondingly diverse, ranging from ion-molecule reactions at 10K to ionization by energetic cosmic rays and γ rays. Because of the extreme physical conditions encountered in space, it is necessary to address the fundamentals of the molecular processes in depth so that they may be appropriately translated to the space environment.

As testified by the diversity of material in this book, Kazuo Takayanagi has done original research on a broad array of subjects involving the entire domain of atomic and molecular processes. He has sustained the development of the field and substantially influenced its directions over the past several decades. Although much of his research has dealt with fundamental aspects of atomic and molecular collision theory, Kazuo Takayanagi has also been concerned with the procedures for the calculation of collision cross sections and with the applications of atomic and molecular collision physics to other branches of physics and chemistry, notably plasma physics, atmospheric science and astronomy.

For my introductory chapter I have made an arbitrary selection of topics of recent interest relating to astronomical phenomena. My selection of topics, though limited and reflecting my personal interests, will, I hope, help to exhibit Kazuo Takayanagi's profound originality and remarkable versatility in developing our understanding of molecular processes in space.

2. INFRARED EMISSION FROM VIBRATIONALLY EXCITED MOLECULAR HYDROGEN

Infrared emission from vibrationally excited molecular hydrogen was first detected in the interstellar medium[1] towards the Kleinmann-Low nebula in Orion. Several rotational lines of the S and Q branches of the $v=1\rightarrow0$ vibrational transition of the ground electronic state were detected. Because of the presence in the region of high velocity motions, the source of the excitation was immediately interpreted as collisional excitation in a heated shocked gas. It appears that the shock is powered by a high-velocity outflow from an embedded protostar, interacting with and shocking the ambient molecular gas.

Many more emission lines have been detected subsequently[2-6]. It is convenient to characterize the relative populations $N(v,J)$ of the individual vibration-rotation levels with vibrational quantum number v and rotational quantum number J by a Boltzmann excitation temperature T_{vJ}. Measurements of emission from low-lying levels with $v\leqq3$ and $J\leqq5$[7,8] yield an excitation temperature varying between 3500 K for the higher-lying vJ levels and 1600 K for the lower-lying levels. There are no systematic differences between the vibrational temperatures inferred from the relative populations of vibrational levels and the rotational temperatures inferred from the relative populations of rotational levels and the population are broadly consistent with thermal excitation in a hot gas, with a range of temperatures.

However, care is needed in interpreting the excitation temperature as a kinetic temperature. Thermal equilibrium holds if the radiative lifetimes are longer than the mean time for collisional quenching. The radiative lifetimes have been reliably determined to vary from 1.3×10^6 s for the $v=1$ level to 2×10^5 s as v increases up to $v=9$[9,10]. The major uncertainty in the analysis stems from the collisional quenching of the vibrational levels of H_2. The approximate estimates give values between 10^{-12} cm^3s^{-1} and 10^{-13} cm^3s^{-1} at 2000 K so that a density of at least 10^6 cm^{-3} in the shocked gas is required for thermal equilibrium to be attained.

The existence of thermal equilibrium does not identify the nature of the source of heating of the gas, though shocks are clearly the probable cause in OMC-1[11-13]. Early attempts to model the Orion observations with a hydrodynamic J-type shock failed but more elaborate magnetohydrodynamic (MHD) models[14-16] met with more success. In the MHD shocks the magnetic field plays a crucial role not only in its effect on energy balance between the pre-shock and the post-shock gas but also in creating a magnetic precursor that heats and compresses the gas before the shock arrives so that the shock becomes a continuous C-type shock[17]. The Orion data are consistent with a shock velocity of 38 km s^{-1}, a shock velocity which would, in the absence of the magnetic field, lead to dissociation of H_2.

Questions about the appropriateness of the C-type shock models have been raised by further measurements[6] of vibration-rotation lines including the 4-3 S(3) and 3-2 S(2) lines and pure rotation lines 0-0 S(15) and 0-0 S(16). In their present form, the C-type models do not reproduce the measured intensities of the lines originating in high v and J levels. Brand et al.[6] argue that a more satisfactory explanation is obtained by postulating gas, emitting in the cooling zone of a hydrodynamic shock propagating with a shock velocity near 25 km s^{-1}.

Ultraviolet fluorescence is inadequate because it has difficulty in populating levels with high rotational angular momenta. However, with sufficient gas density, ultraviolet fluorescence followed by exchange collisions,

$$H_2(v,J) + H \rightarrow H + H_2(v',J')$$

in which $v' < v$ and $J' > J$ such that the amount of energy transfer is minimized may be rapid enough to contribute significantly.

Following the detection in OMC-1, infrared emission lines from H_2 have been observed in many objects in our galaxy including other molecular clouds, planetary nebulae, supernova remnants, Herbig-Haro objects, the Galactic Center, and reflection and bipolar nebulae and in the Magellanic Clouds, Seyfert galaxies and interacting galaxies. Many of the objects were seen only in the 1-0 S(1) line and in analogy with the Orion shocked region it was often assumed that shocked gas was the source of the emission.

Observations of many lines in the reflection nebulae Parsamyan 18[18] and NGC 2023[8,19] and the compact planetary nebula HB 12[20], reveal a pattern of intensities of different lines in the spectrum that cannot be attributed to collisional excitation but which is generally consistent with theoretical predictions of the spectra resulting from ultraviolet pumping[9,21-23].

Detailed theoretical studies of NGC 2023 have been carried out by Takayanagi, Sakimoto and Onda[24] and Black and van Dishoeck[21]. Takayanagi et al.[24] addressed the question of the ortho/para ratio in H_2, which differs in NGC 2023 from the statistical 1:3 value. Fluorescence does not change the ortho/para ratio but exchange reactions with H^+,

$$H_2(\text{even } J) + H^+ \rightleftarrows H_2(\text{odd } J) + H^+$$

tend to drive it towards the value appropriate to thermal equilibrium[25]. Takayanagi et al. showed that the initial H_2 formation temperature, assumed to be the grain surface temperature, is 60-70 K. Black and van Dishoeck[21] drew attention to the importance of the dependence on depth into the cloud of the various parameters and also argued that the photoionization of vibrationally excited H_2 in vibrational levels $v \geqq 4$,

$$H_2(v \geqq 4) + h\nu \rightarrow H_2^+ + e$$

followed by photodissociation of H_2^+

$$H_2^+ + h\nu \rightarrow H + H^+$$

is an important additional source of H^+ ions. With these modifications, no certain conclusion about the initial temperature can be drawn. However, the removal process

$$H^+ + H_2(v \geqq 4) \rightarrow H + H_2^+$$

appears to have been omitted in the analysis. It is presumably a rapid reaction. It remains an open question whether the observed ortho/para ratio retains any information on the formation temperature.

The agreement between the theoretical calculations[21] and the measured intensities of the numerous lines observed in NGC 2023[8,19] is remarkable, and clearly establishes the ultraviolet fluorescence of a low density gas as the excitation mechanism. A high density gas excited by ultraviolet radiation or X rays would have a characteristic emission pattern similar to a shocked gas[22-23].

In observations of external galaxies, few lines are detected and the identification of the excitation mechanism is less certain. Many of the galaxies are observed only in the 1-0 S(1) line. Other evidence is used to argue that a shock is the probable mechanism. Starburst galaxies presumably contain many Orion-like sources; merging galaxies are likely sites of massive shocks; Seyfert galaxies contain compact objects in their centers which drive powerful winds that could produce shocks.

The line intensity data for many of the galaxies found to contain vibrationally excited H_2 are sufficient to establish that the levels are thermally populated. Thermal excitation requires a gas density in excess of 10^5 cm^{-3} and a temperature of about 2000 K. The heat source may be a shock or conceivably an intense ultraviolet radiation[23] or X-ray radiation field[36].

Recent studies[33,35] have discovered galaxies in which the emission is consistent with ultraviolet fluorescence in a gas with a density that must be less than about 10^5 cm^{-3}. Puxley et al. have argued that intense ultraviolet radiation is provided during episodes of vigorous star formation.

In some galaxies a contribution may arise from high energy X rays.

X rays have large optical depths and create fast electrons within a substantial volume. The fast electrons slow down by exciting and ionizing hydrogen molecules. The excited states decay by cascading to lower states and the singlet states terminate in the vibration-rotation levels of the ground state. The resulting infrared emission is similar to that which occurs for ultraviolet fluorescence, but differing in detail[37]. Emission of some of the 1→0 lines is enhanced by the direct excitation of the v=1 level, but higher levels are populated mostly by cascading from electronically excited states.

Similar processes may be at work in the atmospheres of the outer planets. The ultraviolet emission in the dayglow of Jupiter has been interpreted as ultraviolet fluorescence of solar radiation by H_2[38,39] and fast electrons are present in the auroral regions. A possible detection of the 1-0 S(1) line in the Jupiter infrared spectrum has been reported[40].

3. ION-MOLECULE REACTIONS IN THE INTERSTELLAR GAS

A diverse array of molecules has been found in dense interstellar clouds where star formation occurs. The kinetic temperatures are low and the gas phase chemistry is largely an ion-molecule chemistry driven by cosmic rays[41]. The behavior of ion-molecule reactions at low temperatures is critical to the attempts to reproduce the observed abundances by constructing chemical reaction schemes.

For reactions involving neutral molecular hydrogen, the Langevin formula is usually adopted. The Langevin formula assumes that the collision is dominated by the long-range polarization attraction and that reaction occurs in close collisions with a probability of unity. These assumptions lead to a rate coefficient that is independent of temperature. It has been common practice to assume that reactions of ions with heteronuclear molecules behave similarly despite the additional longer range force that arises from the interaction of the ion with the permanent dipole.

Takayanagi[42] introduced an approximation, the perturbed rotational state method, which took explicit account of the orientation effects of the ion-dipole interaction and in 1980 in collaboration with Sakimoto[43] pointed out that in interstellar clouds enhancements in the rate coefficients by an order of magnitude over the Langevin value would occur for some systems. The modifications of the Langevin formula by the interaction between the ion and the permanent quadrupole moment of the molecule were explored by Takayanagi[43]. These studies also demonstrated that the rate coefficient depends upon the rotational state of the molecule. In most laboratory experiments, the rotational states are thermally populated, whereas in astrophysical environments, the rotational populations of heteronuclear molecules are often far from thermal equilibrium. These papers presaged considerable theoretical activity[45-53]. The measurements at very low temperatures[54,55] show clearly the expected increase over the Langevin rate coefficients.

The influence of the enhanced rate coefficients on the composition of interstellar clouds has been explored[56-58]. It appears that the abundances of polar molecules are reduced and of nonpolar molecules increased though the changes are less than an order of magnitude. Of greater significance is the time scale for forming complex molecules which is reduced by the faster ion-polar molecule reactions.

The influence of fine structure is seen in the reaction

$$N^+ + H_2 \rightarrow NH^+ + H$$

Experiments has shown that for normal H_2 the reaction is endoergic by several meV[59-62], but the role of rotational energy and of spin-orbit energy in driving the reaction was unclear. If the internal energy present in the reactant systems is used to overcome the endoergicity, the process will be much slower in the cold interstellar gas where the N^+ ions are in the lowest fine structure $^3P_0$ level and the molecular hydrogen may be mostly in the J=0 rotational level. From the recent work of Marquette, Rebrion and Rowe[63] who made measurements at temperatures as low as 8 K, it appears that the spin-orbit and rotational energies do participate and the endoergicity for the systems in their lowest energy states is 18±2 meV.

At the temperatures of dense clouds, the N^+ ions will react preferentially with CO and the formation of NH_3 which is initiated by the reaction of N^+ with H_2 is severely inhibited.

The difficulty may be partially overcome by recognizing that the nitrogen ions are initially hot[64]. They are produced by charge transfer of He^+ with N_2,

$$He^+ + N_2 \rightarrow He + N + N^+$$

with an initial energy of 0.14 eV. The hot ions lose energy in elastic collisions with H_2 but some fraction of them will react to form NH^+[65,66]. Fine structure transitions may occur during the slowing down of the N^+ ions and during the ion-molecule reactions, and it remains uncertain that a sufficient abundance of ammonia can be formed in the gas phase by reactions of N^+ with H_2[67]. Galloway and Herbst[67] have pointed out that energetic N^+ ions are also produced by the reaction

$$He^+ + CN \rightarrow N^+ + C + He$$

and for it the initial kinetic energy of N^+ is 1.05 eV. An additional source of NH^+ which does not require hot atom chemistry is

$$H_2{}^+ + N \rightarrow NH^+ + H$$

but it is probably insufficient to provide the measured abundances.

Neutral particle reactions also depend on the internal energy modes of the reactants, including the distribution amongst fine structure levels and the conventional assumption that the rate coefficients diminish as temperature $T^{1/2}$ is usually incorrect[68-70]. The variations with temperature measured in laboratory experiments reflect the changing rotational populations.

In regions heated by shocks, X rays or intense ultraviolet radiation fields, the chemistry is dominated by reactions with H and H_2, both exothermic and endothermic, of the kind

$$X + H_2 \rightleftarrows XH + H$$

The composition depends on the H/H_2 ratio and on the temperature. The chemistry may be sensitive to the internal energy distributions.

Vibrationally excited H_2 molecules may be created by ultraviolet pumping or by photoelectron pumping. If so, the reaction

$$C^+ + H_2(v \geqq 1) \rightarrow CH^+ + H$$

which is rapid[71] may be a significant source of CH^+ molecules and

$$He^+ + H_2(v \geqq 2) \rightarrow \text{products}$$

also rapid[71] will lead to the removal of He^+ ions and consequently limit the destruction of CO by

$$He^+ + CO \rightarrow He + C^+ + O$$

4. PHOTODISSOCIATION IN INTERSTELLAR CLOUDS

Photodissociation processes play a major role in the chemistry of diffuse interstellar clouds. The photodissociation of the two most abundant molecules H_2 and CO is of special significance. The photodissociation of H_2 is now well understood. The mechanism is that suggested by Solomon[72] in which vibrational levels of the $B^1\Sigma_u^+$ and $C^1\Pi_u$ electronic states, populated by absorption of the interstellar radiation field, undergo spontaneous radiative transitions into the vibrational continuum of the ground electronic state. The quantitative demonstration of its efficiency was presented by Stecher and Williams[73] and by Nishimura and Takayanagi[74]. A continuum emission is associated with the process which has been seen in laboratory experiments[75,76]. The emission lines that also occur following the initial absorption have sufficient energy to photodissociate and in some cases photoionize other interstellar molecules.

In the interior of dense clouds the interstellar field is excluded by grain absorption and it penetrates only the outer layers unless it is very intense. There exists nevertheless a source of ultraviolet photons produced by the excitation of molecular hydrogen by the energetic electrons ejected during cosmic ray ionization or X ray ionization[77]. Detailed calculations of the energy degradation of fast electrons moving in a gas of H_2 have been carried out and the resulting ultraviolet spectrum has been determined[58].

The effects of the internally generated photons on the equilibrium abundances of complex molecules are substantial. The reaction sequence in which carbon atoms are added sequentially contains systems such as CH_4, C_2H_2 and C_3H_2 that are stable against chemical reactions with abundant neutral gases. Their lifetimes can be shortened by photodissociation so that they are less available to serve as a building block for the addition of another carbon atom. Because the destructive effects of photodissociation amplify along the chemical sequences, the abundances of complex hydrocarbons are drastically diminished[58]. The time scales for photodestruction are long so that the abundances of complex species predicted at early times in a cloud evolving from an initial state in which the carbon exists mostly as neutral atoms are not changed[78].

Carbon monoxide is also photodissociated through the absorption of emission lines of H_2. A quantitative description of the mechanisms by which CO may undergo photodissociation has emerged recently through laboratory studies of absorption and fluorescence at high spectral resolution[79]. Continuum absorption is negligible and dissociation occurs through several excited singlet states which predissociate. It happens that many of the H_2 emission lines overlap the absorption lines of CO that populate the predissociating states. The resulting photodissociation rates of CO in interstellar clouds depend upon the abundance of CO, its rotational temperature and its velocity distribution[80].

It also happens that many of the absorption lines of H_2 and of H overlap the absorption lines of CO so that CO can be shielded by H_2 and H[81-83] from the interstellar ultraviolet radiation field. There remains nevertheless a problem in reproducing the large abundances of CO seen in diffuse clouds, and a reduction in the interstellar radiation field below 100 nm may be indicated[83].

5. MOLECULES IN SUPERNOVA 1987A

On February 23, 1987, a blue supergiant star in the Large Magellanic Cloud underwent core collapse and the rebound shock resulted in an explosion which ejected the stellar envelope into the interstellar medium of the galaxy. The optical and infrared emission is powered by the decay of radioactive ^{56}Co. The decay produces γ-ray lines at 847 keV and 1238 keV which diffuse outward and lose energy in Compton collisions with the free and bound electrons. The resulting X rays are absorbed in K-shell ionizations of the heavy elements. The fast photoelectrons and recoil electrons excite and ionize the atoms and ions. The density is diminishing as the ejecta expand homologously and the material is cooling rapidly by adiabatic expansion and by radiation losses. The supernova can be regarded as a dynamically evolving experiment in atomic and molecular physics and many atomic and molecular processes involving the interactions of heavy elements are occurring.

Many features have appeared in the spectrum which are produced in the inner core region where the material is composed mostly of helium with lesser amounts of oxygen, carbon, silicon and sulphur.

The ionization stages seen in the spectrum are singly-ionized and neutral systems. It is likely that charge transfer is the main mechanism for reducing the ionization stages once some neutral atoms are formed by radiative and dielectronic recombination. Thus the charge transfer reaction

$$Fe^{2+} + O \rightarrow Fe^{+} + O^{+}$$

which is almost certainly a rapid process, controls the conversion of Fe^{2+} into Fe^{+}.

Of special interest is the appearance in the spectrum of the $\Delta v=2$ overtone emission of carbon monoxide near 2.3 μm and the $\Delta v=1$ fundamental near 4.6 μm[84,85]. The broadening implies a velocity of the order of 1000 km s^{-1} and hence an origin for the emitting molecule in the interior of the supernova.

The vibrationally excited levels could be populated by ultraviolet pumping or infrared pumping, but collisional excitation by helium atoms or by electrons is a more likely mechanism. The collisional excitation of CO by helium and by electrons has been discussed by Takayanagi[86]. The rate coefficient

$$He + CO(v=0) \rightarrow He + CO(v=1)$$

is a rapidly increasing function of temperature[87,88]. The kinetic temperature is not well determined. It cannot be too high because then CO would be destroyed by thermal dissociation, but it is higher than the temperatures derived from the relative line intensities. A value at 500 days after the explosion of 6000 K is plausible. At 6000 K, the rate coefficient may be of the order of 10^{-13} cm^3s^{-1} but this estimate is very uncertain.

The electron impact excitation of the vibrational levels has been studied by Itikawa and Takayanagi[89], by Haddad and Milloy[90] and by Sohn et al.[91]. The rate coefficient for vibrational excitation is about 3×10^{-10} cm^3s^{-1} at 6000 K. The fractional ionization in the interior is a few percent so that electron impacts are more effective than helium impacts in exciting CO.

If the electron density is large enough, vibrational excitation and de-excitation in electron collisions will bring the levels into thermal equilibrium. The temperature can then be derived from the relative band intensities. At 92 days after collapse it was 3000 K and after 255 days 1800 K[85]. From the absolute intensities a total mass of CO of 10^{-4} $M_\odot$ is inferred[85].

The assumption of local thermodynamic equilibrium is valid only early in the appearance of CO emission when the temperature and density are high. The radiative lifetimes of the vibrational levels of CO are 0.032 s for the v=1 vibrational level and 0.017 s for the v=2 vibrational level[91]. These lifetimes are shorter than the collisional lifetimes throughout most of the period in which CO has been detected in the infrared.

Because there are no dust particles in SN 1987A, the molecular formation and destruction mechanisms must be occurring in the gas phase. The CO molecules can be destroyed by direct collision-induced dissociation

$$e + CO \rightarrow e + C + O$$

$$He + CO \rightarrow He + C + O$$

$$O + CO \rightarrow O + C + O$$

or by chemical reactions

$$O + CO \rightarrow O_2 + C$$

Collision induced dissociation of CO by heavy particle impact is diminished in efficiency by radiative stabilization[93] but dissociation by electron impact, an electronic excitation process, is unaffected. The efficiency of electron impact dissociation places a lower limit on the temperature above which CO and other molecules cannot survive in an ionized plasma.

Carbon monoxide is also destroyed by photodissociation, as discussed in Section 3. A major source of photons is the two-photon decay of $He(2^1S)$ metastable atoms

$$He(2^1S) \rightarrow He(1^1S) + h\nu_1 + h\nu_2$$

Its efficiency in limiting the abundance of CO in the post-shock gas following the propagation of a dissociative shock has been worked out[94]. It may be diminished in effectiveness in the supernova ejecta by electron impact

$$e + He(2^1S) \rightarrow e + He(2^1P)$$

$$He(2^1P) \rightarrow He(1^1S) + h\nu$$

which converts the two-photon decays to the single photons emitted by the 2^1P level of helium. These photons can photoionize CO

$$CO + h\nu \rightarrow CO^+ + e$$

The photoionization process is only partly destructive because the reaction

$$CO^+ + O \rightarrow CO + O^+$$

restores CO.

An estimate of the radiation field that photoionizes CO can be obtained from the distribution of ionization between Fe^+ and Fe^{2+}. We modify an argument of Petuchowski et al.[95] and assume that the balance is controlled by photoionization

$$Fe^+ + h\nu \rightarrow Fe^{2+} + e$$

and charge transfer

$$Fe^{2+} + O \rightarrow Fe^+ + O^+$$

Then, using the data of Petuchowski et al., we derive an ultraviolet flux of about 10^{10} $cm^{-2}s^{-1}$ at 260 days after the explosion.

The value is an upper limit because Fe^{2+} may be produced by other mechanisms, such as

$$e + Fe^{+} \rightarrow e + Fe^{2+} + e$$

However, the effective ionization rate of CO may not be much altered because

$$e + CO \rightarrow e + CO^{+} + e$$

is also a source of CO^{+}.

A tentative detection of CO^{+} emission from the supernova has been reported[85]. Because of the Coulomb attraction, the collision excitation and de-excitation of CO^{+} by electrons

$$e + CO^{+}(v) \rightarrow e + CO^{+}(v')$$

is much more rapid than the similar process for CO, and the CO^{+} vibrational level populations will remain in local thermodynamic equilibrium for a longer period than CO. The CO^{+} is removed by dissociative recombination

$$e + CO^{+} \rightarrow C + O$$

The rate coefficient of dissociative recombination has been measured to be $2\times10^{-7}(300/T)^{0.48}$ $cm^{3}s^{-1}$ in a merged-beam experiment at energies up to 5 eV[96].

The CO^{+} ions can be removed also by photodissociation

$$CO^{+} + h\nu \rightarrow C^{+} + O$$
$$\rightarrow C + O^{+}$$

and by chemical reactions, but dissociative recombination is probably the most rapid destruction mechanism. The relative abundance of CO^{+} and CO that results from a balance between photoionization of CO and dissociative recombination of CO^{+} is of the order of 0.1, a ratio not inconsistent with the measurements.

Several possible channels exist for the formation of CO, of which the most promising is

$$C^{+} + O \rightarrow CO^{+} + h\nu$$

which takes place when C^{+} and O approach along the excited $A^{2}\Pi_{i}$ state, which may radiate to the $X^{2}\Sigma^{+}$ state in an allowed transition. The neutral molecule is then formed by charge transfer processes

$$CO^{+} + O \rightarrow CO + O^{+}$$

Negative ion chemistry may also contribute. Radiative attachment

$$O + e \rightarrow O^- + h\nu$$

forms O^- which leads directly to CO by associative detachment

$$C + O^- \rightarrow CO + e$$

The supply of neutral carbon atoms may be limited and the alternative route

$$O + O^- \rightarrow O_2 + e$$

$$C^+ + O_2 \rightarrow CO + O^+$$

may be more efficient.

The molecule SiO has also been detected. It can be formed and destroyed by processes similar to those that affect the abundance of CO. It may also be destroyed by

$$C^+ + SiO \rightarrow Si^+ + CO$$

and by charge transfer

$$C^+ + SiO \rightarrow C + SiO^+$$

Most of this discussion of molecular processes in SN 1987A is adapted from a joint study, still in progress, by S. Lepp, R. McCray and myself. I present it here because, in my view, it demonstrates the basic importance of a fundamental understanding of atomic and molecular processes in the development of quantitative descriptions of astronomical phenomena, an area of research to which Kazuo Takayanagi has made enduring contributions.

Acknowledgments This work has been supported by the National Science Foundation, Division of Astronomical Sciences under Grant AST-86-17675.

REFERENCES

1. T.N. Gautier III, U. Fink, R.R. Treffers, and H.P. Larson, Detection of molecular hydrogen quadrupole emission in the Orion nebula, Ap. J. Lett. 207, L129-L133 (1986).
2. R.F. Knacke and E.T. Young, Detection of the S(8), S(12), S(13), and S(15), v=0→0 rotation lines of molecular hydrogen in Orion, Ap. J. Lett. 249, L65-L69 (1981).
3. N.Z. Scoville, D.N.B. Hall, S.G. Kleinmann and S.T. Ridgeway, Velocity, reddening and temperature structure of the H_2 emission in Orion Ap. J. 253, 136-148 (1982).
4. D. Nadeau, T.R. Geballe, and G. Neugebauer, The motion and distribution of the vibrationally excited H_2 in the Orion molecular cloud, Ap. J. 253, 154-166 (1982).
5. T.R. Geballe, S.E. Persson, T. Simon, C.J. Lansdale, and P.J. McGregor, comparison of 2.1 and 3.8 micron line profiles of shocked H_2 in the Orion molecular cloud, Ap. J. 302, 693-700 (1986).

6. P.W.J.L. Brand, A. Moorhouse, M.G. Burton, T.R. Geballe, M. Bird, and R. Wade, Ratios of molecular hydrogen line intensities in shocked gas: Evidence for cooling zones, Ap. J. Lett. 334, L103-L106 (1988).
7. S. Beckwith, N.J. Evans II, I. Gatley, G. Gull, and R.W. Russell, Observations of the extinction and excitation of the molecular hydrogen emission in Orion, Ap. J. 264, 152-160 (1983).
8. T. Hasegawa, I. Gatley, R.P. Garden, P.W.J.L. Brand, M. Ohishi, M. Hayashi, and N. Kaifu, Level population and para/ortho ratio of fluorescent H_2 in NGC 2023, Ap. J. Lett. 318, L77-L80 (1987).
9. J.H. Black and A. Dalgarno, Interstellar H_2: The population of excited rotational states and the infrared response to ultraviolet radiation, Ap. J. 203, 132-142 (1976).
10. J. Turner, K. Kirby-Docken, and A. Dalgarno, The quadrupole vibration-rotation transition probabilities of molecular hydrogen, Ap. J. Suppl. 35, 281-292 (1977).
11. J. Kwan, On the molecular hydrogen emission at the Orion nebula, Ap. J. 216, 713-723 (1977).
12. D.J. Hollenbach and J.M. Shull, Vibrationally excited molecular hydrogen in Orion, Ap. J. 216, 419-426 (1977).
13. R. London, R. McCray, and S-I. Chu, A shock model for infrared line emission from H_2 molecules, Ap. J. 217, 442-447 (1977).
14. B.T. Draine and W.G. Roberge, A model for the intense molecular line emission from OMC-1, Ap. J. Lett. 259, L91-L96 (1982).
15. D.F. Chernoff, D.J. Hollenbach, and C.F. McKee, Molecular shock waves in the BN-KL region of Orion, Ap. J. 259, L97-L101 (1982).
16. B.T. Draine, W.G. Roberge, and A. Dalgarno, Magnetohydrodynamic shock waves in molecular clouds, Ap. J. 264, 485-507 (1983).
17. B.T. Draine, Interstellar shock waves with magnetic precursors, Ap. J. 241, 1021-1038 (1980).
18. K. Sellgren, Ultraviolet-pumped infrared fluorescent molecular hydrogen emission in reflection nebulae, Ap. J. 305, 399-404 (1986).
19. I. Gatley, T. Hasegawa, H. Suzuki, R. Garden, P. Brand, J. Lightfoot, W. Glencross, H. Okuda, and T. Nagata, Fluorescent molecular hydrogen emission from the reflection nebula NGC 2023, Ap. J. Lett. 318, L73-L76 (1987).
20. H.L. Dinerstein, D.F. Lester, J.S. Carr, and P.M. Harvey, Detection of fluorescent molecular hydrogen emission in the planetary nebula Hubble 12, Ap. J. Lett. 327, L27-L30 (1988).
21. J.H. Black and E.F. van Dishoeck, Fluorescent excitation of interstellar H_2, Ap. J. 322, 412-449 (1987).
22. A. Sternberg, The infrared response of molecular hydrogen gas to ultraviolet radiation: A scaling law, Ap. J. 332, 400-409 (1988).
23. A. Sternberg and A. Dalgarno, The infrared response of molecular hydrogen gas to ultraviolet radiation: High density regions, Ap. J. 338, 197-233 (1989).
24. K. Takayanagi, K. Sakimoto, and K. Onda, Para/ortho ratio of molecular hydrogen in NGC 2023, Ap. J. Lett. 318, L81-L84 (1987).
25. A. Dalgarno, J.H. Black, and J.C. Weisheit, Ortho-para transitions in H_2 and the fractionation of HD, Astrophys. Lett. 14, 77-79 (1973).
26. R.I. Thompson, M.J. Lebofsky, and G.H. Rieke, The 2-2.5 micron spectrum of NGC 1068: A detection of extragalactic molecular hydrogen, Ap. J. Lett. 222, L49-L53 (1978).
27. D.N.B. Hall, S.G. Kleinmann, N.Z. Scoville, and S.T. Ridgway, 2 micron spectroscopy of the nucleus of NGC 1068, Ap. J. 248, 898-905 (1981).
28. J. Koornneef and F.P. Israel, Detection of molecular hydrogen in the small Magellanic cloud H II region N81, Ap. J. 291, 156-159 (1985).
29. R.D. Joseph, G.S. Wright, and R. Wade, Detection of molecular hydrogen in two merging galaxies, Nature 311, 132-133 (1984).
30. N.K. Reay, P.D. Atherton, and N.A. Walton, Extra-nuclear molecular hydrogen in NGC 1068, Mon. Not. Roy. Astron. Soc. 218, 13P-17P (1986).

31. T.M. Heckman, S. Beckwith, L. Blitz, M. Skrutskie, and A.S. Wilson, Molecular gas in the type 1 Seyfert galaxy NGC 7469: Implications for nuclear activity, Ap. J. 305, 157-166 (1986).
32. K. Kawara, M. Nishida, and B. Gregory, Two micron spectroscopy of IRAS galaxies, Ap. J. 321, L35-L40 (1987).
33. J. Fischer, T.R. Geballe, H.A. Smith, M. Simon, and J.W.V. Storey, Molecular hydrogen line emission in Seyfert galactic nuclei, Ap. J. 320, 667-675 (1987).
34. T.R. Geballe, Infrared spectroscopy of the ultraluminous IRAS galaxy 14348-1447: A distant detection of the H_2 line emission, Mon. Not. Roy. Astron. Soc. 234, 1P-4P (1988).
35. P.J. Puxley, T.G. Hawarden, and C.M. Mountain, Fluorescent molecular hydrogen in galaxies, Mon. Not. Roy. Astron. Soc. 234, 29P-40P (1988).
36. S. Lepp and R. McCray, X-ray sources in molecular clouds, Ap. J. 269, 560-567 (1983).
37. R. Gredel and A. Dalgarno, in preparation (1989).
38. R.V. Yelle, J.C. McConnell, B.R. Sandel, and A.L. Broadfoot, The dependence of electroglow on the solar flux, J. Geophys. Res. 92, 15110-15124 (1987).
39. R.V. Yelle, H_2 emissions from the outer planets, Geophys. Res. Lett. 15, 1145-1148 (1988).
40. L. Trafton, J. Carr, D. Lester, and P. Harvey, A possible detection of Jupiter's northern auroral $S_1(1)$ H_2 quadrupole line emission, Icarus 74, 351-356 (1988).
41. E. Herbst and W. Klemperer, The formation and depletion of molecules in dense interstellar clouds, Ap. J. 185, 505-533 (1973).
42. K. Takayanagi, Low-energy ion-polar molecule collision -- The perturbed rotational state approach, J. Phys. Soc. Japan 45, 976-985 (1978).
43. K. Sakimoto and K. Takayanagi, Influence of the dipole interaction on the low-energy ion-molecule reactions, J. Phys. Soc. Japan 48, 2076-2083 (1980); J. Phys. Soc. Japan 51, 2036 (1982).
44. K. Takayanagi, Low-velocity ion-molecule collisions with Quadrupole interaction, J. Phys. Soc. Japan 51, 3337-3344 (1982).
45. K. Sakimoto, Ion transport in polar gases, Chem. Phys. 63, 419-436 (1981).
46. T. Su and W.J. Chesnavich, Parametrization of the ion-polar molecule collision rate constant by trajectory calculations, J. Chem. Phys. 76, 5183-5185 (1982).
47. D.R. Bates, Ion-polar molecule encounters, Proc. Roy. Soc. Lond. A384, 289-300 (1982).
48. K. Sakimoto, Orbiting collisions between ions and polar molecules: Semiclassical PRS approaches, Chem. Phys. 85, 273-278 (1984).
49. K. Sakimoto, On the capture rate constant of collisions between ions and symmetric top molecules, Chem. Phys. Lett. 116, 86-88 (1985).
50. D.C. Clary, Calculations of rate constants for ion-molecule Reactions using a combined capture and centrifugal sudden approximation, Mol. Phys. 54, 605-618 (1985).
51. N.G. Adams, D. Smith, and D.C. Clary, Rate coefficients for the reactions of ions with polar molecules at interstellar temperatures, Ap. J. Lett. 296, L31-L34 (1985).
52. W.L. Morgan and D.R. Bates, Ion-dipolar molecule rate coefficients, Ap. J. 314, 817-821 (1987).
53. D.R. Bates and W.L. Morgan, Adiabatic invariance treatment of hitting collisions between ions and symmetrical top dipolar molecules, J. Chem. Phys. 87, 2611-2616 (1987).
54. D.C. Clary, D. Smith, and N.G. Adams, Temperature dependence of rate coefficients for reactions of ions with dipolar molecules, Chem. Phys. Lett. 119, 320-326 (1985).
55. J.B. Marquette, B.R. Rowe, G. Dupeyrat, G. Poissant, and C. Rebrion, ion-polar-molecule reactions: A CRESU study of He^+, C^+, N^+ + H_2O, NH_3 at 27, 68 and 163 K, Chem. Phys. Lett. 122, 431-435 (1985).

56. T.J. Millar, N.G. Adams, D. Smith, and D.C. Clary, The HCS$^+$/CS abundance ratio in interstellar clouds, Mon. Not. Roy. Astron. Soc. 216, 1025-1031 (1985).
57. E. Herbst and C.M. Leung, Effects of large rate coefficients for ion-polar neutral reactions on chemical models of dense interstellar clouds, Ap. J. 310, 378-382 (1986).
58. A. Sternberg, A. Dalgarno, and S. Lepp, Cosmic-ray-induced photodestruction of interstellar molecules in dense clouds, Ap. J. 320, 676-682 (1987).
59. N.G. Adams and D. Smith, A study of the nearly thermoneutral reactions of N^+ with H_2, HD and D_2, Chem. Phys. Lett. 117, 67-70 (1985).
60. J.B. Marquette, B.R. Rowe, G. Dupeyrat, and E. Roueff, CRESU study of the reaction $N^+ + H_2 \rightarrow NH^+ + H$ between 8K and 70K and interstellar chemistry implications, Astron. Ap. 147, 115-120 (1985).
61. J.A. Luine and G.H. Dunn, Ion-molecule reaction probabilities near 10K, Ap. J. 299, L67-L70 (1985).
62. K.M. Ervin and P.B. Armentrout, Energy dependence, kinetic isotope effects, and the thermo-chemistry of the nearly thermoneutral reactions $N^+(^3P) + H_2(HD,D_2)$-$NH^+(ND^+) + H(D)$, J. Chem. Phys. 86, 2659-2673 (1987).
63. J.B. Marquette, C. Rebrion, and B.R. Rowe, Reactions of $N^+(^3P)$ ions with normal, para and deuterated hydrogens at low temperatures, J. Chem. Phys. 89, 2041-2047 (1988).
64. N.G. Adams, D. Smith, and T.J. Millar, The importance of kinetically excited ions in the synthesis of interstellar molecules, Mon. Not. Roy. Astron. Soc. 211, 857-865 (1984).
65. E. Herbst, D.J. DeFrees, and A.D. McLean, A detailed investigation of proposed gas-phase syntheses of ammonia in dense interstellar clouds, Ap. J. 321, 898-906 (1987).
66. J.-H. Yee, S. Lepp, and A. Dalgarno, Energetic N^+ ions in the interstellar medium, Mon. Not. Roy. Astron. Soc. 227, 461-466 (1987).
67. E.T. Galloway and H. Herbst, A refined study of the rate of the $N^+ + H_2 \rightarrow NH^+ + H$ reaction under interstellar conditions: Implication for NH_3 productions, Astron. Ap., in press (1989).
68. D.C. Clary and H.-J. Werner, Quantum calculations of the rate constant for the O + OH reaction, Chem. Phys. Lett. 112, 346-350 (1984).
69. A.F. Wagner and M.M. Graff, Oxygen chemistry of shocked interstellar clouds. I. Rate constants for thermal and non-thermal internal energy distributions, Ap. J. 317, 423-431 (1987).
70. M.M. Graff, Fast neutral reactions in cold interstellar clouds, Ap. J., in press (1989).
71. M.E. Jones, S.E. Barlow, G.B. Ellison, and E.F. Ferguson, Reactions of C^+, He^+ and Ne^+ with vibrationally excited H_2 and D_2, Chem. Phys. Lett. 130, 218-223 (1986).
72. cf. P.M. Solomon and N.C. Wickramasinghe, Molecular and solid hydrogen in dense interstellar clouds, Ap. J. 158, 449-460 (1969).
73. T.P. Stecher and D.A. Williams, Photodestruction of hydrogen Molecules in HI regions, Ap. J. 149, L29-L30 (1967).
74. S. Nishimura and K. Takayanagi, On the photodestruction rate of hydrogen molecules in HI regions, Publ. Astron. Soc. Japan 21, 111-118 (1969).
75. A. Dalgarno, G. Herzberg, and T.L. Stephens, A new continuous emission spectrum of the hydrogen molecule, Ap. J. Lett. 162, L49-L53 (1970).
76. H. Schmoranzer and R. Zeitz, Observation of selectively excited continuous vacuum ultraviolet emission in molecular hydrogen, Phys. Rev. A18, 1472-1475 (1978).
77. S.S. Prasad and S.P. Tarafdar, UV radiation field inside dense clouds: Its possible existence and chemical implications, Ap. J. 267, 603-609 (1983).
78. R. Gredel, S. Lepp, A. Dalgarno, and E. Herbst, Cosmic ray induced photodissociation and photoionization rates of interstellar molecules, Ap. J., in press (1989).

79. C. Letzelter, M. Eidelsberg, F. Rostas, J. Breton, and B. Thieblemont, Photoabsorption and photodissociation cross sections of CO between 88.5 and 115 nm, Chem. Phys. 114, 273-288 (1987).
80. R. Gredel, S. Lepp, and A. Dalgarno, The C/CO ratio in dense interstellar clouds, Ap. J. Lett. 323, L137-L139 (1987).
81. A.E. Glassgold, P.J. Huggins, and W.D. Langer, Shielding of CO from dissociating radiation in dense interstellar clouds, Ap. J. 290, 615-626 (1985).
82. Y.P. Viala, C. Letzelter, M. Eidelsberg, and F. Rostas, The photodissociation of interstellar CO, Astron. Ap. 193, 265-272 (1988).
83. E.F. van Dishoeck and J.H. Black, The photodissociation and chemistry of interstellar CO, Ap. J. 334, 771-802 (1988).
84. E. Oliva, E.A.F.M. Moorwood, and I.J. Danziger, A 1-5 μm infrared spectrum of SN 1987A, ESO Messenger No. 50, 18-21 (1987).
85. J. Spyromilio, W.P.S. Meikle, R.C.M. Learner, and D.A. Allen, Carbon monoxide in Supernova 1987A, Nature 334, 327-329 (1988).
86. K. Takayanagi, Low energy molecular collision processes in space, in: Astrochemistry, IAU Symposium No. 120 (M.S. Vardya and S.P. Tarafdar, eds.), Reidel, Dordrecht, pp.31-42 (1987).
87. M.M. Maricq, E.A. Gregory, C.T. Wickham-Jones, D.J. Cartwright, and C.J.S.M. Simpson, Experimental and theoretical study of the vibrational relaxation of CO(v=1) and N_2(v=1): V-T relaxation by the isotopes of helium, Chem. Phys. 75, 347-370 (1983).
88. R.J. Price, D.C. Clay, and G.D. Billing, Validity of the rotational sudden approximation for vibrational relaxation in CO, Chem. Phys. Lett. 101, 269-273 (1983).
89. Y. Itikawa and K. Takayanagi, Vibrational excitation of CO by slow electron collision, J. Phys. Soc. Japan 27, 1293-1300 (1969).
90. G.N. Haddad and H.B. Milloy, Cross sections for electron-carbon monoxide collisions in the range 1-4 eV, Aust. J. Phys. 36, 473-484 (1983).
91. W. Sohn, K-H. Kochem, K. Jung, H. Ehrhardt, and E.S. Chang, Electron scattering from CO below resonance energy, J. Phys. B. 18, 2049-2055 (1985).
92. K. Kirby-Docken and B. Liu, Absorption oscillator strengths for vibration-rotation transitions of ground-state CO, Ap. J. Suppl. 36, 359-387 (1978).
93. W.G. Roberge and A. Dalgarno, Collision-induced dissociation of H_2 and CO molecules, Ap. J. 255, 176-180 (1982).
94. D. Neufeld and A. Dalgarno, Fast molecular shocks I. Reformation of molecules behind dissociative shock, Ap. J., in press (1989).
95. S.J. Petuchowski, E. Dewk, J.E. Allen, and J.A. Nuth, CO formation in the metal-rich ejecta of SN 1987A, Ap. J., in press (1989).
96. J.B.A. Mitchell and H. Hus, The dissociative Recombination and excitation of CO^+, J. Phys. B. 18, 547-555 (1985).

THEORY OF ELECTRON- AND PHOTON-MOLECULE COLLISIONS

P. G. Burke[†] and I. Shimamura[*]

[†]Department of Applied Mathematics & Theoretical Physics
The Queen's University of Belfast, Belfast BT7 1NN
Northern Ireland

[*]RIKEN, The Institute of Physical and Chemical Research
Wako, Saitama 351-01, Japan

1. INTRODUCTION

The interactions of electrons and photons with molecules lead to a variety of continuum processes that are not only of fundamental interest as a branch of molecular physics but also of immense importance in many applications; knowledge of these processes is indispensable in understanding stellar atmospheres, interstellar matter, planetary nebulae, the earth's ionosphere and aurorae, fusion plasmas, gaseous electronics, radiation physics and chemistry, etc.

The deeper understanding of the dynamics of electron- and photon-molecule collision processes gained through experimental and theoretical studies has resulted in many review articles. Especially, the early article by Takayanagi[1] develops a basic formulation of low-energy electron-molecule collisions, and that by Takayanagi and Itikawa[2] overviews the physics of rotational excitation in electron-molecule collisions at the end of the sixties.

In the absence of techniques of detailed ab-initio calculations, the theoretical studies in those days were based on model potentials having the correct long-range form. Takayanagi's work on rotational and vibrational transitions in the plane-wave and distorted-wave Born approximations revealed the applicability of such perturbative approaches to very low-energy electron-molecule collisions. He and his collaborators then proceeded with close-coupling calculations for many molecules, still using model potentials with parameters changing with the molecule. He also suggested to his student Hara that he should use Slater's free-electron-gas model to approximate the nonlocal exchange potential in e + H_2 collisions by a local potential[3]. This model is still widely used because of its simplicity and accuracy. His group also studied the breakdown of the Born approximation for highly polar molecules using better approximations. These studies led to interesting fundamental facts about the dynamics of electron-molecule interactions and stimulated more accurate theories to be formulated and applied.

The developments made in the seventies and eighties, which are reviewed, for example, in a book edited by Shimamura and Takayanagi[4], have been enormous both quantitatively and qualitatively. These recent developments are due, on one hand, to those of experimental apparatus, such as high-resolution electron spectrometers, intense tunable lasers and synchrotron radiation sources and, on the other hand, to the use of the current generation of computers, together with the developments of computational methods, that have enabled accurate predictions to be made at least on simple molecules.

This review article describes the present status of the theory of elastic scattering and excitation of nuclear and electronic motion in electron-molecule collisions, and molecular photoionization. Some representative examples of recent results are briefly discussed.

2. SUMMARY OF BASIC PROCESSES

In this section we summarize the basic processes that can occur in electron- and photon-molecule collisions. They are:

(i) Rotational and vibrational excitation

$$e^- + AB(vj) \rightarrow e^- + AB(v'j') \tag{1}$$

In this and the following equations we refer for notational simplicity to a diatomic molecule, although the basic processes apply equally well to polyatomic molecules. Also in our later equations (2) to (11) rotational and vibrational excitation may be present in addition to the other processes considered.

(ii) Electronic excitation

$$e^- + AB \rightarrow e^- + AB^* \tag{2}$$

where the star denotes an electronically excited state.

(iii) Dissociative attachment, or in the case of positive ions, dissociative recombination

$$e^- + AB \rightarrow A + B^- \quad (e^- + AB^\top \rightarrow A + B) \tag{3}$$

(iv) Radiative recombination

$$e^- + AB^+ \rightarrow AB + h\nu \tag{4}$$

(v) Dielectronic recombination

$$e^- + AB^+ \rightarrow AB^{**} \rightarrow AB^* + h\nu \tag{5}$$

where the double star denotes an intermediate resonance state where two electrons are in excited electronic orbitals.

(vi) Ionization

$$e^- + AB \rightarrow AB^+ + 2e^- \tag{6}$$

(vii) Dissociation

$$e^- + AB \rightarrow A + B^* + e^- \tag{7}$$

(viii) Dissociative ionization

$$e^- + AB \rightarrow A + B^+ + 2e^- \tag{8}$$

(ix) Photoionization

$$h\nu + AB \rightarrow AB^+ + e^- \tag{9}$$

(x) Photodissociation

$$h\nu + AB \rightarrow A + B^* \tag{10}$$

(xi) Dissociative photoionization

$$h\nu + AB \rightarrow A + B^+ + e^- \tag{11}$$

We see that these processes are far more varied than those which occur in electron and photon collisions with atoms and atomic ions because of the possibility of exciting nuclear as well as electronic degrees of freedom. In addition, the multicentered and nonspherical nature of the electron-molecule interaction considerably complicates the solution of the collision problem by reducing its symmetry.

There are two other aspects of electron and photon collisions with molecules which make them more varied than collisions with atoms and ions. These are firstly the crucial role of resonances in processes involving the excitation of nuclear degrees of freedom, and secondly the importance of the long-range interaction between an electron and a molecule. In particular, in the case of polar molecules the long-range dipole interaction behaving asymptotically as r^{-2} gives rise to bound or virtual states which lead to associated threshold peaks in the cross sections.

3. THEORY OF ELECTRON-MOLECULE COLLISIONS

3.1. Laboratory-Frame Representation

We now turn to the derivation of the basic equations describing electron collisions with a general polyatomic molecule. Following Takayanagi[3] we write the Schrödinger equation for the electron-molecule system as

$$(H_m + T + V)\Psi = E\Psi \tag{12}$$

Here, H_m is the Hamiltonian for the target molecule, and T is the kinetic energy of relative motion of the electron-molecule system

$$T = -\frac{\hbar^2}{2\mu}\nabla^2 , \quad \mu = \frac{M}{1+M} \approx 1 \tag{13}$$

where M is the mass of the molecule. Here and in the following we use atomic units with m_e (the electron mass) = $\hbar$ = e = 1. In eq. (12), V is the electron-molecule interaction potential

$$V = \sum_j \frac{1}{|\underline{r} - \underline{r}_j|} - \sum_{j'} \frac{Z_{j'}}{|\underline{r} - \underline{R}_{j'}|} \equiv V(\underline{R}, \underline{r}_m, \underline{r}) \tag{14}$$

where $\underline{R}$ stands symbolically for the positions $\underline{R}_{j'}$ of all the nuclei of the molecule, $\underline{r}_m$ for the set of coordinates $\underline{r}_j$ of the electrons in the molecule, and $\underline{r}$ for the coordinate of the incident electron. The total energy of the system in the reference frame where the center-of-mass of the whole system is at rest is E.

We expand the total wave function Ψ in the form

$$\Psi = \mathcal{A} \sum_\alpha \Phi_\alpha(\underline{R}, \underline{r}_m)\, \mathcal{F}_\alpha(\underline{r}) + \sum_i \chi_i(\underline{R}, \underline{r}_m, \underline{r})\beta_i \tag{15}$$

The first expansion is a sum over a limited set of target eigenstates and pseudo-states Φ_α satisfying

$$\langle\Phi_\alpha|H_m|\Phi_\beta\rangle = E_\alpha\, \delta_{\alpha\beta} \tag{16}$$

where the pseudo-states may be included to represent additional properties of the target eigenstates such as their polarizability. The second expansion in eq. (15) is a sum over quadratically integrable functions χ_i representing short-range correlation effects. The suffix α represents the rovibrational and electronic state of the molecule as well as the spin state of the scattered electron, and the operator $\mathcal{A}$ antisymmetrizes the first expansion with respect to interchange of the space and spin coordinates of any pair of electrons.

We now derive coupled equations for the functions $\mathcal{F}_\alpha$ and the coefficients β_i by substituting eq.(15) into eq.(12) and projecting onto the functions Φ_α and χ_i. After eliminating the coefficients β_i we obtain the following coupled integrodifferential equations for the $\mathcal{F}_\alpha$:

$$(\nabla^2 + k_\alpha^2)\, \mathcal{F}_\alpha(\underline{r}) = 2 \sum_\beta [V_{\alpha\beta}(\underline{r})\, \mathcal{F}_\beta(\underline{r})$$

$$+ \int \{W_{\alpha\beta}(\underline{r}, \underline{r}') + K_{\alpha\beta}(\underline{r},\underline{r}')\}\, \mathcal{F}_\beta(\underline{r}')d\underline{r}'] \tag{17}$$

where

$$k_\alpha^2 = 2(E - E_\alpha) \tag{18}$$

and where the direct potential

$$V_{\alpha\beta}(\underline{r}) = \iint \Phi_\alpha^*(\underline{R}, \underline{r}_m)V(\underline{R}, \underline{r}_m, \underline{r})\Phi_\beta(\underline{R}, \underline{r}_m)d\underline{R}d\underline{r}_m \tag{19}$$

The nonlocal exchange potential $W_{\alpha\beta}$ arises from the antisymmetrization terms, and the correlation potential $K_{\alpha\beta}$ arises from the quadratically integrable functions. By expanding the $\mathcal{F}_\alpha$ in partial waves we obtain a set of coupled radial equations which can be solved numerically. These were first written down for a diatomic molecule by Arthurs and Dalgarno[5].

The scattering amplitudes and cross sections are obtained by considering the asymptotic form of the total wave function. The asymptotic form corresponding to an electron incident on the target molecule in an initial state Φ_α is

$$\Psi_\alpha \underset{r\to\infty}{\sim} \Phi_\alpha\, e^{ik_\alpha z} + \sum_\beta \Phi_\beta\, f_{\beta\alpha}(\Theta\phi)\, \frac{e^{ik_\beta r}}{r} \tag{20}$$

where $f_{\beta\alpha}(\Theta\phi)$ is the scattering amplitude corresponding to the scattered electron moving after the collision in a direction making polar angles $(\Theta\phi)$ with the incident beam direction and leaving the molecule in a state Φ_β. The differential cross section for the transition from the initial state Φ_α to the final state Φ_β is given by

$$\frac{d\sigma_{\beta\alpha}}{d\Omega} = \frac{k_\beta}{k_\alpha} \left| f_{\beta\alpha}(\Theta\phi) \right|^2 \tag{21}$$

in units of a_0^2/steradian, and the total cross section is obtained by integrating over all scattering angles.

3.2. Molecular-Frame Representation

The theory developed in the previous section is completely general and has been the basis of many calculations for light diatomic molecules. For example, Takayanagi and Geltman[6,7] have used this approach as the basis of pioneering studies of excitation of molecular rotations in H_2 and N_2 by slow electrons. However, a major difficulty arises for complex molecules because of the large number of channels that are coupled after the partial-wave decomposition of eqs. (17) has been carried out.

These difficulties can to a large extent be overcome by adopting an approximation similar to that used in molecular bound-state calculations.

In this case the Born-Oppenheimer separation of the electronic and nuclear motion is adopted. The electronic motion is first determined with the nuclear positions fixed. This is referred to as the fixed-nuclei approximation. The molecular rotational and vibrational motion is then included in a second step of the calculation. This whole procedure is called the adiabatic-nuclei approximation. It owes its validity to the large ratio of the nuclear mass to the electronic mass, and can be adopted in electron-molecule collision processes when the collision time is much shorter than the period of molecular rotation and/or vibration. The approximation is thus expected to be valid when the scattered electron energy is not close to the threshold of any channel, or when the energy is out of the region of any narrow resonances. As we shall see, however, it can be extended to apply to these situations as well.

The fixed-nuclei approximation is not new, having been used to describe scattering of electrons by homonuclear diatomic molecules by Stier[8] and Fisk[9] in the 1930's. However, in the last few years it has gained general acceptance as the basis of ab-initio computational methods that are yielding the most accurate low-energy cross sections for complex molecules.

3.2.1. Basic formulation

In order to formulate the collision in this approximation, we adopt a molecular frame of reference that is rigidly attached to the molecule as illustrated in Fig. 1 for a diatomic molecule. In this figure, A and B are the two nuclei, G is the center of gravity which is taken as the origin of coordinates, and the z axis is chosen to lie along the internuclear axis. The molecular frame of reference can be related to the laboratory frame of reference, discussed in the previous section, by a unitary transformation defined by

$$Y_{\ell m}(\hat{\underline{r}}') = \sum_{m'} D^{\ell}_{mm'}(\alpha\beta\gamma)\, Y_{\ell m'}(\hat{\underline{r}}) \tag{22}$$

where the primed coordinates refer to the laboratory frame, the unprimed coordinates to the molecular frame, and where the molecular frame is oriented in a direction defined by the Euler angles $(\alpha\beta\gamma)$ with respect to the laboratory frame. Also, we have adopted the notation of Rose[10] for the Wigner rotation matrices $D^{\ell}_{mm'}(\alpha\beta\gamma)$.

The fixed-nuclei approximation starts from the Schrödinger equation

$$(H_{e\ell} + T + V)\psi = E\psi \tag{23}$$

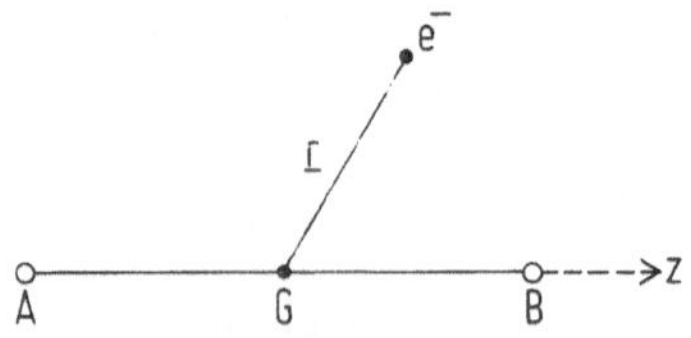

Fig. 1. Molecular frame for a diatomic molecule.

where $H_{e\ell}$ is the electronic part of the Hamiltonian for the target molecule defined by assuming that the target nuclei have fixed coordinates denoted by $\underline{R}$. It follows that $H_{e\ell}$ is related to H_m in eq. (12) by the equation

$$H_m = H_{e\ell} + T_{\underline{R}} \tag{24}$$

where $T_{\underline{R}}$ is the kinetic energy of rotational and vibrational motion of the nuclei. The remaining quantities T and V are defined by eqs. (13) and (14), respectively.

The solution of eq. (23) proceeds in an analogous way to the solution of eq. (12) except that now the nuclear coordinates only appear parametrically in eq. (23) and we have to re-solve the equations for each combination of coordinates $\underline{R}$ of interest. In analogy with eq. (15) we expand the wave function ψ in the form

$$\psi = \mathcal{A} \sum_i \phi_i(\underline{R}, \underline{r}_m)\, F_i(\underline{r}) + \sum_{i'} \chi_{i'}(\underline{R}, \underline{r}_m, \underline{r})\, b_{i'} \tag{25}$$

where the first expansion is a sum over target eigenstates and pseudostates ϕ_i satisfying

$$\langle\phi_i \mid H_{e\ell} \mid \phi_j\rangle = \varepsilon_i \delta_{ij} \tag{26}$$

while the second expansion is a sum over quadratically integrable functions representing short-range correlation effects. The suffix i now represents only the electronic state of the molecule as well as the spin state of the scattered electron. As before $\mathcal{A}$ antisymmetrizes the first expansion with respect to interchange of the space and spin coordinates of any pair of electrons.

We can derive coupled equations for the functions F_i and the coefficients $b_{i'}$ by substituting eq. (25) into eq. (23) and projecting onto the functions ϕ_i and $\chi_{i'}$. After eliminating the coefficients $b_{i'}$ we then obtain the following coupled integrodifferential equations for the functions F_i.

$$(\nabla^2 + k_i^2)\, F_i(\underline{r}) = 2 \sum_j [V_{ij}^{FN}(\underline{r})\, F_j(\underline{r})$$
$$+ \int \{W_{ij}^{FN}(\underline{r}, \underline{r}') + K_{ij}^{FN}(\underline{r}, \underline{r}')\}\, F_j(\underline{r}')d\underline{r}'] \tag{27}$$

where

$$k_i^2 = 2\,(E - \varepsilon_i) \tag{28}$$

and where the fixed-nuclei direct potential

$$V_{ij}^{FN}(\underline{r}) = \int \phi_i^*(\underline{R}, \underline{r}_m)\, V(\underline{R}, \underline{r}_m, \underline{r})\, \phi_j(\underline{R}, \underline{r}_m)d\underline{r}_m \tag{29}$$

which depends parametrically on the nuclear coordinates $\underline{R}$. The nonlocal exchange potential W_{ij}^{FN} and the correlation potential K_{ij}^{FN} also depend parametrically on $\underline{R}$. However, we will not write down explicit expressions for these potentials here.

The final step in order to reduce eqs. (27) to a form that is numerically tractable is to separate out the angular variables. We expand the

$F_i(\underline{r})$ in symmetry-adapted angular functions $X^{p\mu}_{h\ell}(\hat{\underline{r}})$ which transform as appropriate irreducible representations (IR) of the molecular point group[11]

$$F_i(\underline{r}) = \sum_{h\ell} r^{-1} f_{h\ell i}(r) X^{p_i\mu_i}_{h\ell}(\hat{\underline{r}}) \tag{30}$$

where p_i denotes the particular IR, μ_i distinguishes each component of the basis if the IR has dimension greater than one, and h distinguishes different bases of the same IR that corresponds to the same angular momentum ℓ. These angular functions can be expanded in terms of spherical harmonics. Equations (27) then reduce to the coupled radial integrodifferential equations

$$\left(\frac{d^2}{dr^2} - \frac{\ell(\ell+1)}{r^2} + k_i^2\right) f_s(r) = 2 \sum_{s'} [v^{FN}_{ss'}(r) f_{s'}(r)$$

$$+ \int_0^\infty \{w^{FN}_{ss'}(r, r') + k^{FN}_{ss'}(r, r')\} f_{s'}(r')dr'] \tag{31}$$

where, for notational convenience, we have combined the electronic index i and the component $h\ell$ of the IR into one index s. Also the potentials $v^{FN}_{ss'}(r)$, $w^{FN}_{ss'}(r,r')$ and $k^{FN}_{ss'}(r,r')$ are obtained by projecting the corresponding potentials in eqs. (27) onto the angular functions $X^{p\mu}_{h\ell}(\hat{\underline{r}})$.

We now look for solutions of eqs. (31) satisfying the boundary conditions

$$f_{ss'}(0) = 0 \tag{32a}$$

and

$$f_{ss'}(r) \underset{r\to\infty}{\sim} k_i^{-1/2} (\sin\theta_s\, \delta_{ss'} + \cos\theta_s\, K_{ss'}) \tag{32b}$$

where for neutral molecules $\theta_s = k_i r - (1/2)\ell\pi$ and the second index s' on $f_{ss'}$ labels the independent solutions of eqs. (31). These equations define the real symmetric K-matrix $K_{ss'}$ from which the S-matrix can be obtained from the matrix equation

$$\underline{S} = \frac{\underline{1} + i\,\underline{K}}{\underline{1} - i\,\underline{K}} \tag{33}$$

The scattering amplitude in the molecular frame of reference is then given by

$$A_{ii'}(\hat{\underline{k}}\cdot\hat{\underline{r}}) = \frac{2\pi}{ik_i} \sum_{\substack{h\ell h'\ell'\\ p\mu}} X^{p\mu}_{h\ell}(\hat{\underline{k}}) X^{p\mu}_{h'\ell'}(\hat{\underline{r}}) i^{\ell-\ell'}$$

$$\times (S^{p\mu}_{h\ell i,h'\ell' i'} - \delta_{hh'}\, \delta_{\ell\ell'}\, \delta_{ii'}) \tag{34}$$

where $\hat{\underline{k}}$ and $\hat{\underline{r}}$ are the initial and final directions of the scattered electron. The corresponding expression for the scattering amplitude in the laboratory frame of reference can be obtained using the transformation given by eq. (22).

3.2.2. Methods of solution of the fixed-nuclei equations

In this subsection we review recent computational methods used for solving the coupled integrodifferential equations (31). We commence with a brief discussion of local exchange and separable exchange methods. We then consider the linear algebraic equations method, the Schwinger variational method, the complex Kohn variational method, and the R-matrix method which are now giving the most accurate results for small molecules.

3.2.2a. Local exchange potentials If the nonlocal exchange and correlation potentials can be replaced by local potentials, then eqs. (31) reduce to a set of coupled differential equations which are much easier to solve. Most early work on diatomic molecules and nearly all work on polyatomic molecules have used such approximations[2].

One of the most successful and still widely used methods is the Hara free-electron gas exchange (HFEGE) method[12]. In this approach, the total wave function is assumed to be composed of plane wave states which are antisymmetrized in accordance with the Pauli exclusion principle, and the exchange energy is then obtained by summing all states up to the Fermi level. The resultant form of this potential is

$$V_{ex}^{HFEGE}(\underline{r}) = -\frac{2}{\pi} k_F(\underline{r}) \left[\frac{1}{2} + \frac{1-\eta^2}{4\eta} \ln \left| \frac{1+\eta}{1-\eta} \right| \right] \tag{35a}$$

where

$$k_F(\underline{r}) = [3\pi^2 \rho(\underline{r})]^{1/3} \tag{35b}$$

and

$$\eta(\underline{r}) = \frac{[k^2 + 2I + k_F^2(\underline{r})]^{1/2}}{k_F(\underline{r})} \tag{35c}$$

Also, I is the ionization potential and $\rho(\underline{r})$ is the molecular charge distribution. As $r \to \infty$, the numerator of η should be k and not $(k^2 + 2I)^{1/2}$ as in eq. (35c). This has led to the introduction of the asymptotically adjusted form of the HFEGE potential where

$$\eta(\underline{r}) = \frac{[k^2 + k_F^2(\underline{r})]^{1/2}}{k_F(\underline{r})} \tag{36}$$

which is referred to as the AAFEGE model. Alternatively, I is sometimes taken to be an adjustable parameter.

Another approach for defining a local exchange potential is the semiclassical exchange (SCE) approximation. If we carry out a Taylor series expansion about $\underline{r}$, the point in space where the exchange integral is computed, we find that[13]

$$I = \int \phi^*(\underline{r}_1) \mid \underline{r} - \underline{r}_1 \mid^{-1} F(\underline{r}_1)d\underline{r}_1$$

$$= \int \mid \underline{r}' \mid^{-1} \exp[(\nabla_\phi + \nabla_F)\cdot\underline{r}'] \, \phi^*(\underline{r}) \, F(\underline{r})d\underline{r}' \tag{37}$$

where $\underline{r}' = \underline{r}_1 - \underline{r}$ and the ∇ operators act either on the bound-electron wave function ϕ or on the continuum function F depending on their subscripts. Using spherical polar coordinates we can evaluate the integral giving

$$I = - \frac{4\pi}{|\nabla_\phi + \nabla_F|^2} \phi^*(\underline{r}) \, F(\underline{r}) \tag{38}$$

For high-energy collisions, the bound functions are slowly varying compared with the continuum functions, and hence we can disregard the ∇_ϕ operator compared with the ∇_F operator. We then obtain the following expression for the SCE potential.

$$V_{ex}^{SCE}(\underline{r}) = \frac{1}{2} [2k^2 - V_{st}(\underline{r})] - \frac{1}{2} \{[2k^2 - V_{st}(\underline{r})]^{1/2} + 4\pi\rho(\underline{r})\}^{1/2} \tag{39}$$

where we have summed over all bound orbitals and we note that $V_{st}(\underline{r})$ is the static potential and $\rho(\underline{r})$ the charge distribution of the molecule. The above SCE potential is strictly valid only for high-energy collisions. However, some work has been carried out by Gianturco and Scialla[14] to extend its validity to lower energies by including more correctly the effects of the bound electron momenta on the scattered electron starting from the full form of eq. (38). The SCE potential has been successfully used for collisions involving diatomic and polyatomic molecules.

Finally, we remark that in order to obtain accurate results using local exchange potentials it is necessary to ensure that the scattered functions in eqs. (31) are made orthogonal to the fully occupied bound orbitals of the same symmetry. This can be achieved by adding additional Lagrange orthogonalization terms into the coupled equations as suggested by Burke and Chandra[15].

3.2.2b. Separable exchange potentials. The large computational effort necessary to include the exchange and correlation terms in eqs. (31) is mainly a result of their non-separable character rather than their non-locality. It was pointed out by Rescigno and Orel[16] and by Schneider and Collins[17] that representing the exchange kernel $W_{ij}(\underline{r}, \underline{r}')$ by the expansion

$$W_{ij}(\underline{r}, \underline{r}') = \sum_\alpha \sum_\beta \chi_\alpha(\underline{r}) \, U_{\alpha\beta} \, \chi_\beta(\underline{r}') \tag{40}$$

where the $\{\chi_\alpha\}$ is an orthonormal basis set of Gaussian or Slater functions, enables the equations to be solved very much more rapidly. Such approaches have been widely used in recent years.

3.2.2c. Linear algebraic equations method. The linear algebraic (LA) method is the first of several methods that we now discuss which are yielding accurate ab-initio results. The LA method, which is similar in spirit to the approach used in nuclear collisions by Robertson[18] and electron-atom collisions by Seaton[19], was introduced for electron-molecule collisions by Collins and Schneider[17,20]. It starts from eqs. (31) which are written in integral form as the matrix equation

$$\underline{f}(r) = \underline{G}^1(r) - \int_0^\infty \underline{G}^0(r,r') \int_0^\infty \underline{U}(r',r'') \underline{f}(r'')dr'dr'' \tag{41}$$

where the potential $\underline{U}$ includes the direct, exchange, and correlation terms in eqs. (31) and the Green's functions are defined by

$$\underline{G}^0(r,r') = \begin{cases} \underline{G}^1(r)\, \underline{G}^2(r') & r < r' \\ \underline{G}^2(r)\, \underline{G}^1(r') & r > r' \end{cases} \tag{42}$$

with the diagonal matrices

$$\underline{G}^1(r) = \underline{k}r\, j_{\underline{\ell}}\,(\underline{k}r) \underset{r\to\infty}{\sim} \sin(\underline{k}r - \tfrac{1}{2}\,\underline{\ell}\pi) \tag{43a}$$

and

$$\underline{G}^2(r) = -\, r\, n_{\underline{\ell}}\,(\underline{k}r) \underset{r\to\infty}{\sim} \underline{k}^{-1} \cos(\underline{k}r - \tfrac{1}{2}\,\underline{\ell}\pi) \tag{43b}$$

The functions $j_{\underline{\ell}}(\underline{k}r)$ and $n_{\underline{\ell}}(\underline{k}r)$ are spherical Bessel and Neumann functions. The integrals in eq. (41) are replaced by quadratures out to a radius r = a beyond which the direct potential achieves its asymptotic form and the exchange and correlation potentials can be neglected. Equation (41) then reduces to a set of linear algebraic equations for the values of the functions $\underline{f}(r)$ at the quadrature points where the order of the matrix is equal to the number of channels times the number of quadrature points. These equations can be solved by an iteration-variation method which avoids storage of the full matrix. The method is very fast and in particular maps efficiently onto the current generation of vector supercomputers.

3.2.2d. The Schwinger variational method. This is the second of several methods now yielding accurate ab-initio results. It has been developed and applied to multichannel electron-molecule collisions by McKoy and co-workers[21,22]. The method starts from the Lippmann-Schwinger integral equations obtained from eq. (23), viz.,

$$\mathcal{T} = V + V\, G_0^+\, \mathcal{T} \tag{44}$$

where V is the total electron-molecule interaction defined by eq. (14) and $\mathcal{T}$ is the transition operator. The potential is first written in the separable form

$$V = \sum_{\alpha\beta} V\,|\,\chi_\alpha > [<\chi_\alpha\,|\,V\,|\,\chi_\beta>]^{-1} < \chi_\beta\,|\,V \tag{45}$$

where the $\{\chi_\alpha\}$ are discrete basis functions. The Schwinger variational functional for the T-matrix is then given by[23]

$$< \underline{k}_f | T | \underline{k}_i > = \sum_{\alpha\beta} < \underline{k}_f | V | \chi_\alpha > [D^{-1}]_{\alpha\beta} < \chi_\beta | V | \underline{k}_i > \quad (46)$$

where $[D^{-1}]_{\alpha\beta}$ is the inverse of the matrix with elements

$$D_{\alpha\beta} = < \chi_\alpha | V - V G_0^+ V | \chi_\beta > \quad (47)$$

and where $|\underline{k}_i>$ and $|\underline{k}_f>$ are plane wave states. An important feature of the method is that the discrete basis functions $\{\chi_\alpha\}$ only appear in conjunction with the potential V. Hence these functions need only have the range of the potential.

The method has been refined through an iterative prescription in which numerical oscillating functions are used to augment the discrete basis. This gives better convergence when long-range potentials are important in the collision.

3.2.2e. The complex Kohn variational method. For many years the Kohn variational method for the K-matrix has been used to describe electron-atom collisions[24]. However, a major difficulty was the appearance of spurious singularities in the solution. Within the last two years it has been shown by Miller and Jansen op de Haar[25] that these singularities do not occur if the Kohn variational method for the T-matrix is used instead. This has opened up the possibility of using it as an accurate and efficient ab-initio method to calculate electron-molecule collision cross sections, and already applications to low-energy electron-molecule collisions have been made[26].

The method starts by writing the solution of eqs. (31) in the form

$$f_{ss'} = [s_s(r)\, \delta_{ss'} + e_s(r)\, T_{ss'}] + \sum_i \chi_i(r)\, b_{iss'} \quad (48)$$

where as in eq. (32) the second index s' on $f_{ss'}$ labels the independent solutions of eqs. (31), χ_i are square-integrable functions, and s_s and e_s are chosen to go to zero at the origin as $r^{\ell+1}$ and to behave asymptotically as

$$s_s(r) \underset{r\to\infty}{\sim} k_i^{-1/2} \sin(k_i r - \tfrac{1}{2}\ell\pi)$$

$$e_s(r) \underset{r\to\infty}{\sim} k_i^{-1/2} \exp[i(k_i r - \tfrac{1}{2}\ell\pi)] \quad (49)$$

The Kohn variational principle then requires that the following functional is stationary

$$[\, \underline{T}\,] = \underline{T} + \int \underline{f}^T \underline{\mathcal{L}}\, \underline{f}\, dr \quad (50)$$

where $\underline{\mathcal{L}}$ is the operator in eqs. (31) in which all the terms are taken onto the left-hand side of the equation. Variations of eq. (50) with respect to the variables $T_{ss'}$ and $b_{iss'}$ then yield linear algebraic equations from which the stationary value of the T-matrix can be obtained.

These equations do not involve any spurious singularities and can again be efficiently solved on the current generation of vector supercomputers.

3.2.2f. The R-matrix method. This is the fourth approach that is giving accurate results. It was developed as an ab-initio method for electron-atom collisions by Burke et al.[27,28] and for electron-molecule collisions by Schneider[29,30] and by Burke et al.[31]. The method proceeds by partitioning the interaction region into an internal and an external region by a sphere of radius r = a which is chosen to just envelope the charge distribution of the molecular states of interest. This is illustrated in Fig. 2.

In the internal region the interaction between the electron and the molecule is strong, multicentered, and nonlocal, and a CI expansion of the wave function analogous to eq. (25) is used, viz.,

$$\psi_k = \mathcal{A} \sum_{si} \bar{\Phi}_s u_i(r) a_{sik} + \sum_{i'} \chi_{i'} b_{i'k} \tag{51}$$

The channel functions $\bar{\phi}_s$ are obtained by coupling the target states ϕ_i with the angular functions of the scattered electron. The u_i are radial basis functions defined over the range $0 \leq r \leq a$ and the $\chi_{i'}$ are quadratically integrable functions. The coefficients a_{sik} and $b_{i'k}$ are obtained by diagonalizing

$$\langle\psi_k|H_{e\ell} + T + V + L_b|\psi_{k'}\rangle = E_k \, \delta_{kk'} \tag{52}$$

where the integral is over the internal region and L_b is a surface operator introduced by Bloch[32] and defined by

$$L_b = \frac{1}{2} \sum_s \bar{\Phi}_s > \delta(r - a)\left(\frac{d}{dr} - \frac{b}{r}\right) < \bar{\Phi}_s| \tag{53}$$

The parameter b is arbitrary and ensures that $H_{e\ell} + T + V + L_b$ is Hermitian. The Schrödinger equation (23) can now be solved in the internal region by expanding the solution in terms of the basis ψ_k. We obtain the formal solution

$$\psi = (H_{e\ell} + T + V + L_b)^{-1} L_b\psi \tag{54}$$

which can be written as

$$|\psi\rangle = \sum_k \frac{|\psi_k\rangle\langle\psi_k|L_b|\psi\rangle}{E_k - E} \tag{55}$$

Projecting this equation onto the channel functions $\bar{\phi}_s$ and evaluating on the boundary of the internal region then yields the matrix equation

$$\underline{f}(a) = \underline{R}\left(r \frac{d\,\underline{f}}{dr} - b\,\underline{f}\right)_{r=a} \tag{56}$$

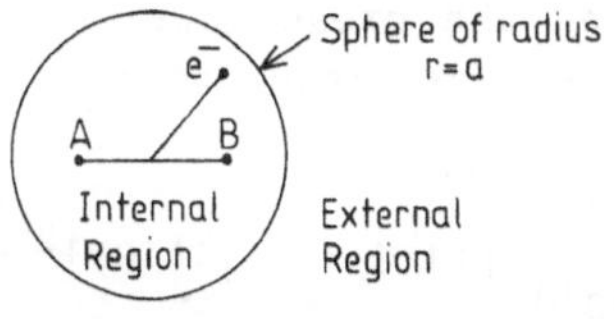

Fig. 2. Partitioning of the interaction region in the R-matrix method.

where the R-matrix, which has the dimension of the number of channels retained in the expansion, is defined by

$$R_{ss'} = \frac{1}{2a} \sum_k \frac{w_{sk}\, w_{s'k}}{E_k - E} \tag{57}$$

with the surface amplitudes w_{sk} given by

$$w_{sk} = \sum_i u_i(a)\, a_{sik} \tag{58}$$

Equations (56) and (57) are the basic equations of the R-matrix method. They give the logarithmic derivatives of the radial channel functions f_s on the surface of the internal region at any energy E in terms of the surface amplitudes w_{sk} and eigenenergies E_k. These in turn are obtained by a single diagonalization defined by eq. (52). The K-matrix and S-matrix are then easily obtained by matching directly to the asymptotic solutions given by eq. (32b) in the external region, if the potential in this region is negligible. If it is not, then eqs. (31) are solved in the external region with the boundary condition (56) out to a point where the solutions have attained the asymptotic form (32b). This can easily be done because in the external region the potential is local and can be represented by a simple asymptotic expansion form.

3.2.3. Inclusion of the Nuclear Motion

We discuss briefly in this section how observables involving the nuclear motion, such as rotational and vibrational excitation cross sections and dissociative attachment cross sections, can be extracted from solutions of the fixed-nuclei equations.

The most widely used approach is the adiabatic-nuclei approximation proposed by Drozdov[33] and by Chase[34] in studies of neutron scattering by nuclei. In the case of diatomic molecules in a $^1\Sigma$ state the relevant scattering amplitude can be written as

$$A_{ivjm,i'v'j'm'}(\hat{\underline{k}}\cdot\hat{\underline{r}}) = \langle \chi_{iv}(R)\, Y_{jm}(\hat{\underline{R}}) | A_{ii'}(\hat{\underline{k}}\cdot\hat{\underline{r}};\, \underline{R}) | \chi_{i'v'}(R)\, Y_{j'm'}(\hat{\underline{R}}) \rangle \tag{59}$$

where $A_{ii'}(\hat{\underline{k}}\cdot\hat{\underline{r}};\underline{R})$ is the fixed-nuclei scattering amplitude defined by eq. (34), which depends parametrically on the nuclear coordinates $\underline{R}$, and $\chi_{iv}(R)$ and $Y_{jm}(\hat{\underline{R}})$ are the molecular vibrational and rotational eigenfunctions. This approximation is valid provided that the collision time is short compared with the vibration and/or rotation time and is widely used in such situations.

The cross section is usually averaged over the degenerate sublevels m and summed over m' giving the cross section for the transition ivj to i'v'j'. The form of eq. (59) then leads to the relation[33,35]

$$\frac{d\sigma_{ivj,i'v'j'}}{d\Omega} = \sum_{j_t=|j-j'|}^{j+j'} [(j\ 0\ j_t\ 0|j'\ 0)]^2 \frac{d\sigma_{iv0,i'v'j_t}}{d\Omega} \tag{60}$$

regardless of the form of the interaction potential provided we can neglect the small differences in the wavenumbers for different rotational channels. The quantities $(j\ 0\ j_t0|j'0)$ in eq. (60) are the usual Clebsch-Gordan coefficients[36]. A similar relation follows for symmetric-top molecules such as NH_3. For spherical-top molecules such as CH_4 the relation is[37]

$$\frac{d\sigma_{ivj,i'v'j'}}{d\Omega} = \frac{2j'+1}{2j+1} \sum_{j_t=|j-j'|}^{j+j'} \frac{1}{2j_t+1} \frac{d\sigma_{iv0,i'v'j_t}}{d\Omega} \tag{61}$$

The sum in eq. (60) or in eq. (61) over the final rotational state j' is independent of the initial state j. Also, if we multiply the cross section by the transition energy and then sum over j', the result, which is the mean energy loss by the incident electron, is still independent of j. These sum rules may be proved by using a relation for the Clebsch-Gordan coefficients[37]. A more general operator technique for deriving them also applies to sum rules for asymmetric-top molecules[38].

The adiabatic-nuclei approximation breaks down close to threshold or in the neighborhood of narrow resonances as discussed by Morrison[39]. One of the most straightforward ways of including nonadiabatic effects that arise in vibrational excitation is to retain the vibrational term in the Hamiltonian but to still treat the rotational motion adiabatically. Hence, instead of eq. (23) we now solve the equation

$$(H_{e\ell} + T_{vib} + T + V)\, \tilde{\Psi} = E\, \tilde{\Psi} \tag{62}$$

where T_{vib} is the kinetic energy operator for the nuclear motion and where the other quantities have the same meaning as in eq. (23). Instead of using eq. (25) we now expand the wave function $\tilde{\Psi}$ in the form

$$\tilde{\Psi} = \mathcal{A} \sum_{iv} \phi_i(\underline{R}, \underline{r}_m)\, \chi_{iv}(R)\, G_{iv}(r) + \sum_{i'} \eta_{i'}(\underline{R}, \underline{r}_m, \underline{r})\, c_{i'} \tag{63}$$

where we adopt a frame of reference in which the molecule has a fixed spatial orientation and where $\chi_{iv}(R)$ are the molecular vibrational eigenfunctions. We can then derive coupled integrodifferential equations for the functions G_{iv}, where the potentials now couple the target vibrational as well as the electronic states. This approach has been adopted by Chandra and Temkin[40,41] but it is computationally demanding since the number of coupled channels can become very large.

Other approaches that are particularly appropriate for describing vibrational excitation and dissociative attachment are based on resonance theories of electron-molecule collisions. Such approaches have been pursued particularly by Herzenberg and collaborators[42-44], by Bardsley[45], and more recently by Domcke et al.[46-48]. The basic idea is that we introduce a series of fixed-nuclei resonance states ψ_n^r as functions of $\underline{R}$ either by imposing resonance boundary conditions[49] or by introducing Feshbach projection operators[50]. The amplitude for a transition from an initial electronic-vibrational state iv to a final state i'v' is then given by

$$T_{iv,i'v'} = \sum_n \langle \chi_{iv}(\underline{R})\ \zeta_{ni}(\underline{R}) | G_n^r(\underline{R}, \underline{R}') | \zeta_{ni'}(\underline{R}')\chi_{i'v'}(\underline{R}')\rangle \qquad (64)$$

where $\chi_{iv}(\underline{R})$ are the vibrational eigenfunctions, $\zeta_{ni}(\underline{R})$ are the "entry amplitudes" from the resonant states ψ_n^r into the initial or final electronic states and $G_n^r(\underline{R}, \underline{R}')$ is the Green's function which describes the propagation in the intermediate resonant state ψ_n^r. Dissociative attachment cross sections can be obtained by a straightforward extension of the above theory. The incident electron is captured as before into the intermediate resonance state. The nuclei then move apart along the resonance potential curve. If they separate to the point where the resonance curve crosses the neutral molecule curve before the resonance decays into the final state, then dissociative attachment occurs. Otherwise vibrational excitation occurs.

The fixed-nuclei R-matrix method has also been extended to treat vibrational excitation and dissociative attachment by Schneider et al.[51]. A generalized R-matrix is introduced by an equation, which can be written in analogy with eq. (64), as

$$\mathcal{R}_{iv,i'v'} = \frac{1}{2a} \sum_k \langle \chi_{iv}(\underline{R})\ w_{ik}(\underline{R}) | G_k^{RM}(\underline{R}, \underline{R}') | w_{i'k}(\underline{R}')\ \chi_{i'v'}(\underline{R}')\rangle \qquad (65)$$

where the surface amplitudes $w_{ik}(\underline{R})$ are defined by eq. (58) and the Green's function $G_k^{RM}(\underline{R}, \underline{R}')$ now describes the propagation in the intermediate R-matrix state ψ_k defined by eq. (51). Once the generalized R-matrix has been determined, the final step in the calculation is to solve the collision problem in the external region including all relevant electronic and ro-vibrational channels.

Finally we mention an energy-modified adiabatic approximation introduced by Nesbet[52]. In this approach the S-matrix elements connecting the vibrational states are defined by

$$S_{iv,i'v'} = \langle \chi_{iv} | S_{ii'}(E - H_i;R) | \chi_{i'v'}\rangle \qquad (66)$$

where $S_{ii'}(E - H_i;R)$ is the S-matrix calculated in the fixed-nuclei approximation at the internuclear separation R at an energy defined by the operator $H_i = E_i(R) + T_{vib}$. This has the effect of including the internal energy of the target into the S-matrix elements giving threshold energies in the correct locations.

4. THEORY OF MOLECULAR PHOTOIONIZATION

In this section we briefly review the theory of molecular photoionization describing processes denoted by eqs. (9), (10), and (11). We adopt the fixed-nuclei approximation in which we assume that the molecule is fixed in space during the photoionization process. We introduce a laboratory frame of reference in which the z' axis is along the photon polarization direction which is assumed linear. We then assume that the molecular frame, in which the electron-molecule calculation is carried out, is oriented with respect to the laboratory frame by the Euler angles ($\alpha\ \beta\ \gamma$). We illustrate this in Fig. 3.

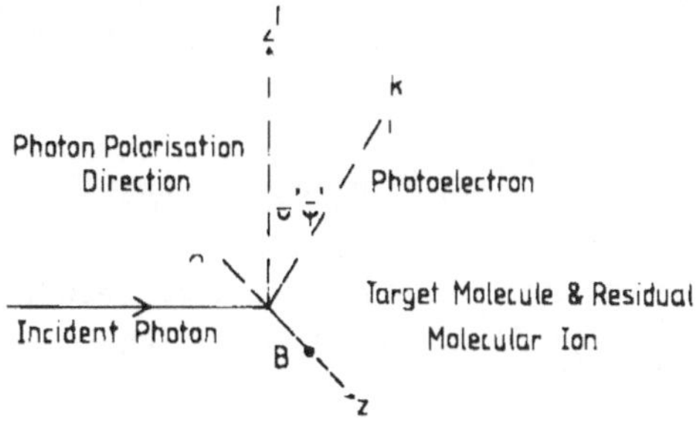

Fig. 3. Frames for molecular photoionization process.

The cross section for photoionization leaving the molecular ion in a final electronic state denoted by j with the photoelectron ejected in the direction $\hat{\underline{k}}'$ is given in the dipole length approximation by

$$\frac{d\sigma_j^L}{d\Omega'} = 4\pi^2 \alpha a_0^2 \omega |<\psi_{jE}^{(-)}(\hat{\underline{k}}')| \hat{\underline{\varepsilon}} \cdot \underline{D}^L | \psi_0 >|^2 \tag{67}$$

In addition, calculations can be carried out in the dipole velocity or dipole acceleration approximation (e.g., see Burke[53]). In eq. (67), α is the fine structure constant, a_0 is the Bohr radius of the hydrogen atom, ω is the incident photon energy in atomic units, $\hat{\underline{\varepsilon}}$ is the photon polarization direction, and $\underline{D}^L$ is the dipole length operator which has the spherical components

$$D_\mu^L = \left(\frac{4\pi}{3}\right)^{1/2} \sum_i r_i Y_{1\mu}(\hat{\underline{r}}_i') \tag{68}$$

where the summation is over all the electronic coordinates of the target molecule. Finally, ψ_0 and $\psi_{jE}^{(-)}$ are the initial bound state and final continuum state of the molecule satisfying the normalization conditions

$$<\psi_0|\psi_0> = 1 \tag{69}$$

and

$$<\psi_{jE}^{(-)}|\psi_{j'E'}^{(-)}> = \delta_{jj'} \delta(E - E') \tag{70}$$

where the boundary condition satisfied by $\psi_{jE}^{(-)}$ corresponds to a Coulomb-modified plane wave in direction $\hat{\underline{k}}'$ incident on the residual ion in an electronic state denoted by j together with ingoing waves in all open channels.

We now relate the final states in eq. (67) to the wave function describing the corresponding electron-molecular ion collision state defined by eq. (25). We find in the molecular frame of reference that

$$\psi_E^{(-)}(\underline{k}) = \sum_{\substack{h\ell h'\ell' \\ p\mu}} i^{\ell+1} \exp(-i\sigma_{\ell'}) X_{h'\ell'}^{p\mu}(\underline{k}) \psi_{ih\ell}^{p\mu} \tag{71}$$

where $\psi_{ih\ell}^{p\mu}$ is defined by eqs. (25), (30), and (31) and where the radial functions $f_{ss'}$ satisfy the boundary conditions

$$f_{ss'}(0) = 0 \tag{72a}$$

and

$$f_{ss'}(r) \underset{r\to\infty}{\sim} (2\pi k_i)^{-1/2} [\exp(-i\theta_s)S^*_{ss'} - \exp(i\theta_s)\delta_{ss'}] \tag{72b}$$

As in eqs. (31), the quantum numbers $ih\ell$ have been denoted by the single index s, the quantity $S^*_{ss'}$ is the complex conjugate of an S-matrix element $S_{ss'}$, and

$$\begin{aligned} \theta_s &= k_i r - \frac{1}{2}\ell\pi - \eta_i \ln 2k_i r + \sigma_s \\ \eta_i &= -z/k_i \\ \sigma_s &= \arg \Gamma(\ell + 1 + i\,\eta_i) \end{aligned} \tag{73}$$

where z is the charge on the residual molecular ion.

Our final step in the evaluation of eq. (67) is to transform the dipole length operator from the laboratory frame to the molecular frame using eq. (22). We find after some algebra[54] that after averaging over all orientations of the molecular axis

$$\left(\frac{d\sigma_j}{d\Omega}\right)_{Av} = \frac{\sigma_j}{4\pi}[1 + \beta_j\, P_2(\cos\Theta')] \tag{74}$$

where σ_j is the total photoionization cross section and β_j is the asymmetry parameter. Explicit expressions for these quantities have been given for linear molecules by Burke[54].

Once we have determined the solution of the fixed-nuclei equations using one of the methods described in Section 3.2.2 we can determine the matrix element in eq. (67) and hence the photoionization cross section. In practice it is often convenient and more accurate to determine both ψ_0 and $\psi^{(-)}_{jE}$ using the same approximation. In other words ψ_0 is a bound state solution of eqs. (31) where all channels are closed, while $\psi^{(-)}_{jE}$ corresponds to a scattering solution of these equations as discussed above.

If the time for emitting the photoelectron is much shorter than the vibrational/rotational time, an adiabatic-nuclei approximation analogous to eq. (59) may be used to calculate the cross section for photoionization in which the vibrational/rotational states of the initial neutral molecule and of the final molecular ion are specified. The asymmetry parameter is different for different vibrational and rotational transitions. Because the amplitude for photoionization is expressible in the form of eq. (59) in this approximation, the relations (60) and (61) and the sum rules mentioned below these equations apply equally well to photoionization[37,38]. Hence, not all the asymmetry parameters for different rotational transitions are independent.

5. ILLUSTRATIVE EXAMPLES

In this section we present two examples of recent work which illustrate the theory presented in the previous sections.

(i) $e^- - H_2^+$ Collisions

Careful calculations of $e^- - H_2^+$ collisions with a fine mesh of collision energies yield series of resonances. The R-matrix method is particularly suited for such calculations, because of the extremely short computation time at each energy even for very accurate and detailed treatment of the process. Thus, Shimamura et al.[55] have recently found many resonances in the $^1\Sigma_g^+$ symmetry at each of many fixed internuclear distances R. They are classified into two doubly excited Rydberg series $(1\sigma_u)(np\sigma)$ and $(1\sigma_u)(nf\sigma)$. The complex resonance energies $E_r - i\Gamma/2$ may be written in terms of the quantum defect $\mu(n)$ and the threshold energy E_{th} for excitation as $E_{th} - E_r = (1/2)(n-\mu)^{-2}$ and $\Gamma = 2\gamma(n-\mu)^{-3}$. The quantity γ may be regarded as the imaginary part of the complex quantum defect $\mu + i\gamma$ when $(\gamma/n)^2 << 1$. Figure 4 shows examples of the real and imaginary parts of the quantum defect as functions of n. Singly excited bound Rydberg series have also been calculated to a good accuracy using the same R-matrix method but with decaying asymptotic boundary conditions.

These results are now being used to predict the cross section for dissociative recombination

$$e^- + H_2^+ \rightarrow H + H^* \tag{75}$$

which proceeds through these intermediate resonance states at low impact energies.

(ii) e^- - HF Collisions

Sharp threshold peaks were observed in the vibrational excitation cross sections for electron scattering by hydrogen halide molecules by Rohr and Linder[56]. These peaks have been the subject of numerous theoretical papers (see, for example, the recent review by Morrison[39]).

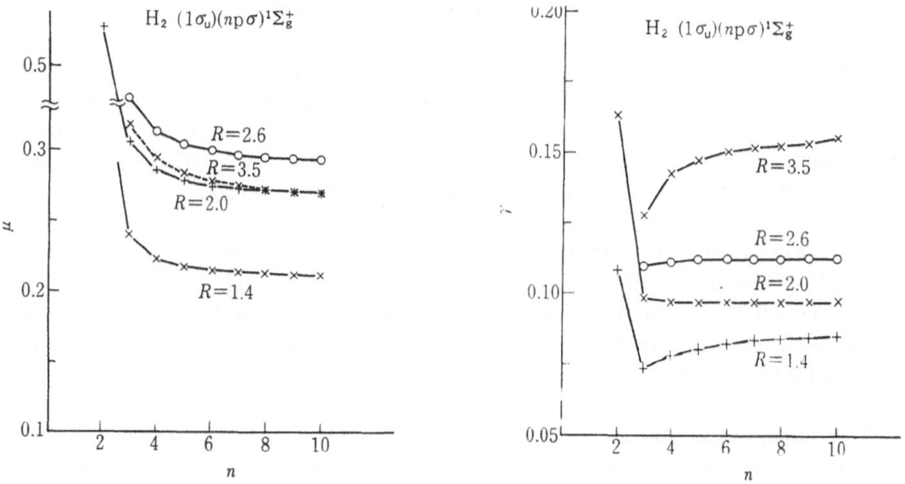

Fig. 4. Real (μ) and imaginary (γ) parts of the quantum defect for the H_2 resonance series $(1\sigma_u)(np\sigma)$ $^1\Sigma_g^+$ as a function of n.

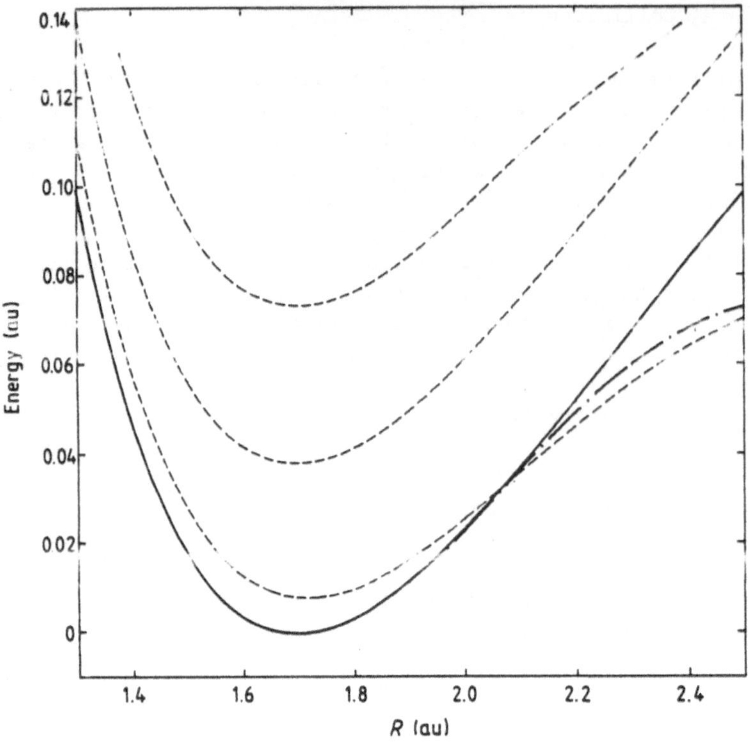

Fig. 5. Potential energy curves of HF and HF^-. The energies are given in a.u. relative to the energy of HF at equilibrium separation. Full curve, SCF $^1\Sigma$ ground state of HF; chain curve, $^2\Sigma$ bound state of HF^-; broken curves, R-matrix pole positions for $^2\Sigma$ symmetry (Figure 1 from Morgan and Burke[57]).

In a recent calculation by Morgan and Burke[57] based on the R-matrix method the collision wave function was represented by eq. (51) where the SCF target wave function given by McClean and Yoshimine[58] was included in the first expansion and 2-particle 1-hole (2p-1h) configurations generated by excitation of the target electrons into the virtual orbitals obtained in the SCF diagonalization were included in the second expansion. The inclusion of the 2p-1h configurations allow for short-range polarization effects in the collision. The vibrational matrix is then calculated using the nonadiabatic R-matrix theory given by eq. (65).

The fixed-nuclei electronic R-matrix poles for $^2\Sigma$ symmetry are plotted in Fig. 5 which also shows the SCF $^1\Sigma$ ground state and the HF^- bound state curves. It is found that the HF^- state is bound at all internuclear separations although for $R \leq 1.65$ a.u. the HF and HF^- curves are almost indistinguishable.

The integrated cross sections for the $v = 0 \rightarrow 1$ transition given by this theory is presented in Fig. 6 where it is compared with model calculations of Gauyacq[59] and the experiments of Rohr and Linder[56]. The peaks in the measurements of Rohr and Linder are broader than the R-matrix theory but the overall shapes of the cross sections and the

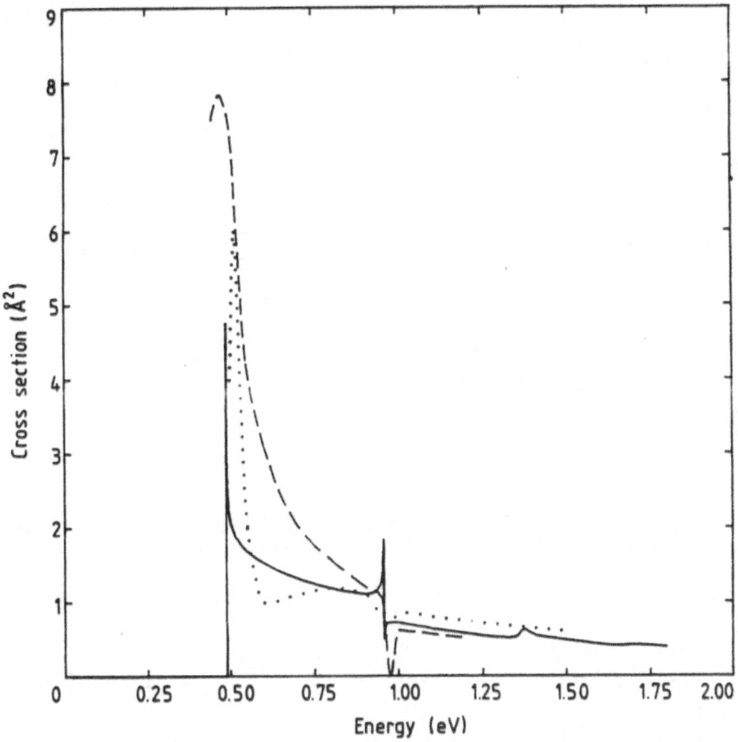

Fig. 6. Integrated cross sections for excitation of the v = 1 level in e^- - HF collisions. Full curve, R-matrix theory; broken curve, Gauyacq[59]; dotted curve, Rohr and Linder[56]. (Figure 6 from Morgan and Burke[57]).

magnitude of the background are in reasonable agreement. The v = 0 → 1 cross sections obtained from the R-matrix method also show marked anisotropy, but only at energies very close to threshold, which is in agreement with recent measurements of Knoth et al.[60]. In general the results are consistent with a "nuclear-excited Feshbach resonance" model proposed by Gauyacq and Herzenberg[61].

6. CONCLUSIONS

The theory of electron- and photon-molecule collisions, coupled with modern computational methods for solving the fixed-nuclei equations, is now capable of yielding accurate cross sections for diatomic molecules. Increasingly, the emphasis is now turning to collisions involving poly-atomic molecules where there is still much work to be done before reliable calculations can be carried out. This field, which owes so much to Takayanagi's pioneering work, is now one of the most active, stimulating and important areas of atomic and molecular collisions.

REFERENCES

1. K. Takayanagi, Prog. Theor. Phys. Suppl. 40, 216 (1967).
2. K. Takayanagi and Y. Itikawa, Adv. Atom. Mol. Phys. 6, 105 (1970).
3. K. Takayanagi, in Electron-Molecule Collisions (I. Shimamura and K. Takayanagi, eds.), Plenum Press, New York and London, Chap.1, Sec.3.2 (1984).
4. I. Shimamura and K. Takayanagi, Electron-Molecule Collisions, Plenum Press, New York and London (1984).
5. A.M. Arthurs and A. Dalgarno, Proc. Roy. Soc. A304, 465 (1960).
6. K. Takayanagi and S. Geltman, Phys. Rev. 138, A1003 (1965).
7. S. Geltman and K. Takayanagi, Phys. Rev. 143, 25 (1966).
8. H.C. Stier, Z. Phys. 76, 439 (1932).
9. J.B. Fisk, Phys. Rev. 49, 167 (1936).
10. M.E. Rose, Elementary Theory of Angular Momentum, John Wiley, New York (1957).
11. P.G. Burke, N. Chandra, and F.A. Gianturco, J. Phys. B: (Atom. Molec. Phys.) 5, 2212 (1972).
12. S. Hara, J. Phys. Soc. Japan 22, 710 (1967).
13. J.B. Furness and I.E. McCarthy, J. Phys. B: (Atom. Molec. Phys.) 6, 2280 (1973).
14. F.A. Gianturco and S. Scialla, in Electron Molecule Scattering and Photoionization (P.G. Burke and J.B. West, eds.), Plenum Press, New York (1988).
15. P.G. Burke and N. Chandra, J. Phys. B: (Atom. Molec. Phys.) 5, 1696 (1972).
16. T.N. Rescigno and A.E. Orel, Phys. Rev. A23, 1134 (1981).
17. B.I. Schneider and L.A. Collins, Phys. Rev. A24, 1264 (1981).
18. H.H. Robertson, Proc. Camb. Phil. Soc. 52, 538 (1956).
19. M.J. Seaton, J. Phys. B: (Atom. Molec. Phys.) 7, 1817 (1974).
20. L.A. Collins and B.I. Schneider, Phys. Rev. A24, 2387 (1981).
21. K. Takatsuka and V. McKoy, Phys. Rev. A24, 2473 (1981).
22. K. Takatsuka and V. McKoy, Phys. Rev. A30, 1734 (1984).
23. S.K. Adhikari and I.H. Sloan, Phys. Rev. C11, 1133 (1975).
24. R.K. Nesbet, Variational Methods in Electron-Atom Scattering Theory, Plenum Press, New York and London (1980).
25. W.H. Miller and B.M.D.D. Jansen op de Haar, J. Chem. Phys. 86, 6213 (1987).
26. B.I. Schneider and T.N. Rescigno, Phys. Rev. A37, 3749 (1988).
27. P.G. Burke, A. Hibbert and W.D. Robb, J. Phys. B: (Atom. Molec. Phys.) 4, 153 (1971).
28. P.G. Burke and W.D. Robb, Adv. Atom. Molec. Phys. 11, 143 (1975).
29. B.I. Schneider, Chem. Phys. Lett. 31, 237 (1975).
30. B.I. Schneider, Phys. Rev. A11, 1957 (1975).
31. P.G. Burke, I. Mackey, and I. Shimamura, J. Phys. B: (Atom. Molec. Phys.) 10, 2497 (1977).
32. C. Bloch, Nucl. Phys. 4, 503 (1957).
33. S.I. Drozdov, Sov. Phys. JETP 1, 591 (1955); Sov. Phys. JETP 3, 759 (1956).
34. D.M. Chase, Phys. Rev. 104, 838 (1956).
35. I. Shimamura, in Electron-Molecule Collisions (I. Shimamura and K. Takayanagi, eds.), Plenum Press, New York and London, Chap. 2 (1984).
36. A.R. Edmonds, Angular Momentum in Quantum Mechanics, Princeton University Press (1957).
37. I. Shimamura, Chem. Phys. Lett. 73, 328 (1980); J. Phys. B: (Atom. Molec. Phys.) 15, 93 (1982).
38. I. Shimamura, Phys. Rev. A23, 3350 (1981).
39. M.A. Morrison, Adv. Atom. Molec. Phys. 24, 51 (1987).
40. N. Chandra and A. Temkin, Phys. Rev. A13, 188 (1976).
41. N. Chandra and A. Temkin, Phys. Rev. A14, 507 (1976).
42. A. Herzenberg and F. Mandl, Proc. Roy. Soc. A270, 48 (1962).

43. A. Herzenberg, J. Phys. B: (Atom. Molec. Phys.) 1, 548 (1968).
44. L. Dubé and A. Herzenberg, Phys. Rev. A20, 194 (1979).
45. J.N. Bardsley, J. Phys. B: (Atom. Molec. Phys.) 1, 349, 365 (1968).
46. W. Domcke and L.S. Cederbaum, J. Phys. B: (Atom. Molec. Phys.) 13, 2829 (1980).
47. C. Münsdel and W. Domcke, J. Phys. B: (Atom. Molec. Phys.) 17, 3593 (1984).
48. M. Berman and W. Domcke, Phys. Rev. A29, 2485 (1984).
49. A.J.F. Siegert, Phys. Rev. 56, 750 (1939).
50. H. Feshbach, Ann. Phys. (N.Y.) 19, 287 (1962).
51. B.I. Schneider, M. Le Dourneuf, and P.G. Burke, J. Phys. B: (Atom. Molec. Phys.) 12, L365 (1979).
52. R.K. Nesbet, Phys. Rev. A19, 551 (1979).
53. P.G. Burke, in Atomic Processes and Applications (P.G. Burke and B.L. Moiseiwitsch, eds.), North-Holland, Amsterdam, p.199 (1976).
54. P.G. Burke, in Atomic and Molecular Collision Theory (F.A. Gianturco, ed.), Plenum Press, New York, p.69 (1982).
55. I. Shimamura, C.J. Noble, and P.G. Burke, to be published (1989).
56. K. Rohr and F. Linder, J. Phys. B: (Atom. Molec. Phys.) 9, 2521 (1976).
57. L.A. Morgan and P.G. Burke, J. Phys. B: (Atom. Molec. Opt. Phys.) 21, 2091 (1988).
58. A.D. McLean and M. Yoshimine, J. Chem. Phys. 47, 3256 (1967).
59. J.P. Gauyacq, J. Phys. B: (Atom. Molec. Phys.) 16, 4049 (1983).
60. G. Knoth, M. Rädle, H. Ehrhardt, and K. Jung, Europhys. Lett. 4, 805 (1987).
61. J.P. Gauyacq and A. Herzenberg, Phys. Rev. A25, 2959 (1982).

EXPERIMENTS ON LOW-ENERGY ELECTRON-MOLECULE COLLISIONS

H. Ehrhardt

Fachbereich Physik der Universität Kaiserslautern
D - 6750 Kaiserslautern
West Germany

1. INTRODUCTION

During the last 40 years an important stimulus for the study of electronic and atomic collision physics was the necessity to understand the atmospheres of the earth and other planets as well as the composition and physical properties of interstellar matter. This knowledge is of importance for flights of rockets, satellites and space vehicles and also for the interpretation of astronomical observations.

Although experimental achievements in measuring cross sections for a large variety of collision processes have been rather impressing, only theoreticians could judge the role of different particles and processes in such atmospheres, calculate model atmospheres and compare these with observations.

Professor Takayanagi certainly is one of the very active and successful theoreticians in this field of physics. The author remembers very well one typical example in connection with the work of Takayanagi. In the early 1960's the rotational excitation of molecules by electrons could not be measured in the laboratory since the energy resolution of electron impact spectrometers was by far insufficient and the analysis of swarm data was partly based on rather simple theoretical calculations. It was believed at this time that electron impact rotational excitation cross sections would be very small, possibly of the order of 10^{-18} cm^2, and therefore not interesting for the energy transfer from free electrons in the ionosphere into rotational motion of molecules, heating up the neutral gas.

In 1964, Takayanagi together with Geltman[1] showed, by distorted wave approximations including the quadrupole moment of the hydrogen molecule and its long-range polarizability by the colliding slow electron, that the pure rotational excitation cross sections could very well be of the order of 10^{-16} cm^2 for electrons of a few eV, i.e., one hundred times larger than estimated earlier.

The first experimental determination of the rotational excitation of a molecule in a crossed beam experiment was achieved in 1968 by Ehrhardt and Linder[2]. For H_2 the transition $J = 1 \rightarrow 3$ with and without a simultaneous vibrational excitation was measured in the energy range from 1 to 10 eV. The results agreed quite well with the predictions of Takayanagi and Geltman and other calculations[3] which had been made in the meantime.

Although the electron impact rotational excitation of molecules in their electronic ground states is now better understood due to the work of Takayanagi and his school and the work of several other theoreticians, experiments are still scarce. Most of the crossed beam data for other molecules than H_2 have been obtained for the first time only during the last six years. In Sections 3 and 4 such experimental results will be presented.

In addition to fundamental quantum mechanical many-particle problems, electron impact spectrometry has become quite important to many fields of applications. They reach from laboratory and astrophysical plasma, gaseous laser media, radiation chemistry to electron impact induced chemical reactions on surfaces. For such applications low impact energies are of importance because of the rapid thermalization of the electrons in the plasma or in the first 4 or 5 layers of the surface. Therefore the most interesting range of collision energies for such applications extend from a few tenth of eV to about 10 eV or 20 eV. Having this in mind the author will restrict the scope of the article mostly to this energy range.

During the last decade, an increasing effort is going on to supplement the classical electron impact spectrometer with additional techniques in order to prepare the target particle before the collision with the electron into a specific state and also to identify the target state after the collision as precisely as possible. These techniques are often called state-to-state or state-specific spectrometry. In Section 2 the most important techniques of this kind are mentioned together with their intrinsic intensity problems.

2. ELECTRON IMPACT SPECTROMETERS

Only such experimental arrangements are called electron impact spectrometers which allow for single collision conditions, in which the products of the collision do not anymore interact with other particles on their way to the detectors. This definition excludes those experimental arrangements working at pressures higher than about 10^{-14} mbar.

Modern electron impact spectrometers are suitable to work at impact energies as low as 0.1 eV and therefore overlap with swarm experiments in the rather large range from $0.1 \leqq E_0 \leqq 1$ eV corresponding to an electron temperature in an ionized gas between 1000 K to 10,000 K. This temperature range is most important in planetary atmospheres and there for the rotational and vibrational excitation of molecules.

2.1. Classical Instruments

Several experimental methods exist for special measurements. Total cross sections can be measured very sensitively with trochoidal spectrometers developed by Stamatovic and Schulz[4] and time-of-flight spectrometers brought to great perfection by Ferch et al.[5]. Bederson et al.[6] have used the beam recoil method, i.e., the deflection of a molecular beam induced by the momentum transfer from the colliding electron to the target particle for low energy elastic collisions and for rotational excitations of highly polar alkali halogen molecules.

By far the most flexible electron impact spectrometer has been developed in its essential parts during the time between 1960 and 1967. Most important was the development of small 127° electrostatic analyzers by Marmet and Kervin[7] and the 180° analyzers by Simpson and Kuyatt and by Simpson[8]. A modern version of such a spectrometer is shown in Fig. 1. It is used in the laboratory of the author.

The electrons are produced by thermal emission from a heated tungsten or iridium cathode and are focussed to the entrance slit of the first tandem energy selector. This tandem system has in principle the same

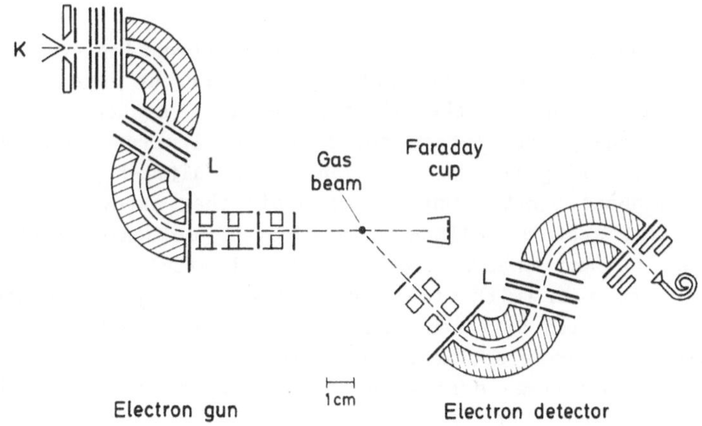

Fig. 1. Schematic representation of a double tandem electron impact spectrometer with rotatable electron detector in the angular range from 0° to 150°. The energy resolution can be tuned electrically.

energy resolution as a single system, but it reduces the background due to stray electrons which are still present after passing through the first 127° selector. The second system reduces the stray intensity in the wings of a spectral line on the high and low energy side so that small signals close to large peaks can be measured. The two cylindrical selectors in Fig. 1 are electrically separated by a small lens system L. This lens can be used to shift the transmission curves of the two selectors slightly with respect to each other so that the overall resolution is electrically tunable from outside the vacuum chamber. The exit lens system of the gun is quite open in order to reduce electron impact contaminations on its metal surfaces. The shape of the electron beam is visualized bv multichannel plates followed by a phosphor (not shown in the figure) and its intensity is measured by a Faraday cup.

The electron detector is built and operated similarly to the electron gun. It can be rotated from 0° to 150° with respect to the primary electron beam. For operation in the forward direction the Faraday cup is pushed away by the detector system and it moves back into the 0° position if the detector moves to higher scattering angles.

The gas beam is produced by a single channel tube of 10 mm length and 0.5 mm diameter. The target gas pressure just above the tube is about 10^{-3} mbar.

All systems, the gun, the detector and the gas inlet tube are heated during operation up to 250°C in order to reduce potential disturbances on the metal surfaces. The residual magnetic field in the region of the gun and detector is less than 1 mG and the residual electric field in the scattering area is less than 1 mV if surface charging is not considered.

The operational data of such a spectrometer are as follows: Energy of the electron beam tunable from 0.05 eV to about 10 eV or higher, energy inhomogeneity of about 10 meV FWHM for all four selectors together, primary current of about 10^{-10} A or higher depending on the desired energy resolution of the gun system. More and more experimental groups[9] use the double monochromators and analyzers, often with 180° electrostatic energy selectors.

The final energy resolution of such an electron impact spectrometer, the lowest impact energy and the primary beam current depend very much on the surface conditions in the electron optical systems. This is the reason why the energy resolution could only be lowered from about 50 meV to about 10 meV during the last 20 years. In ultra-high vacuum systems one may reach about 5 meV. One cannot avoid that stray electrons in the electron optical systems collide with adsorbed molecules such as water, hydrocarbons, O_2 and others and initiate chemical reactions producing surface layers or even (with time) thin films of badly conducting material or at least films with different contact potentials compared to the clean metal. This leads to electrical field inhomogeneities and therefore to chromatic and other image errors in the lens systems. Since the surface contaminations depend very much on the gas introduced for analytical purposes, there is little hope that electron impact spectrometers can be developed with better operational data than given above. It seems as if other methods must be employed to increase the resolution of such spectrometers. On the other hand we know from electron spectroscopy and other fields in physics that improvements in energy resolution always lead to new questions and a better understanding of physics.

2.2. New Developments

Of course, the so-called classical electron impact spectrometer is quite versatile. Threshold energies of many processes can be measured, cross sections which are optically forbidden, angular dependencies which give information on the symmetry of a short-living negative ion state or on the partial waves which are involved in the scattering and therefore one obtains information on the type of process which is occuring, i.e., whether it is resonant or a direct process proceeding mainly via the dipole or quadrupole moment or the long- and/or short-range polarizability of the molecule.

Therefore many attempts have been made in the last years to improve such spectrometers, but keeping their versatility. One step into the right direction for major improvements is the production of photoelectrons by lasers or synchrotron radiation as a substitute for the thermionic cathode and the complicated gun system which introduces the major troubles in spectrometers. Thermally produced electrons have an energy distribution which normally is about 300 meV wide, which means that if one wants an energy inhomogeneity of only one meV, one has to eliminate about 99 % of all produced electrons and use only about 1 % in order to form a decent electron beam. Of course the 99 % electrons produce space charge and surface contaminations in the whole gun system. Therefore it is reasonable to produce only electrons which are already monochromatic and the classical gun system is reduced to an electron optical system which forms a well collimated electron beam of those photoelectrons.

Such attempts have been made by several groups. Kennerly et al.[10] have photoionized metastable barium atoms ($^1D_2$) with 3.8 eV photons from a He-Cd ultraviolet laser and obtained electrons with a kinetic energy of 17 meV. The photoelectrons are shaped to a beam and then accelerated to the proper impact energy. A supersonic target beam was used in order to reduce the Doppler broadening of the scattered electrons. The spectrometer was tested by measuring the width of the well-known Feshbach resonance in helium below the 2^3S state. The energy inhomogeneity was 5 meV and the electron beam current about 10^{-12} A.

More intensity was obtained by Field et al.[11] by producing very slow electrons using synchrotron radiation for the photoionization of argon. The photoelectrons have been shaped to a beam with kinetic energy ranging from 0.1 to 1.3 eV and directed to a supersonic beam of molecular oxygen. The authors measured the energy dependence of the vibrational excitation of the molecule. They were able to measure fine structure splitting and saw effects of rotational excitations on the width of the O_2^--resonance peaks. The final energy resolution was 5.5 meV.

It might very well be possible to obtain with those photoionization guns an energy inhomogeneity as low as 1 meV, but the space charge from the remaining ions, the different Doppler effects, the remaining angular divergence of the electron beam, and the surface contaminations will always be limiting factors.

Another rather new development is the introduction of position sensitive detectors in electron spectrometers which was achieved first by Comer et al.[12]. Such a device reduces the time for measurements of energy loss spectra by a factor up to 1000, which means that one obtains a much higher reproducibility and therefore reliability of the data and an enormous increase in sensitivity. These authors were able to obtain a lowest overall energy spread of 10 to 25 meV in the range of impact energies from about 1.5 eV to some 100 eV. Energy loss measurements have been performed using N_2, CO, CO_2, HCl and other molecules.

A somewhat changed version of such an instrument has been built in the author's laboratory (Fig. 2). This spctrometer has a tandem selector

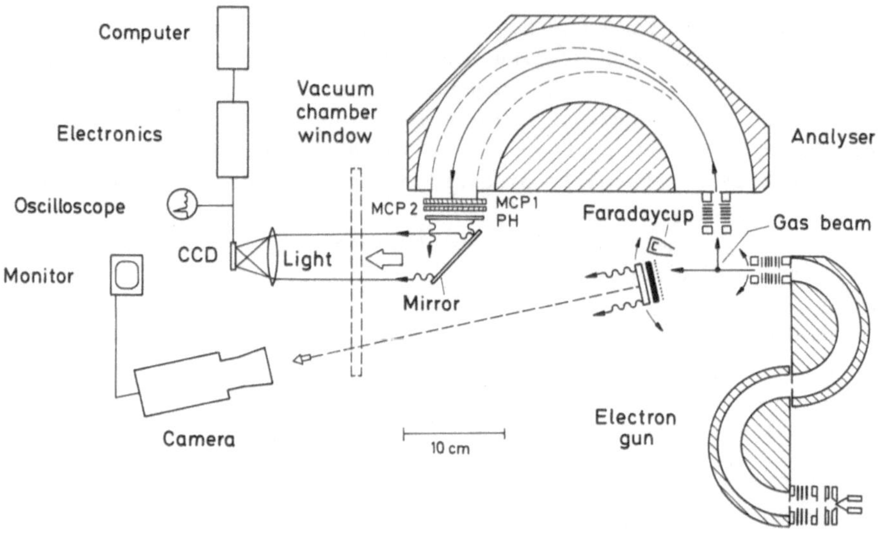

Fig. 2. Electron impact spectrometer with rotatable gun from 0° to 125° and position sensitive detection. The gun system can be electrically adjusted by using either the Faraday cup or a multichannel plate with phosphorus, video camera and monitor.

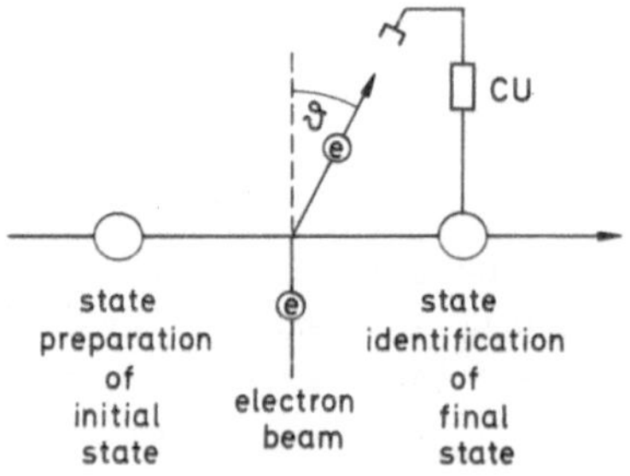

Fig. 3. Schematic representation of an electron impact spectrometer with state preparation and state identification. Even if the final state is identified, the information carried by the scattered electron (energy transfer in the case of a break up channel, the momentum transfer and the spin) is of interest. In this case a coincidence unit CU must connect the signals of the scattered electrons and the state identification.

in the electron gun in order to reduce effects on the electron beam which derive from space charge and stray electrons in the gun system. The electron gun can be rotated in the range from 0° to 125°. The primary electron beam is controlled with respect to intensity and shape either by a Faraday cup or a multi-channel plate connected to a phosphorus, a video camera and a monitor. The mean radius of the detector system is 13 cm.

There are always interesting new developments coming up for any part of the classical spectrometer, the electron optics, the dispersion systems, spectrum stabilization with little time delay (control devices) and the data handling which tends towards high speed data acquisition systems with 2-dimensional CCD's[13]. One can be sure that all these developments contribute to better spectrometers especially for industrial (mostly analytical) purposes, but it seems that the principal limitations discussed above favour other techniques to be used together with some elements of a classical spectrometer.

2.3. State Preparation and State Identification in Scattering Experiments

Occasionally the expressions "state-to-state" or "state-specific" cross sections are used indicating that the highest perfection in an experiment is reached. And indeed, for the reaction

$$e^-(E_0) + M(i) \rightarrow M(f) + e^-(E_0-\Delta E)$$

$$\rightarrow A(f_1) + B(f_2) + e^-(E_0-\Delta E)$$

it would be the most detailed information if we could prepare an atom or molecule in a specific state with a set of quantum numbers i and if we could find out how many particles after the collisions are in a specific final state f or, in the case of a break-up channel, how many particles A and B are produced with quantum numbers f_1 and f_2 (see Fig. 3). The scattered electron carries the information about the process energy, the momentum transfer and the spin.

Of course, such an experiment is an old dream which in a very few cases is already a reality. Experiments with polarized electrons[14], polarized target particles and the analysis of the outgoing electrons with respect to energy, scattering angle and spin represent already a perfect situation. But also in such beautiful experiments the target particles are mostly in the electronic ground state, and cannot be prepared (with some degree of saturation) in any excited state. We are still far away from such ideal conditions. Nevertheless, during the last few years the technical feasibilities to prepare the initial state have increased quite rapidly, especially by the use of lasers, supersonic beams and by sputtering. The following list contains the most important technical means, the particles which can be prepared and the particle fluxes which can be obtained in the scattering center:

(1) Gaseous matter or easy-to-evaporate target material,
particles in the electronic ground state,
particle flux of about 10^{16} p/s

(2) Supersonic beams,
low rotational and vibrational excitations, dimers and clusters,
particle flux 10^8-10^{12} p/s

(3) Gas discharge, microwave and radio frequency sources,
long living metastable particles, atoms like H, O, N, Cl, and radicals,
particle flux 10^8-10^{15} p/s
(4) Sputtering (mostly with argon ions on solid surfaces),
atoms, dimers and clusters, which cannot be produced by evaporation,
particle flux 10^8-10^{13} p/s
(5) Lasers, single and multiphoton excitation,
particles in well-defined rotational, vibrational and electronic states, Rydberg states, fine and hyperfine states,
particle flux $\leqq 10^{14}$ p/s
(6) Polarized atoms 10^{14} p/s

This is of course a very coarse list with rough orders of magnitudes. A more detailed information would be far beyond the scope of this article.

The final states can be identified by:

- energy loss of the scattered electron, its scattering angle and spin state
- detection of emitted photons and its polarization
- detection of metastable particles on a metal surface
- detection of ions in a mass spectrometer
- laser induced fluorescence
- photoionization with a laser, spin state of the photoelectron with a Mott detector
- cluster identification by scattering with neutral helium, photoionization or electron impact.

From the two lists one can easily see the future trends in experimental electron collision physics. Nearly any method of the first list can be combined with any method of the second list. Of course, we are still far away from a perfect preparation and a perfect identification in any collision process. But we are close to it in a few cases.

In Section 3.2.1 the measurement of rotational rainbow scattering[15] will be described. This experiment represents the combination of a supersonic beam with laser induced preparation (or marking) of Na_2 in a certain rotational state and identification of the final state after the electron collision by laser induced fluorescence. It therefore is a perfect state-to-state experiment.

3. ROTATIONAL EXCITATION OF MOLECULES

3.1. Low-Energy Electron-Molecule Collisions: Resonant and Nonresonant Interactions

Pure rotational excitation of molecules by low energy electrons (0.1-10 eV) can have cross sections which differ by approximately five orders of magnitude depending on the interaction mechanism between the electron and the molecule. For a molecule with weak polarizability and a weak permanent quadrupole moment like N_2 molecules[16,17], rotational excitation cross sections are of the order of 10^{-18} cm^2, whereas for the same molecule the rotational excitation via a shape resonance may have cross sections of about 10^{-16} cm^2. Molecules with strong dipole moments have rotational excitation cross sections of the order of 10^{-13} cm^2[18-22]. In the last six years the laboratory of the author has therefore made a systematic study of rotational excitations by low energy electron impact with consideration of the different interaction mechanisms.

This has been done by selecting molecules with different permanent multipole configurations and by energy loss measurements in an electron spectrometer at different scattering energies and angles. The angular dependence of the scattered electrons gives information on the interaction mechanism.

On the other hand, the energy resolution even of the best electron spectrometer is still too low to separate state-to-state transitions except for H_2, HD and D_2[2,23,24]. Therefore, the only measurable quantity with information on the cross sections of rotational transitions is the line width and line shape of the "elastic" scattering peak. The spectrometer must of course have a good energy resolution ($\Delta E \sim$ 10-20 meV) and must be very stable in order to measure the wings of a line quite

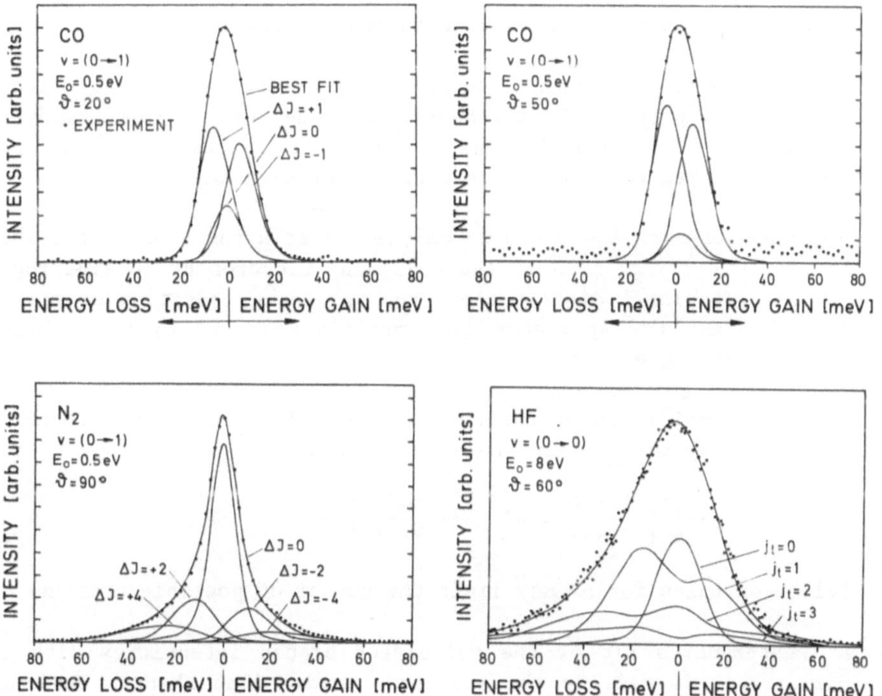

Fig. 4. Rotational excitation of CO, N_2 and HF using the branch fitting and the high-j approximation or the Shimamura formula. The true line width of the apparatus is measured using neon and argon (see ΔJ = 0 and j_t = 0 curves) and is always considerably smaller than the measured line shape (dots). The measured lines are unsymmetric, since the energy transfer from the electron to the molecule (energy loss) is preferred to the energy transfer from the molecule to the scattered electron (energy gain).

accurately[17]. Experimentally one can see (see Fig. 4) that the lines are always broader (up to 4 times) than the line width measured using helium, neon or argon. With rare gases the apparatus profile (true elastic scattering) is obtained. The line shape contains information on the relative transitions $\Delta J = 0,1,2,...$ and whether the electron transfers energy to the molecule (energy loss) or rotational energy is transferred from the molecule to the free electron (energy gain).

The evaluation of line shapes in terms of rotational cross sections can proceed via one of the two methods:

i) The branch construction method, which depends on the high-J approximation[25].
ii) A cross section formula for transitions of a spherical top or a linear rotator molecule by Shimamura[26].

The branch construction method assumes that all cross sections for transitions from J_i to $J_f = J_i + \Delta J$ are the same and independent of the value of J_i. In other words, a branch exists for each ΔJ if within one ΔJ-branch all transition probabilities are equal. For high J-values ($J \geq 4$) this assumption is very good because the associated Clebsch-Gordan coefficients are equal within a few percent.

A line shape is analyzed via the following steps:

i) Calculation of the initial rotational distribution in the molecular beam according to the beam temperature. The beam is in thermodynamical equilibrium after flowing through a thin heated (500 K) metal tube. The high temperature ensures rather high initial J's.

ii) The shape of each ΔJ-branch is calculated as a function of the energy loss or the energy gain of the colliding electron by considering the initial rotational distribution, the new energy position of the line $J_i + \Delta J$, and the apparatus line profile measured by using helium, neon and/or argon.

iii) The individual lines of the energy gain and energy loss branches are multiplied by the detailed balance factors

$$\frac{2(J_i+\Delta J) + 1}{2J_i + 1} \quad \text{and} \quad \frac{2(J_i-\Delta J) + 1}{2J_i + 1}$$

Selection rules for ΔJ may limit the number of possible branches.

iv) A least-squares fit of the calculated branch intensities with $\Delta J = 0,\pm1,\pm2,...$ is made to the measured overall line shape. The number of free parameters is equal to half the number of branches plus one, since the truly elastic peak and only the loss or the gain branches have to be considered (detailed balance factors). For example, in the case of the rotational excitation of N_2 in the region of the shape resonance (see Fig. 4) only $\Delta J = 0,\pm2,\pm4$ have to be considered because of selection rules (symmetry of the resonance). Therefore only three numbers are the outcome of the experiment (for one value of the impact energy), namely the maximum height of the truly elastic peak, the maximum of the branch distribution $\Delta J = +2$ and the maximum of the distribution $\Delta J = +4$. The method loses its accuracy with increasing numbers of rotational branches.

The Shimamura formula[26] is

$$\sigma(J' \leftarrow J) = \sum_{J_t=|J-J'|}^{J+J'} D\, \sigma(J_t \leftarrow 0)$$

with $D = \frac{2J' + 1}{(2J+1)(2J_t+1)}$ for spherical top molecules and $D = C^2(J,J_t,J';0,0,0)$ for linear-linear transitions, where C are the Clebsch-Gordan coefficients. The formula contains only the assumption that the rotational transitions are adiabatic (high-energy approximation), which is the case in the whole energy range of the experiment. The cross sections $\sigma(J_t \leftarrow 0)$ are the (few) free parameters, which have to be determined from the experiment by a line shape analysis. If these numbers have been determined, state-to-state transition cross sections $\sigma(J' \leftarrow J)$ can be calculated.

Table 1. Order-of-magnitude values of integrated cross sections for rotational excitation in most important ΔJ-branches. For non-resonant, i.e., direct interactions between the colliding electron and the molecule the cross sections refer to energy regions in which the influence of a shape resonance is small.

Integrated Rotational Cross Sections						
Mol.	Transitions via Direct Interactions			Transitions via Shape Resonances		Refs.
	Main interaction	ΔJ	σ (10^{-16}cm^2)	ΔJ	σ (10^{-16}cm^2)	
H_2	Weak quadrupole polarizability	0,±2	0.1	0,±2	0.8	2,23 24,27
N_2	Weak quadrupole polarizability	0,±2	0.015	0,±2,±4	2,8	16,17 28
CO	Weak dipole	0,±1	0.17	0,±1 ...±4	1	17,29
H_2O	Strong dipole	0,±1	2.5	0,±1	>0.3	17
HCl	Strong dipole	0,±1,±2	1	0,±1,±2	1	32,33
CH_4	Octupole polarizability	0,±3,±4	0.06	0,±1, ...±4	1	28,30
CO_2	Strong quadrupole	0,±2,±4	0.1	0,±2,...	0.6	31
HF	Strong dipole	0,±1,±2	1	No shape resonance visible		32

A rather large number of representative molecules have been measured during the last years using the line shape analysis giving electron impact rotational excitation cross sections, which otherwise cannot be obtained. In this paper only a rather short overview can be given. For more details the reader should see the original publications (see Table 1). For all experiments, the impact energies range from 0.1 eV to about 10 eV and only such rotational and vibrational transitions have been measured which start and end in the molecular ground state.

In most molecules there are energy ranges in which transitions occur via a shape resonance, and other energy ranges in which nonresonant, i.e., direct interactions between the colliding electron and the molecule leads to rotational transitions. Therefore it is possible to quote integrated (with respect to scattering angles) rotational cross sections for different interactions (see Table 1). From the table it can be concluded that the most important mechanisms for rotational excitations in the molecular ground states are shape resonances, strong permanent dipole moments and the polarizabilities of the molecules by the colliding electrons.

Several experiments of electrons colliding with highly polar molecules such as CsF, CsCl, CsBr, LiF, KI have been performed by Stern et al.[18,19], Bederson et al.[22] and Trajmar et al.[20,21]. From the data rotational excitation cross sections for $\Delta J = 0,\pm 1$ have been deduced, which are as large as $10^{-13}\mathrm{cm}^2$.

3.2. High-Energy Impact Rotational Excitation

The physics of the rotational excitation of molecules by high energy electrons is in principle not different from the mechanisms at low impact energies. A torque is needed to transfer angular momentum from the impinging electron to the molecule. This torque results from the interaction of the electron with the permanent dipole or quadrupole moment of the molecule or with the induced multipole moment due to the polarization of the molecular electron cloud by the incoming electron. An incoming beam of electrons contains (at higher energies) several partial waves, i.e., waves with higher angular momentum quanta, and these quanta can be transferred to the molecule. For distant collisions, i.e., in forward direction the simple first Born approximation and the selection rules according to the interaction mechanisms are valid. In this case only small ΔJ-values (mostly $\Delta J = 0,\pm 1,\pm 2$ and ± 3) are expected.

For close collisions an increase of large angle scattering is expected and accordingly the first Born approximation and the corresponding selection rules will break down. One also can assume that higher angular momentum quanta are transferred. Unfortunately there is only one experimental paper published[34] for impact energies as high as 100 eV and scattering angles between 10° and 135°. The pure rotational excitation of H_2 was measured mostly for $J = 1 \rightarrow 3$. It was found that for large scattering angles rotational excitation can even exceed pure elastic scattering. Certainly more experimental information is needed for those collisions.

3.2.1. Rotational rainbow scattering

Recently, an attempt to measure large angle scattering was undertaken

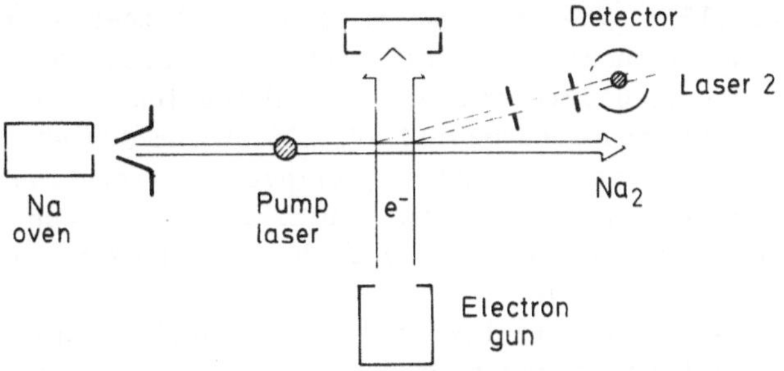

Fig. 5. Experimental set-up for the measurement of state-to-state rotational excitation in the electronic ground state of Na_2 by electron impact. The population of a rotational state j_i in the Na_2 supersonic beam is modulated by a pump laser. As a consequence of the electron collision a modulation signal is detected (by laser fluorescence, Laser 2) in different states j_f, indicating the transitions $j_i \rightarrow j_f$.

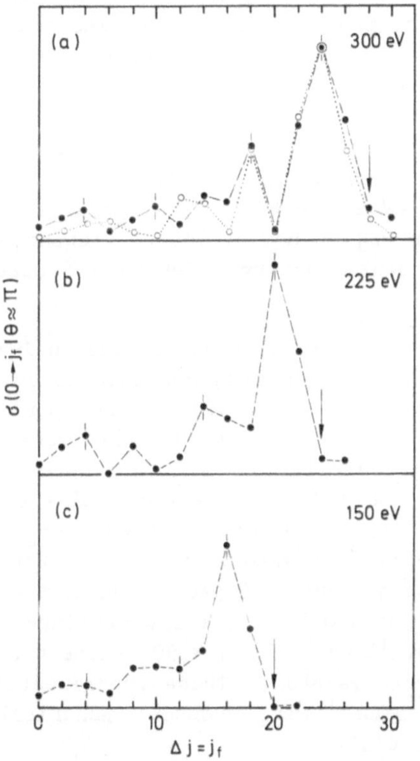

Fig. 6. Experimental relative differential cross sections for rotational transitions $j_i \rightarrow j_f$ at $\theta \approx 180°$ as a function of the number of transferred rotational quanta for E_0 = 300 eV (a), 225 eV (b) and 150 eV (c). The arrows indicate the theoretically possible maximum values of Δj. The main and the supernumerary rainbow structures are visible. The dotted curve in (a) is calculated[37,38] in the positions indicated by the open circles.

by Ziegler et al.[15]. The idea was to measure state-to-state rotational transitions of Na_2 molecules in collisions, by which the electrons are preferentially scattered into the backward direction. The impact energy of the electrons was chosen to be 150 eV, 225 eV and 300 eV, so that short living negative ion states do not contribute to rotational excitation within the electronic ground state. Differential cross sections for scattering angles in the range $90° \leqq \theta \leqq 180°$ have been measured.

Figure 5 shows schematically the experimental setup. A supersonic sodium beam, consisting of about 85 % Na and 15 % Na_2, is crossed with a high-current (about 10 mA) electron beam of energy $150 \leqq E_0 \leqq 300$ eV. Due to the momentum transfer in the scattering process, Na_2 molecules are deflected up to 11° for E_0 = 300 eV. The scattering angle of the deflected electrons is correlated to the deflection angle of the sodium dimers. 11° for Na_2 for 300 eV corresponds to a scattering angle of 180° of the electrons. The angular resolution $\Delta\theta(Na_2) \cong 2°$ is equivalent to $\Delta\theta_e \cong 40°$, which is adequate for the purpose of this study.

If the pump laser is on, a given rotational state J_i of Na_2 is completely depopulated by optical pumping. The number of Na_2 particles in the rotational state J_f is detected by laser induced fluorescence. If the pump laser is switched on and off, then the modulation in the fluorescence signal is a measure of the electron impact induced transition from J_i to J_f. A modulation of the electron beam allows to measure background contributions in the signal.

Figure 6 shows relative cross sections for $J_i = 0$ to J_f induced by electrons, which have been scattered into the backward direction $\theta_e \cong 180°$ for E_0 = 150 eV, 225 eV and 300 eV. The cross sections for odd ΔJ-values are zero, because the Na_2 is homonuclear. Pronounced rotational rainbows as well as resolved supernumerary rainbows are observed with very large ΔJ-values up to 28.

The explanation of the results can be made using classical arguments. The internuclear distance of the Na_2 molecule is quite large (r_e = 0.3079 nm[35]) and larger than the de Broglie wavelength of the colliding electrons. For backward scattering of the electrons the angular momentum transferred to the molecule is $\Delta J = kr_e \sin\gamma$, where k is the magnitude of the wave vector $\underline{k}$ of the incoming electron and γ the angle between $\underline{k}$ and the molecular axis. The maximum value of $\Delta J = kr_e$ occurs for $\gamma = 90°$ and $\theta_e = 180°$ and is indicated by arrows in Fig. 6. Since the molecular positions $\gamma = 90° + \delta$ and $\gamma = 90° - \delta$ produce each a reflected wave with the same angular momentum transfer, the two waves interfere leading to supernumerary rotational rainbows. For $\gamma = 90°$ these two orientations coincide giving rise to the main rainbow. These experimental results are in good agreement with classical and quantum mechanical calculations[36,37] (dotted curve in Fig. 6(a)).

If we try to generalize these results for high energy electron impact rotational excitation of molecules, we come to the following conclusions:

i) For very large angle scattering of the electrons, $120° \leqq \theta_e \leqq 180°$, very large ΔJ-values are to be expected depending on the size of the molecule in comparison to the de Broglie wavelength. The scattering can be understood semiclassically.

ii) For intermediate angle scattering, $60° \leqq \theta_e \leqq 120°$, the colliding electrons penetrate deeply into the electronic cloud of the molecule. Rather large ΔJ-values are to be expected, possibly $6 \leqq \Delta J \leqq 10$, depending on the short-range polarizability of the molecule.

iii) For small angle scattering the first Born approximation is valid and therefore small ΔJ-values are to be expected.

iv) The cross sections for rotational excitations are connected to the elastic cross section.

For the intermediate angular range more experimental and theoretical work has to be made.

4. VIBRATIONAL EXCITATION OF MOLECULES

During the last 25 years numerous experiments applying the crossed beam technique have been made to measure the vibrational excitation of molecules in their electronic ground states[38]. Most of these experiments were connected to the search for new shape resonances, the determination of their energy positions, symmetries, lifetimes, dissociative attachment channels and the population of vibrational states in the molecular ground states after the autoionization of the resonances. A beautiful example of the presentation of the results and of modern electron spectrometry has been published just recently by Comer and his coworkers[39].

On the other hand, vibrationally excited molecules are produced not only by shape resonances, but also through virtual states (see Sections 4.1 and 4.2) and by nonresonant interactions between the colliding electron and the molecule[40].

4.1. Low-Energy Electron-Molecule Collisions: Resonant and Nonresonant Interactions

Different interaction mechanisms can produce vibrationally excited molecules in low energy electron collisions with rather high cross sections. This can be seen easily from Fig. 7[41]. Fig. 7 shows a strong rise of the differential cross section below 0.5 eV incident energy of the electrons colliding with OCS molecules. This rise and its angular dependence is in agreement with a simple Born calculation, in which the permanent dipole moment of the molecule was inserted. In this energy region vibrational excitation mostly of the bending mode (0,n,0) is observed, although its angular dependence is not very well reproduced by the Born calculation.

We therefore conclude that this rise is partly due to a virtual state, which is observable especially in the inelastic channels. A virtual state was first observed in molecules by Rohr and Linder[42] in the electron scattering from HCl. Such a state was explained by a quasi bound state at nearly zero energy and the s-wave playing a dominant role (see Section 4.2).

In Fig. 7 the rise is followed by two maxima at 1.4 eV and 3.5 eV. These maxima are explained by shape resonances. Between the virtual state

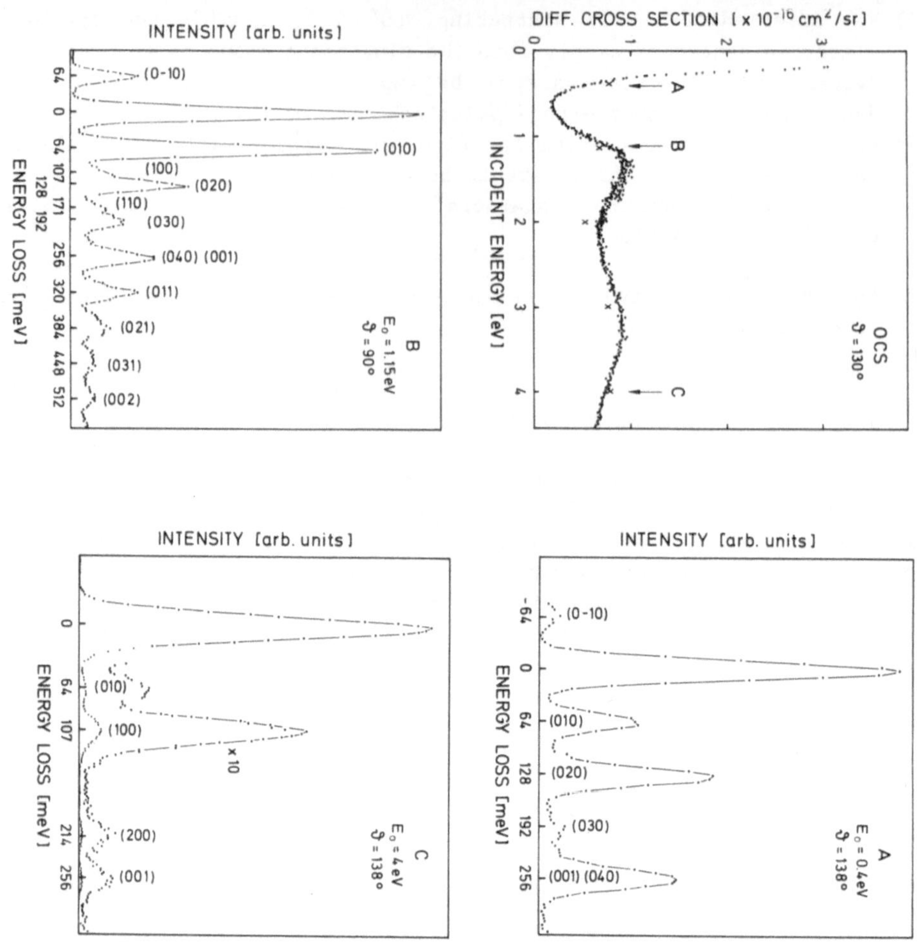

Fig. 7. Energy dependence of the elastic scattering cross section of OCS at a scattering angle of 130°. A low energy peak and two shape resonances (1.3 eV and 3.4 eV) are visible. The arrows and capital letters indicate the energy positions, in which the energy loss spectra of the figures A, B and C have been measured. This is an example of mode selective excitation of vibrations due to different interaction mechanisms.

and the two shape resonances are energy areas in which nonresonant interactions for vibrational excitations play a role. The three figures 7A, B, C show that the vibrational excitation is quite different depending on the impact energy of the electrons. In the virtual state we mostly find excitation of the bending mode, and in the lower shape resonance between 1 and 2 eV the bending and the asymmetric stretch mode is dominant, whereas in the high energy shape resonance (3 to 4 eV) the symmetric stretch mode is quite strongly excited.

Such mode-selective vibrational excitations have been observed in all molecules with more than two atoms. C_2H_2, CO_2 and CS_2 are examples.[43, 44]. Infrared active modes are excited quite effectively by the interaction of the colliding electron with the dipole moment of the molecule.

Many partial waves are involved in the excitation of such a mode, which is evident from the increase of the cross section in forward direction. The Born dipole approximation describes the cross sections quite well at all impact energies and scattering angles from 0° to 30°, at higher angles the Born dipole approximation gives too low cross sections. Vibrational excitation cross sections induced by quadrupole or octupole interaction are very small, e.g., of the order of 10^{-18} cm^2. Test molecules have been H_2 and N_2[45] outside the region of a shape resonance. These findings, however, may be affected by the vibrational excitation through the polarizability. The contributions to vibrational excitation via polarization of the molecule can be quite large. In a molecule with a large long-range polarizability like CS_2[41], the integrated cross section may be as large as 10^{-16} cm^2. In general, the cross sections for the excitation of one or two vibrational quanta are of the order of 10^{-18} to 10^{-16} cm^2 close to the thresholds. The effect of the short range polarizability is well detectable in backward scattering and has been observed for the case of CO, $v = 0 \rightarrow 1$ below the shape resonance at 0.5 eV collision energy[46]. The effect of the polarizability is tested best for the excitation of Raman active vibrational modes.

Shape resonances are very effective in exciting vibrations. The integrated cross sections are mostly of the order of 10^{-16} cm^2. In some molecules quite large numbers of $\Delta\nu$ have been observed, in H_2 up to $\Delta\nu \approx 20$, of course with smaller cross sections[47]. As already mentioned, in polyatomic molecules the shape resonances act often only on selected vibrational modes, namely on those for which the symmetry is in accordance with the symmetry of the resonance. In this way mode-selective vibrational excitation can be achieved quite easily.

4.2. Vibrational Threshold Structures

As already mentioned, Rohr and Linder[42] have discovered in 1976 rather narrow peaks in the vibrational excitation of HCl and HF very close to the thresholds (Fig. 8) of $\Delta\nu = 1$ and $\Delta\nu = 2$. These authors also reported that the angular dependencies of the threshold peaks were constant over the angular range from 30° to 120°. Azria et al.[48] detected rather large steps in the Cl^--formation differential cross sections for dissociative attachment in HCl right at the energy positions of the thresholds of $\nu = 2$, 3, 4 and 5 excitation of HCl. These steps demonstrate large cross sections at the thresholds of the molecular vibrational excitation.

A rather large number of theoretical papers explained these threshold peaks as the so-called virtual states, i.e., quasi-bound states of the very slow electron in the field of the molecule, the outgoing electron being an s-wave electron. It also seems to be clear now that nonadiabatic effects due to the molecular vibration play an important role in the formation of these states.

Although the strongest peaks have been found in molecules with rather strong dipole moments, the dipole does not seem to be the only source for the quasi-binding mechanism, since this effect has been established for many molecules as for example SF_6[49], H_2S[50], HBr[51], H_2O[52], CO_2[44], COS and CS_2[41], and CH_4[53]. For other molecules such as H_2, N_2, CO,

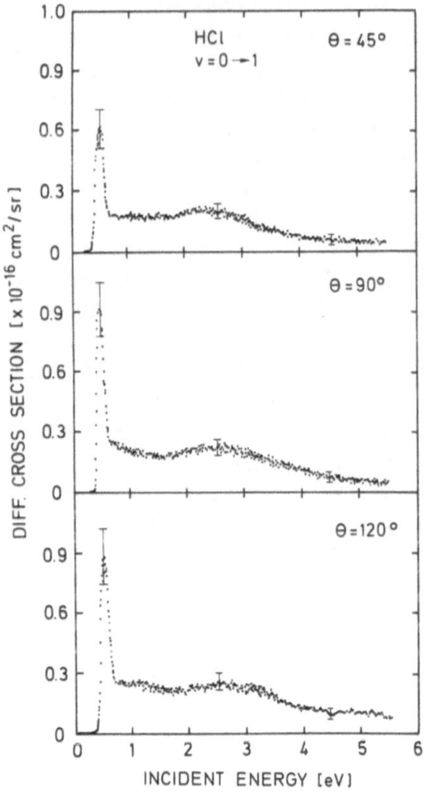

Fig. 8. Energy dependence at three different scattering angles of the excitation of one vibrational quantum of HCl in its electronic ground state[33]. At 2.5 eV is the maximum of a shape resonance.

C_2H_2 such structures have not been detected. In polyatomic molecules it was found (e.g., CO_2[44]) that only certain vibrational modes exhibit a strong threshold peak in the vibrational excitation, whereas other modes do not.

For the symmetric stretch mode of CO_2 theoretical results of Morrison and Lane[54], Whitten and Lane[55] and Estrada and Domcke[56] are in quantitative agreement with the measurements. Because of discrepancies between the experimental results of Rohr and Linder[42] and theoretical attempts to explain the structures quantitatively, it was required to repeat the measurement for HCl and HF with improved techniques and in more detail[33]. Here only the vibrational excitation of HCl, $v = 0 \rightarrow 1$ will be discussed[57]. The most important results are as follows:

i) Absolute cross section and peak shape close to threshold
In Fig. 9 are shown three energy dependencies of the angular integrated cross section for the vibrational excitation $v = 0 \rightarrow 1$ of HCl. The full curve represents a theoretical curve published by Dubé and Herzenberg[58] using an effective range theory including the concept of the virtual state. Other theoretical approaches[59] produce similar peak shapes, namely a vertical onset at the threshold, a cusp

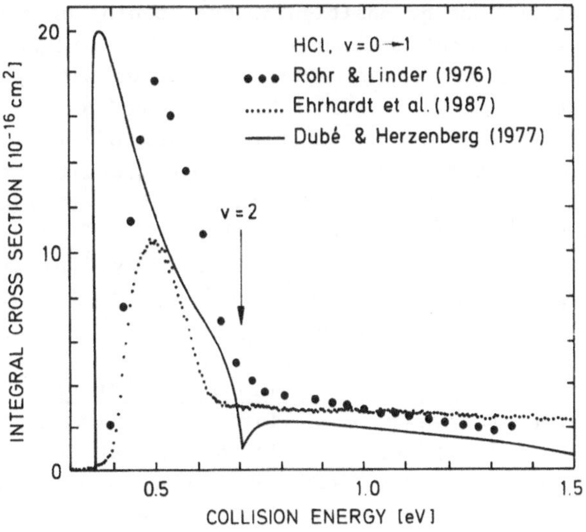

Fig. 9. Comparison of the two existing measurements of the threshold peak in HCl, v = 0 → 1 channel with an effective range approximation of Dubé and Herzenberg[58]. The two experimental curves agree in the shape of the threshold structure (except for the absolute value of the cross section). The sharp onset of the theoretical curve and the cusp structure at the opening of the v = 2 channel were not found experimentally.

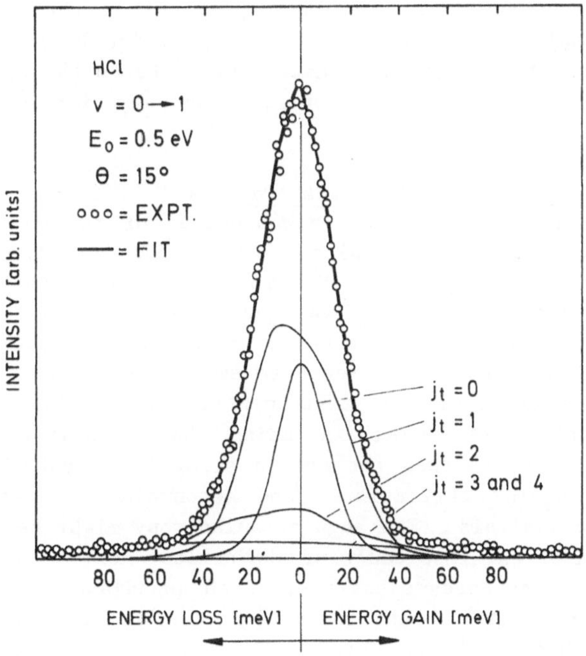

Fig. 10. Rotational excitation of HCl (with simultaneous vibrational excitation v = 0 → 1), measured at the maximum of the threshold peak at E_0 = 0.5 eV. The scattering angle is θ = 15°.

structure at the energy position of the opening of the channel v = 2 and a rather flat contribution of v = 0 → 1 to higher energies. The two other curves are experimental; the curve with the big dots was given by Rohr and Linder[42] and the fine dotted curve by Knoth et al.[33]. Except for the absolute value of the cross section in the peak area, where the Rohr and Linder results are too large by nearly a factor of two, the two curves are very similar in shape; both experimental groups find a gradual rise close to the threshold and the maximum of the peak is shifted away from the threshold to higher energies by about 150 meV. Both groups find no cusp structure at the onset of the v = 2 channel.

ii) Simultaneous rotational and vibrational excitation

Knoth et al.[33] have made a line shape analysis as discussed in Section 3.1 of this paper in order to find out if rotational excitation occurs simultaneously with the vibrational excitation. Figure 10 shows one of the many measurements. The analysis was made for an impact energy of 0.5 eV, i.e., right at the position of the peak maximum. The initial rotational distribution before the collision is equivalent to a temperature of 400 K, and the apparatus line shape was obtained using argon. For the deconvolution of the spectral line the method of Shimamura[26] was used. From Fig. 10 it is seen that the line width is nearly twice the apparatus line shape (curve for $j_t = 0$, i.e., truly elastic scattering), and therefore far outside the experimental error limits. The rotational contributions from $j_t = 1$ and 2 are much larger than for $j_t = 0$. Small contributions are detectable even for $j_t = 3$ and 4. These findings have not yet been discussed theoretically and may explain the rather smooth onset of the threshold peak instead of the theoretically predicted vertical onset, since higher partial waves contribute to the collision and the contributions from these partial waves may depend on the impact energy.

iii) Angular dependence of the threshold peak

Figure 11 shows the angular dependencies of the different rovibrational excitations of HCl with v = 0 → 1 taken at the maximum of the threshold peak at $E_0 = 0.5$ eV. The angular distribution of the sum curve is not constant as was published by Rohr and Linder[42]. Instead, all curves show a rather constant value for angles larger than 60°, but a strong decrease to smaller angles. Also this observation is well outside of instrumental errors. The full line error bars describe the errors including the normalization to absolute cross sections, whereas the dotted error bars contain counting statistics, angular corrections, and deconvolution errors due to the line shape analysis. The angular anisotropy might be an experimental indication of the importance of the nuclear motion during the passage of the slow scattered electrons. Such nonadiabatic coupling effects at threshold have been found for the rotational excitation of molecular hydrogen[27]. This conclusion for nonadiabaticity in the case of HCl is based only on the similar angular dependency of H_2 and has of course to be examined theoretically. A time-dependent model calculation by Domcke[60] shows the strong coupling between the molecular vibration and the quasi-free electron.

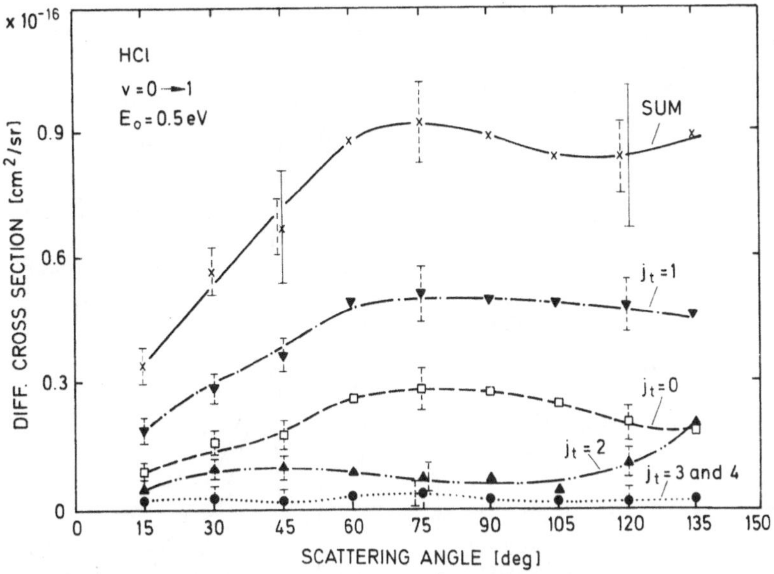

Fig. 11. Differential cross sections for j_t = 0,1,2 and 3 plus 4 and the sum curve as a function of the scattering angle. The rather rapid decrease of all curves below 60° could be a consequence of nonadiabaticity of the collision, i.e., the coupling of the nuclear motion and the motion of the slow outgoing electron. The broken line error bars represent relative errors, and the full line bars include the normalization to absolute values of the cross sections.

Teillet-Billy and Gauyacq[61] have made model calculations to explain the threshold peaks using the picture of nuclear excited Feshbach resonances. Feshbach resonances can be distinguished experimentally from virtual states, since they are positioned energetically slightly below the threshold of the parent state in the open channel, whereas the virtual state has no such structure. The influence of the corresponding pole is only visible by the large cross section right at the threshold of the newly opening channel.

We do not find such Feshbach resonance structures within our energy resolution of approximately 15-20 meV. This has been proved as well in the channel v = 0 → 0 as in the v = 0 → 1 channel. If v = 1 is the parent state for a Feshbach resonance, then below the threshold for v = 0 → 1 a resonance structure should be visible. This is not the case as Fig. 12 shows. The cusp in the v = 0 channel due to the opening of the v = 1 channel is clearly visible but no structure is visible below this energy. The cusp width is approximately equal to the experimental energy resolution. Also no resonance structure is visible in Fig. 13 below the position of the opening of the channel v = 2, which should be the case if v = 2 is the parent state for a resonance in the v = 1 channel. Of course these measurements do not prove that the model of the nuclear excited Feshbach resonance is wrong. They only show, in case this model would be right, that the binding energy of the additional electron would be less than 10 to 20 meV.

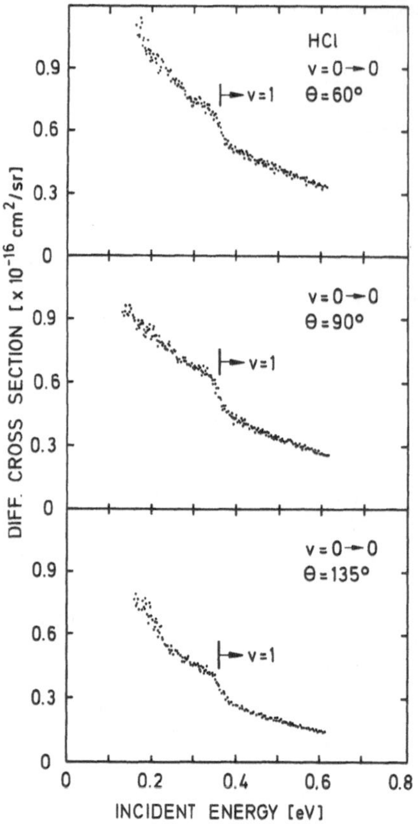

Fig. 12

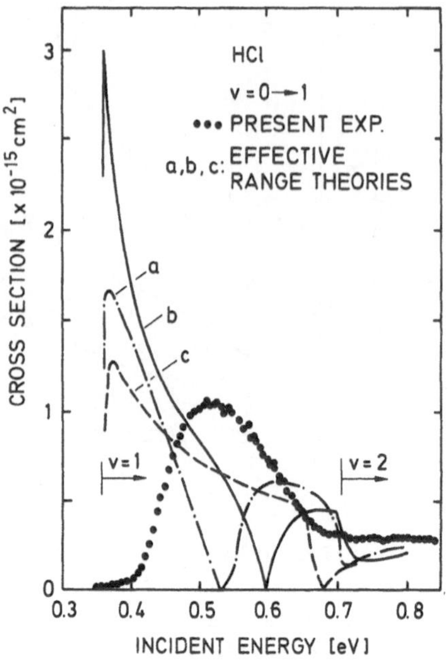

Fig. 13

Fig. 12. Cusp structures in the elastic channel due to the opening of the v = 1 channel at three different scattering angles as a function of impact energy. The energy halfwidth of the cusp structure is nearly equal to the instrumental width.

Fig. 13. Comparison of the experimentally determined shape of the threshold peak of HCl, v = 0 → 1, as a function of impact energy with three theoretical curves calculated by Teillet- Billy and Gauyacq[61]. The theoretical curves show large structures with zero intensity of about 170 meV (a), 100 meV (b) and 20 meV (c) below the opening of the new channel. The experimental curve does not exhibit such a structure.

Acknowledgment The author wants to express his thanks for the financial support of the Deutsche Forschungsgemeinschaft (SFB 91).

REFERENCES

1. K. Takayanagi and S. Geltman, Phys. Lett. 13, 135 (1964); K. Takayanagi and S. Geltman, Phys. Rev. 138, A1003 (1965); S. Geltman and K. Takayanagi, Phys. Rev. 143, 25 (1966).
2. H. Ehrhardt and F. Linder, Phys. Rev. Lett. 21, 419 (1968).

3. N.F. Lane and S. Geltman, Phys. Rev. 160, 53 (1967); E.S. Chang and A. Temkin, Phys. Rev. Lett. 23, 399 (1969).
4. A Stamatovic and G.J. Schulz, Rev. Sci. Instr. 41, 423 (1970).
5. J. Ferch, B. Granitza, C. Masche, and W. Raith, J. Phys. B 18, 967 (1985).
6. B. Jaduszliwer, G.F. Shen, J.L. Cai, and B. Bederson, Phys. Rev. A 31, 1157 (1985).
7. P. Marmet and L. Kerwin, Can. J. Phys. 38, 787 (1960).
8. J.A. Simpson, Rev. Sci. Instr. 35, 1698 (1964); C.E. Kuyatt and J.A. Simpson, Rev. Sci. Instr. 38, 103 (1967).
9. S. Trajmar and D.F. Register, in Electron-Molecule Collisions (I. Shimamura and K. Takayanagi, eds.), Plenum Press, New York (1984).
10. R.E. Kennerly, R.J. van Brunt, and A.C. Gallagher, Phys. Rev. A 23, 2430 (1981).
11. D. Field, G. Mrotzek, D.W. Knight, S. Lunt, and J.P. Ziesel, Proceedings of the XVth International Conference on the Physics of Electronic and Atomic Collisions (ICPEAC), Brighton, p.266 (1987).
12. P.J. Hicks, S. Daviel, B. Wallbank, and J. Comer, J. Phys. E 13, 713 (1980); T.A. York and J. Comer, J. Phys. B 16, 3627 (1983); T.A. York and J. Comer, J. Phys. B 17, 2563 (1984).
13. XVth International Conference on the Physics of Electronic and Atomic Collisions (ICPEAC), Brighton, Book of Abstracts, pp.798-803 (1987).
14. J. Kessler, Polarized Electrons, 2nd ed., Springer-Verlag, Berlin (1985).
15. G. Ziegler, M. Rädle, O. Pütz, K. Jung, H. Ehrhardt, and K. Bergmann, Phys. Rev. Lett. 58, 2642 (1987).
16. S.F. Wong and L. Dubé, Phys. Rev. A 17, 570 (1978).
17. K. Jung, Th. Antoni, R. Müller, K.H. Kochem, and H. Ehrhardt, J. Phys. B 15, 3535 (1982).
18. R.C. Slater, M.G. Fickes, G.W. Becker, and R.C. Stern, J. Chem. Phys. 60, 4697 (1974).
19. R.C. Slater, M.G. Fickes, W.G. Becker, and R.C. Stern, J. Chem. Phys. 61, 2290 (1974).
20. M.R.H. Rudge, S. Trajmar, and W. Williams, Phys. Rev. A 13, 2074 (1976).
21. L. Vuskovic, S.K. Srivastava, and S. Trajmar, J. Phys. B 11, 1643 (1978).
22. B. Jaduszliwer, A. Tino, P. Weiss, and B. Bederson, Phys. Rev. Lett. 18, 1644 (1983).
23. F. Linder and H. Schmidt, Zeit. f. Naturforsch. 26a, 1603 (1971).
24. E.S. Chang and S.F. Wong, Phys. Rev. Lett. 38, 1327 (1977).
25. F.H. Read, J. Phys. B 5, 255 (1972).
26. I. Shimamura, Chem. Phys. Lett. 73, 328 (1980).
27. K. Jung, K.-M. Scheuerlein, W. Sohn, K.-H. Kochem, and H. Ehrhardt, J. Phys. B 20, L327 (1987).
28. H. Tanaka, Symposium on Electron-Molecule Collisions, p.31, Tokyo (1979).
29. E.S. Chang, Th. Antoni, K. Jung, and H. Ehrhardt, Phys. Rev. A 30, 2086 (1984).
30. R. Müller, K. Jung, K.-H. Kochem, W. Sohn, and H. Ehrhardt, J. Phys. B 18, 3971 (1985).
31. Th. Antoni, K. Jung, H. Ehrhardt, and E.S. Chang, J. Phys. B 19, 1377 (1986).
32. G. Knoth, M. Rädle, K. Jung, and H. Ehrhardt, to be published (1989).
33. G. Knoth, M. Rädle, H. Ehrhardt, and K. Jung, Europhys. Lett. 4, 805 (1987).
34. S.K. Srivastava, R.I. Hall, S. Trajmar, and A. Chutjian, Phys. Rev. A 12, 1399 (1975).
35. W. Demtröder, W. Stetzenbach, M. Stock, and J. Witt, J. Mol. Spectrosc. 61, 382 (1976).
36. P. Eckelt, H.-J. Korsch, and V. Philipp, J. Phys. B 7, 1649 (1974).

37. H.-J. Korsch, H. Kutz, and H.D. Meyer, J. Phys. B 20, L433 (1987).
38. G.J. Schulz, Phys. Rev. 125, 229 (1962).
39. T. Reddish, F. Currell, and J. Comer, J. Phys. E 21, 203 (1988).
40. L. Andrick, D. Danner, and H. Ehrhardt, Phys. Lett. 29A, 346 (1969).
41. W. Sohn, K.-H. Kochem, K.-M. Scheuerlein, K. Jung, and H. Ehrhardt, J. Phys. B 20, 3217 (1987).
42. K. Rohr and F. Linder, J. Phys. B 9, 2521 (1976).
43. K.-H. Kochem, W. Sohn, K. Jung, H. Ehrhardt, and E.S. Chang, J. Phys. B 18, 1253 (1985).
44. K.-H. Kochem, W. Sohn, N. Hebel, K. Jung, and H. Ehrhardt, J. Phys. B B 18, 4455 (1985).
45. W. Sohn, K.-H. Kochem, K.-M. Scheuerlein, K. Jung, and H. Ehrhardt, J. Phys. B 19, 4017 (1986).
46. W. Sohn, K.-H. Kochem, K. Jung, H. Ehrhardt, and E.S. Chang, J. Phys. B 18, 2049 (1985).
47. M. Allan, J. Phys. B 18, L451 (1985); M. Allan, J. Phys. B 18, 4511 (1985).
48. R. Azria, Y. LeCoat, D. Simon, and M. Tronc, J. Phys. B 13, 1909 (1980).
49. K. Rohr, J. Phys. B 10, 1175 (1977).
50. K. Rohr, J. Phys. B 11, 4109 (1978).
51. K. Rohr, J. Phys. B 11, 1849 (1978).
52. G. Seng and F. Linder, J. Phys. B 9, 2539 (1976).
53. W. Sohn, K. Jung, and H. Ehrhardt, J. Phys. B 16, 891 (1983).
54. M.A. Morrison and N.F. Lane, Chem. Phys. Lett. 66, 527 (1979).
55. B.L. Witten and N.F. Lane, Phys. Rev. A 26, 3170 (1982).
56. H. Estrada and W. Domcke, J. Phys. B 18, 4469 (1985).
57. H. Eharhardt, M. Gote, K. Jung, G. Knoth, and M. Rädle, More experimental results on HCl and HF have been measured and will be published.
58. L. Dubé and A. Herzenberg, Phys. Rev. Lett. 38, 820 (1977).
59. For example: R.K. Nesbet, J. Phys. B 10, L739 (1977); J.P. Gauyacq and A. Herzenberg, Phys. Rev. A 25, 2959 (1982); W. Domcke and C. Mündel, J. Phys. B 18, 4491 (1985).
60. W. Domcke, Seminar presented at the Universität Kaiserslautern (May 20, 1988).
61. D. Teillet-Billy and J.P. Gauyacq, J. Phys. B 17, 4041 (1984).

SUBEXCITATION ELECTRONS IN GASES*

Mitio Inokuti

Argonne National Laboratory
Argonne, Illinois 60439
U.S.A.

1. INTRODUCTION

The absorption of ionizing radiation in matter always leads to the production of numerous electrons having a wide range of kinetic energies. For simplicity, let us consider first a gas composed of a single species of molecules. Electrons with kinetic energies T exceeding the first electronic-excitation threshold E_1 degrade rapidly through collisions with molecules, resulting in electronic excitation or in ionization. In contrast, electrons with $T < E_1$ degrade much more gradually because they transfer to molecules much smaller quanta of energy per collision; the mechanisms involved are mainly momentum transfer on elastic scattering. rotational excitation, and vibrational excitation of molecules. The rate of energy loss at $T < E_1$ is much smaller than that at $T > E_1$; the ratio

* Work supported by the U.S. Department of Energy, Office of Health and Environmental Research, under Contract W-31-109-Eng-38.

of the rates is 10^{-3} or even less for monatomic gases such as rare gases (in which only elastic scattering occurs at $T < E_1$), and it is about 10^{-2} for H_2, for example. Because of the low rates of energy loss, the electrons with $T < E_1$ warrant special consideration as <u>a transient species</u> in the analysis of the action of ionizing radiations. This point was first made by Weiss[1] and independently by Platzman[2], who also named the electrons with $T < E_1$ <u>subexcitation electrons</u>.

Platzman recognized that subexcitation electrons play a role in the influence of impurities on the total ionization of rare gases resulting from the absorption of α particles. This phenomenon was observed by Jesse and Sadauskis[3] and was later called the Jesse effect[4]. The total ionization of He, for instance, is greatly enhanced by a minute impurity of any other species, say A. The enhancement is primarily due to the Penning ionization, i.e., thermal collisions of excited states He^* with A that result in ionization of A, viz.,

$$He^* + A \rightarrow He + A^+ + e \quad (1)$$

This process is energetically possible with almost any impurity, because the excitation energy of He^* (at least 19.8 eV) is greater than the first ionization threshold of almost any atom or molecule.

Platzman's crucial observation was that the amount of enhancement of the total ionization depends on the species of the impurities; according to the experiments of Jesse and Sadauskis[3] the enhancement increases in the order Ar, Kr, Xe. This order is precisely the order of decreasing ionization thresholds. The difference in the enhancement arises because of the subexcitation electrons in He having kinetic energies T that are greater than the impurity ionization threshold I_A, i.e., those electrons with T in the interval $I_A < T < E_1(He) = 20$ eV. Because the collision processes above E_1 give rise to subexcitation electron energies T distributed over the entire interval $0 < T < E_1$, it is obvious that an impurity with lower I_A necessarily causes a greater enhancement than does an impurity with higher I_A.

The idea of subexcitation electrons is useful in many contexts related to the interactions of charged particles with matter in various forms, and it is also pertinent to some problems in gaseous discharges, astrophysics, and upper-atmospheric physics. What follows is a review of our current knowledge about the behavior of subexcitation electrons.

It is a great pleasure for me to dedicate the present chapter to Professor Takayanagi, who is a pioneer in research on subexcitation electrons and who taught me much about the subject. In this respect I cannot help recalling the summer of 1965, when Professor Takayanagi was a guest of Platzman and me at Argonne. We three wrote a short conference paper[5] and planned for a comprehensive article on subexcitation electrons; however, the intent has not been realized, in part because of Platzman's premature death and in part because of my laziness. I sincerely hope that Professor Takayanagi will generously accept the present chapter as a partial realization of our 1965 plans.

2. THE ENTRY SPECTRUM OF SUBEXCITATION ELECTRONS

Let us consider the energy distribution of subexcitation electrons at the earliest time, immediately following the rapid degradation of electrons at $T > E_1$. To be specific, let $\phi(T)dT$ be the number of electrons that reach energies between T and T + dT in the subexcitation region ($0 < T < E_1$). The function $\phi(T)$ has been often referred to as "the spectrum of subexcitation electrons". This term is confusing because the subexcitation electrons further slow down, and hence their energy distribution changes with time. The function $\phi(T)$ represents the energy distribution before the moderation in the subexcitation region, and in this sense one may be tempted to call $\phi(T)$ the "initial" spectrum. However, the adjective "initial" is used in radiation physics and chemistry to refer to widely different time scales in different contexts. Therefore, it seems appropriate to depart from the tradition and call $\phi(T)$ the entry spectrum of subexcitation electrons[6].

In general, an electron enters the subexcitation region in two ways. In the first way, it enters directly, i.e., it is produced as a secondary electron in an ionizing collision of energetic electrons or other charged particles. This direct way contributes to $\phi(T)$ an amount proportional to the secondary-electron spectrum (or the singly differential ionization cross section), which has been a subject of many studies[7]. Let us designate the secondary-electron spectrum as $d\sigma(T)/dT$, following the tradition of these studies. In the second way, an electron may enter the subexcitation region indirectly, i.e., as a result of many collisions of energetic electrons. This indirect way should contribute to $\phi(T)$ an amount roughly independent of T. Consequently, it is sensible to write

$$\phi(T) = a\, d\sigma(T)/dT + b \tag{2}$$

where a and b are constants. The constants are not independent; they are subject to the obvious requirement that

$$\int_0^{E_1} \phi(T)dT = \bar{N}_i \tag{3}$$

must be equal to the total ionization $\bar{N}_i$ in an originally neutral gas. The above idea is Platzman's, although I have presented it in a slightly more general way.

For He and H_2, it is sensible to set $d\sigma(T)/dT$ as proportional to $(T + I)^{-3}$, on the basis of both theoretical results and experimental data[7]. Thus, one obtains the Platzman form[2],

$$\sigma(T) = a(T + I)^{-3} + b \tag{4}$$

where I is the ionization threshold of the gas.

The precise determination of $\phi(T)$ is possible by the solution of the Spencer-Fano equation[8] at $T > E_1$, as was first carried out by Douthat[9, 10] for He and H_2; indeed, Douthat obtained[11] reasonable fits of his results with eq. (4), except for a small region near $E_1 = 0$. [Alternatively, one may carry out a Monte Carlo simulation of the electron

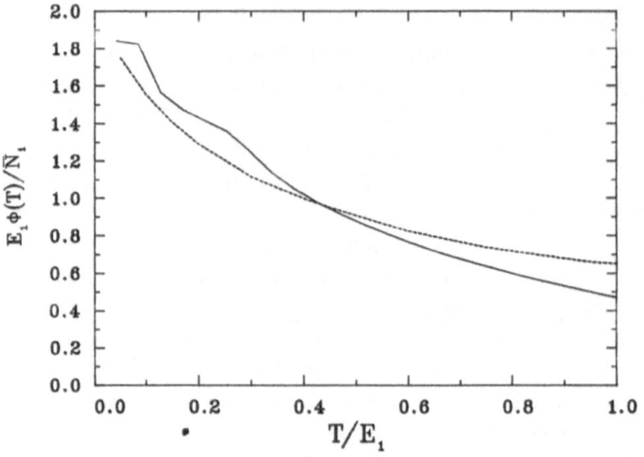

Fig. 1. The entry spectrum of subexcitation electrons in argon. The horizontal axis represents the subexcitation electron energy T measured in units of the first electronic-excitation threshold E_1. The vertical axis represents the normalized spectrum $E_1\phi(T)/\bar{N}_i$, so that the area under each curve is unity. The solid curve shows the result of solution of the Spencer-Fano equation. The broken curve represents the Platzman form, eq. (4). This figure is reproduced with publisher's permission from Inokuti, Kimura, and Kowari[6].

degradation process to obtain $\phi(T)$. An example is seen in the work by Kaplan et al.[12] on H_2O in gas and liquid. However, these authors present the numbers of subexcitation electrons in only four energy intervals, and it is difficult to determine how closely their results fit eq. (4)].

It is important to recognize that eq. (4) is not always appropriate. For Ar, for instance, the solution[6] of the Spencer-Fano equation with the use of realistic cross-section data leads to a result qualitatively different from eq. (4), as is seen in Fig. 1. One reason is that $d\sigma(T)/dT$ for Ar cannot be represented by the form $(T + I)^{-3}$. Further, Ar is not a special case; qualitatively the same results are expected for Kr and Xe and for other molecules for which data on $d\sigma(T)/dT$ are known[7]. In such cases, eq. (2) remains sensible.

In addition, Fig. 1 exhibits wavy structures in $\phi(T)$ at the lowest energies, i.e., $T/E_1 < 0.3$. These structures arise from an interplay of discrete excitation and ionization in the last few collisions of electrons before they enter the subexcitation domain.

3. MODERATION MECHANISMS

Subexcitation electrons slow down by three classes of processes: momentum transfer (to the translation of molecules) upon elastic scattering, rotational excitation, and vibrational excitation (the latter two of which apply to diatomic and polyatomic molecules). The knowledge about these individual processes is extensive, as is seen in the standard literature[13] and in the articles by Burke and Shimamura[14] and by Ehrhardt[15] in the present volume. Indeed, Professor Takayanagi made lasting contributions to the theory of rotational and vibrational excitation. What follows is only a brief summary of points essential to the moderation kinetics of subexcitation electrons, to be treated in Sections 4 and 5.

The elastic scattering by definition leaves the internal state of the struck molecule unchanged. However, the kinetic energy of the electron is changed by an amount depending upon the scattering angle θ. The mean energy loss E_e per elastic scattering of an electron by a molecule of mass M is given by

$$E_e = (2m/M) < 1 - \cos\theta > T \qquad (5)$$

where the brackets signify the mean over the angular distribution of scattering. Thus, the mean energy loss per unit pathlength in a gas containing one molecule per unit volume (i.e., the stopping cross section) is given by

$$E_e\sigma_e = (2m/M)T < 1 - \cos\theta > \sigma_e = (2m/M)T\sigma_m \qquad (6)$$

where σ_e is the total cross section for elastic scattering and σ_m is the momentum transfer cross section. Data on σ_m for many molecules have been extensively reviewed by Itikawa[16]. Inspection of the data shows that σ_m has substantial values of 10^{-16} cm^2 or larger. The energy dependence of σ_m is mostly smooth over a large energy range, except for some small regions. In general, the smooth part arises from long-range interactions, both electrostatic and polarization-induced, between an electron and a molecule. The strongly energy-dependent part arises from short-range interactions that cause a resonance (the temporary capture of an electron to a molecule), which is a subject of extensive studies pioneered by Schulz[17]. (Most often, the resonance is easily understandable as an electron temporarily residing in an unoccupied molecular orbital.) Returning to eq. (6), I note that the stopping cross section for elastic scattering is generally small because the ratio m/M is small, except for the lightest targets, such as H_2.

The rotational excitation is characterized by its cross section σ_r and associated energy loss E_r. The value of σ_r is often comparable to σ_m, and the energy dependence is generally governed by long-range interactions, especially when the molecule has a dipole moment. To cause a molecule to become rotationally excited, an incoming electron must exert a torque; however, it need not approach very closely if there are long-range interactions such as dipole or even quadrupole interactions (either permanent or temporarily induced).

The stopping cross section is given as the sum of $E_r\sigma_r$ over all possible individual processes. The value of E_r is a few multiples of the rotational energy level spacing and is generally of the order of magnitude of meV. Because of the small spacing, some points are relevant to the moderation kinetics. When one treats a gas at room temperature (as is often the case in radiation chemistry), molecules in the gas are in general rotationally excited at the outset. The moderation of subexcitation electrons in the gas is indeed due to the net energy loss, because a rotationally excited molecule can give energy to subexcitation electrons just as well as it can take energy from them. To account for this in full detail, one must use, in principle, a huge volume of cross-section data that describe both rotational excitation and de-excitation for various initial rotational states. Fortunately, much simplification of such a treatment is made possible by a simple scaling of cross sections for different initial rotational states of molecules, provided the electron energy is substantially higher than the thermal energy of the gas. According to the scaling, the cross section for any initial rotational state is readily calculable from the cross section for a particular initial rotational state, e.g., the ground rotational state. This result stems from the treatments of neutron scattering with nonspherical nuclei by Drozdov[18] and by Chase[19], and was given specifically for electron scattering with molecules by Shimamura[20].

The vibrational excitation is likewise characterized by the cross section σ_v and associated energy loss E_v. The value of σ_v is in general smaller than the values of σ_m or σ_r. This is so because an electron must approach close to a nucleus to excite its vibrational motion. An exception arises at a resonance, i.e., when an electron of the correct energy enters into an orbit around the molecule, at least temporarily. Then the forces acting on nuclei change appreciably; as a result, when the electron eventually leaves the molecule, it is highly likely to leave the nuclear motion excited, often to a high vibrational state[17]. When a sufficiently strong resonance thus occurs, σ_v becomes notably large and structured in the neighborhood of the resonance. The energy loss E_v is of the order of 0.1 eV, which is appreciable compared with typical subexcitation electron kinetic energies of several eV. Consequently, the stopping cross section due to vibrational excitation is almost always appreciable and is often dominant.

A quantitative analysis of the moderation of subexcitation electrons must be based on sound cross-section data concerning all the major energy loss processes. An analysis with unrealistic cross-section data is meaningless and often misleading. To adopt a comprehensive set of cross sections for a given molecule, even in the limited energy range of subexcitation electrons, one must not only survey all data in the literature, but also assess their reliability in the light of current knowledge, both experimental and theoretical. A classic example of such a study is seen in the work on N_2 by Itikawa et al.[21]. An article by Takayanagi[22] describes the need and the availability of cross section data for molecules, and includes incisive remarks on theoretical studies. In an article by Hayashi[23] one finds an extensive survey of cross-section data for many molecules of interest to applications ranging from gas discharges to radiation science.

4. MODERATION KINETICS

4.1. General Theory

A powerful method for analytically treating the electron moderation was given by Spencer and Fano[8]. Although they specifically discussed high-energy electrons, their idea applies to subexcitation electrons as well. The basic quantity is the electron degradation spectrum $y(T)$, or the tracklength distribution, as a function of electron energy T, which is defined as follows: $y(T)dT$ represents the total pathlength of electrons that have energies between T and T + dT. Suppose that the medium consists of a single molecular species at the number density n, and that $u(T)dT$ represents the number of source electrons having energies between T and T + dT. Then $y(T)$ satisfies the Spencer-Fano equation

$$nK_T y(T) + u(T) = 0 \tag{7}$$

provided that the source is stationary, i.e., $u(T)$ is independent of time. In eq. (7), K_T is a linear integral operator having the dimension of the cross section, defined as follows: $K_T y(T)$ represents the net gain and loss electrons of energy T in a medium of unit density. More specifically, it is expressed as

$$K_T y(T) = \int dT'\, y(T')\, \sigma(T' \to T) - y(T) \int dT''\, \sigma(T \to T'') \tag{8}$$

where $\sigma(T_1 \to T_2)$ represents the cross section for all the collision processes in which an electron of energy T_1 collides with a molecule and an electron of energy T_2 emerges. The operator K_T depends parametrically on T; the subscript T is meant to indicate this.

Once the electron degradation spectrum is determined through solution of eq. (7), it is straightforward to calculate the yield of any product s that is generated in the course of the electron slowing-down process; for instance, we may wish to calculate the mean number of vibrationally excited states. Let the cross section for the process be $\sigma_s(T)$; then the mean number $N_s(T_0)$ of the product s due to the incidence of an electron of initial energy T_0 is given by

$$N_s(T_0) = n \int dT\, y(T_0, T)\, \sigma_s(T), \tag{9}$$

where $y(T_0, T)$ is the solution of eq. (7) with the source term set as $u(T) = \delta(T - T_0)$.

Recent developments in the Spencer-Fano theory are summarized elsewhere[24]. A generalization to time-dependent cases[25-28] was stimulated largely by recent experiments[29-31] bearing on the moderation of subexcitaion electrons.

In the time-dependent theory, the basic quantity is the incremental electron degradation spectrum $z(T; t)$, which is defined as follows: $z(T; t)dTdt$ represents the increment, during the period between time t and t + dt, of the total pathlength of all the electrons having kinetic energies between T and T + dT. When $u(T; t)dtdT$ represents the number of source electrons that are introduced into the medium in the period between t and t + dt and that have kinetic energies between T and T + dT, $z(T; t)$ obeys the equation

$$[v(T)]^{-1} \partial z(T; t)/\partial t = nK_T\, z(T; t) + u(T; t) \tag{10}$$

where $v(T) = (2T/m)^{1/2}$ is the speed of an electron at kinetic energy T.

With a time-dependent source, the yield of any product is time dependent, and it is calculated from the incremental spectrum. For instance, consider a monoenergetic and instantaneous source as given by $u(T; t) = \delta(T - T_0)\ \delta(t)$. Let the solution of eq. (10) with this source term be $z(T_0, T; t)$. Then, the increment $\nu_s(T_0; t)$ of the number of product s during the period between t and t + dt is given by an expression formally the same as eq. (9), with $y(T_0, T)$ replaced with $z(T_0; T; t)$.

4.2. The Yield of Product Species Expressed in Terms of the G Value

It is customary in radiation chemistry to express the yield of a product in terms of the G value, which is the number of that product formed for 100 eV of the radiation energy absorbed in the medium. Therefore, it may be useful to show how $N_s(T_0)$ of eq. (9) is converted to G_s, which is the G value for the formation of product s. Suppose that energy Q is absorbed from a given ionizing radiation, which may be charged particles, photons, or neutrons with specified fluence. The subexcitation electrons formed as a result are characterized by their entry spectrum $\phi(T)$, as is discussed in Section 2. The total number N_s of product s formed is then given by

$$\bar{N}_s = \int_0^{E_1} dT_0 \phi(T_0)\ N_s(T_0) \tag{11}$$

where $N_s(T_0)$ is given by eq. (10). The G value is then

$$G_s = (100/Q)\bar{N}_s = (100/Q) \int_0^{E_1} dT_0 \phi(T_0)\ N_s(T_0) \tag{12}$$

where Q is measured in units of eV.

It is convenient to rewrite Q in terms of the total ionization $\bar{N}_i$, which is the best studied quantity in radiation physics and dosimetry, both experimentally and theoretically, and is expressed in terms of the W value[32]; thus, one may write $Q = W\bar{N}_i$ and insert this into eq. (12). The result is

$$G_s = (100/W\bar{N}_i) \int_0^{E_1} dT_0 \phi(T_0) N_s(T_0) \tag{13}$$

By use of eq. (3), one can further rewrite this as

$$G_s = (100/W) \int_0^{E_1} dT_0 \Phi(T_0) N_s(T_0) \tag{14}$$

where $\Phi(T_0)$ is the <u>normalized entry spectrum</u> of subexcitation electrons, or

$$\Phi(T_0) = \phi(T_0)/\bar{N}_i = \phi(T_0)/ \int_0^{E_1} dT_0 \phi(T_0) \tag{15}$$

In summary, eq. (14) enables one to convert $N_s(T_0)$ to G_s from the knowledge of the W value and $\Phi(T_0)$.

4.3. The Continuous-Slowing-Down Approximation (CSDA)

For treating the moderation of subexcitation electrons, it is often sensible to simplify the operator K_T. As we saw in Section 3, the fractional energy loss in each collision of a subexcitation electron is often small, until the kinetic energy becomes comparable to the thermal energy of the medium gas. Under these conditions, the operator K_T takes a simple form

$$K_T y(T) = \partial/\partial T[s(T)y(T)] \qquad (16)$$

where s(T) is the stopping cross section,

$$s(T) = \int dE\ E\ d\sigma(T;\ E)/dE \qquad (17)$$

and $[d\sigma(T;\ E)/dE]dE$ is the cross section for the process in which an electron of energy T collides with a molecule and loses energy between E and E + dE. This simplification is known as the continuous-slowing-down approximation (CSDA).

Detailed derivation of eq. (16) is given in Section IV of Inokuti et al.[26]. In summary, one treats the energy loss E upon each collision as small compared with the kinetic energy T, uses the Taylor expansion of the K_T operator of eq. (8), and retains only the terms of the first order in E. Then, the integral operator K_T is approximated by the first-order differential operator, as is seen in eq. (16). The validity of the CSDA therefore depends on two criteria. First, the fractional energy loss E/T should be small in the majority of collisions. Second, the neglect of higher-order terms in the Taylor expansion is justifiable when the operand y(T) is smooth as a function of T. This in turn depends on the smoothness of cross sections as functions of T.

Within the CSDA, the solution $y(T_0, T)$ of the time-independent equation, eq. (7), with $u(T) = \delta(T - T_0)$, is simply the reciprocal of the stopping power,

$$y(T_0, T) = 1/[ns(T)] \qquad (18)$$

at every $T < T_0$.

According to the CSDA, the mean time required for an electron of a fixed initial energy T_0 to slow down to a lower energy T is written as

$$\tau(T_0, T) = \int_T^{T_0} [ns(T')\ v(T')]^{-1} dT' \qquad (19)$$

where $v(T') = (2T'/m)^{1/2}$ is the speed of an electron of energy T'. We refer to $\tau(T_0, T)$ as the CSD time. It is often useful for the consideration of time-dependent aspects of the moderation. The time-dependent equation, eq. (10), becomes a linear partial differential equation of the first order and therefore is always analytically solvable[25,26].

5. APPLICATIONS

5.1. Hydrogen: An Example of Well-Studied Gases

For molecular hydrogen, the behavior of subexcitation electrons is more thoroughly understood than for any other gas, because of the work of Douthat[10,11,33]. First, Douthat studied[10] the entry spectrum, the yields of rotational and vibrational excitation, and the yields of negative hydrogen ions resulting from dissociative electron attachment,

$$e + H_2 \rightarrow H + H^- \tag{20}$$

Process (20) is possible at several bands of electron energies, but the total yield is dominantly determined by electrons with 7.0 eV $< T < E_1 =$ 8.8 eV. At $T <$ 7.0 eV, the cross section for the electron attachment is minute. At $T > E_1$, electrons moderate too rapidly because of electronic excitation to contribute appreciably to the yield. This exemplifies the general principle that the yield of any process is determined by the competition of that process with the moderation. Indeed, one interpretation of eq. (9) is precisely this principle. Recall that $y(T_0, T)dT$ represents the pathlength of electrons having energies near T. If the moderation at T is weak, then $y(T_0, T)$ is large, and the integrand of eq. (9) is appreciable at T unless $\sigma_s(T)$ is negligible. If the moderation at T is strong, then $y(T_0, T)$ is small, and the integrand of eq. (9) is small.

Next, Douthat[11] evaluated the CSD time [eq. (19)] required for an electron to moderate to various terminal energies between 0.1 eV and 1.0 eV. His results are in good agreement with experiment[34-36]. Note that the terminal energies he chose are substantially higher than thermal energies at room temperature. To determine the time required for an electron to reach thermal energies, one must fully analyze the kinetics, incorporating the energy gain of the electron from molecules. [This topic is beyond the scope of the present chapter. For fuller discussions on this topic, see Refs. 37-40, for instance.]

Finally, Douthat[33] calculated the yields of rotational and vibrational excitations in molecular hydrogen at various temperatures between 50 K and 1500 K. His motivation was to point out a mechanism for molecular-hydrogen infrared emissions from various astronomical objects.

5.2. Helium or Neon Admixed with Nitrogen

Dillon, Inokuti, and Kimura[25] gave an analysis of an experiment by Cooper, Denison, and Sauer[29], who studied light emitted from helium or neon admixed with a small amount of molecular nitrogen immediately after pulse irradiation with 0.6-MeV electrons. The light observed has a wavelength of 379 nm; it stems from the transition C $^3\Pi_u$, v = 0 → B $^3\Pi_g$, v'=2 of the N_2 molecule, and it serves as a monitor of the number of molecules in the C $^3\Pi_u$ state present at the time of observation. This state has the excitation energy E_{ex} = 11.15 eV.

Cooper et al.[29] interpreted the observed growth of the light in terms of subexcitation electrons in the parent gas, i.e., the electrons

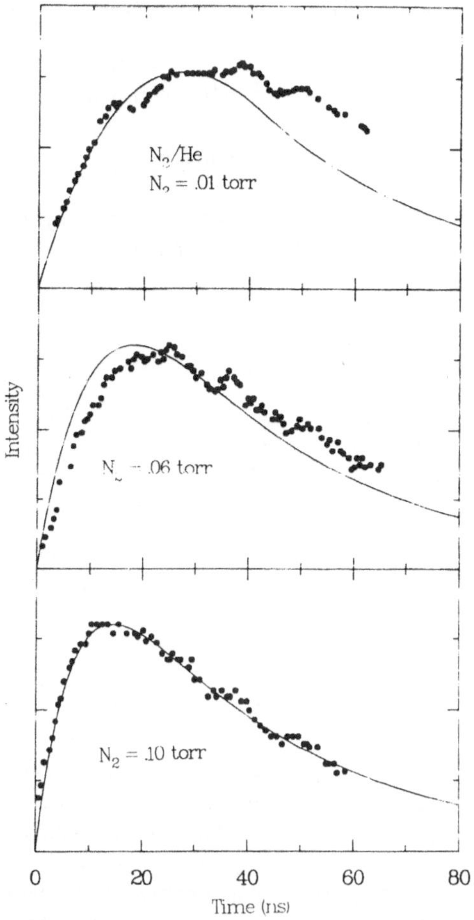

Fig. 2. Time dependence of the nitrogen emission in helium. The curves represent results of calculations[25] and the dots results of measurements[29]. In every case the partial pressure of helium is 20 Torr. This figure is reproduced with publisher's permission from Dillon, Inokuti, and Kimura[25].

with $T < E_1$ (E_1 = 19.8 eV in He and E_1 = 16.6 eV in Ne). The growth of the light is governed by the competition of the excitation of the N_2 C $^3\Pi_u$ state with the degradation of the electrons in the interval $E_{ex} < T < E_1$.

Dillon et al.[25] calculated the time-dependent yield of the C $^3\Pi_u$ by use of the theory outlined in Section 4 and obtained results in good agreement with the experiment[29], as is seen in Fig. 2. In the analysis of Dillon et al., an additional element deserves mention. The degradation of the electrons in the interval $E_{ex} < T < E_1$ occurs not only through elastic collisions with He or Ne, but also through collisions with N_2 that result in several discrete electronic excitations as well as in ionization. The energy loss upon the inelastic collisions with N_2 is greater than 8 eV, and the cross sections for these collisions are appreciable. Therefore, these collisions must be fully taken into account; most often they remove electrons from the interval. This effect was incorporated into the analysis as a negative source term of the form

$$u(T;\, t) = -\, k(T)\, z(T;\, t) \tag{21}$$

in eq. (9), where k(T) is the product of the number density and the total inelastic-collision cross section of nitrogen for an electron of energy T.

5.3. Nitrogen: Effects of a Strong Resonance

Stephens and Robicheaux[41] conducted a survey of cross-section data for H_2, N_2, O_2, and CO in the subexcitation domain and studied the effects

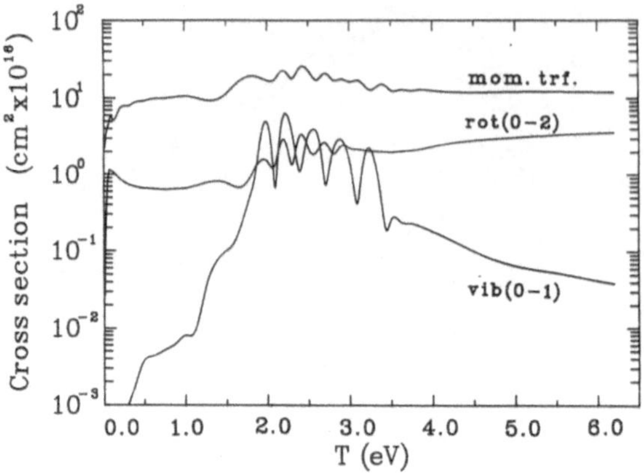

Fig. 3. Cross sections of N_2 for electrons. The curve labeled "mom. trf." represents the momentum transfer cross section $\sigma_m(T)$. The curve labeled "rot(0-2)" represents the cross section for the transition from J = 0 to J = 2, J being the rotational quantum number. The curve labeled "vib(0-1)" represents the cross section for the transition from v = 0 to v = 1, v being the vibrational quantum number. This figure is reproduced with publisher's permission from Kowari, Kimura, and Inokuti[42].

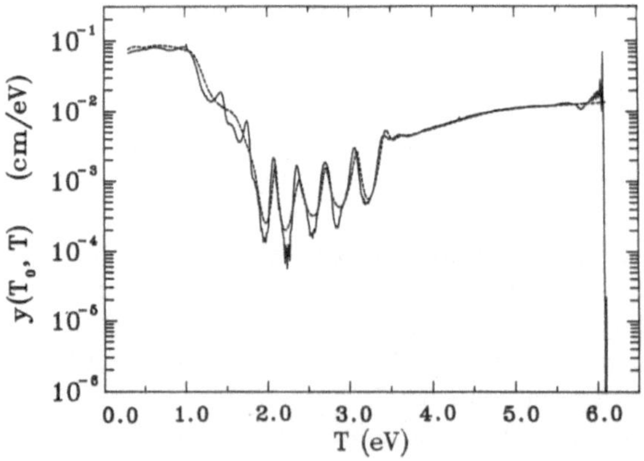

Fig. 4. An example of the degradation spectra in N_2. The spectrum $y(T_0,\, T)$ is shown as a function of electron energy T, for the incident energy T_0 = 6.1 eV. The solid curve represents the solution of the Spencer-Fano equation, and the broken curve the CSDA. This figure is reproduced with publisher's permission from Kowari, Kimura, and Inokuti[42].

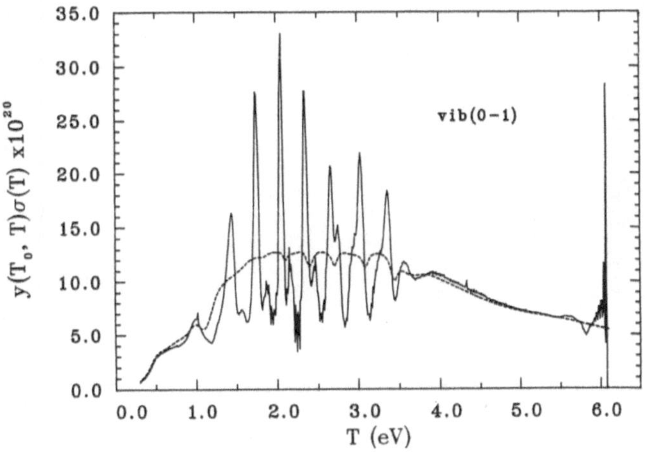

Fig. 5. Yield spectrum for the vibrational excitation in N_2. The integrand of eq. (10) is plotted, with σ here representing the cross section for the excitation from the vibrational state v = 0 to v = 1. The solid curve shows the result from the solution of the Spencer-Fano equation, and the broken curve the result from the CSDA. This figure is reproduced with publisher's permission from Kowari, Kimura, and Inokuti[42].

of resonances on the moderation within the CSDA. Stimulated by that work, I renewed my interest in the resonance effects. Results of work with my co-workers[28,42] on N_2 are summarized below.

The resonance influences all the cross sections at electron energies between 2.0 eV and 4.0 eV, as is seen in Fig. 3. The influence is seen in rich structures, most conspicuous in the vibrational-excitation cross section. As a result of the resonance, the degradation spectra also exhibit rich structures in the same interval of electron energy T (Fig. 4). In this interval the CSDA deviates appreciably from the Spencer-Fano degradation spectrum when one looks at its value at each energy, but the CSDA does represent the average behavior of the Spencer-Fano spectrum. This is an illustration of the point, made in Section 4.3, that the smoothness of the dominant cross-section with respect to the electron energy is a requirement for the validity of the CSDA.

Another major finding of the work on N_2 is the presence of distinct energy regions that contribute to the rotational excitation on one hand and to the vibrational excitation on the other hand. The yield of the vibrational excitation arises from practically the entire subexcitation domain, and most importantly from the resonance region between 2.0 eV and 4.0 eV (Fig. 5). The yield of the rotational excitation, in contrast, arises from the region above 4.0 eV and from the region below 2.0 eV (Fig. 6). This work shows an example of the competitive interplay of the vibrational and the rotational excitation.

5.4. Carbon Dioxide: Evaluation of Negative-Ion Formation

In carbon dioxide, studied by Pagnamenta et al.[43], one sees another

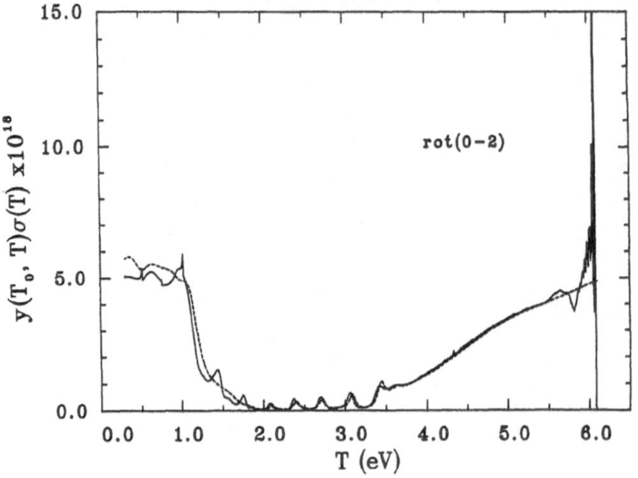

Fig. 6. Yield spectrum for the rotational excitation in N_2. The integrand of eq. (10) is plotted, with σ here representing the cross section for the excitation of the rotational state J=0 to J=2. The solid curve shows the result of the solution of the Spencer-Fano equation, and the broken curve the result of the CSDA. This figure is reproduced with publisher's permission from Kowari, Kimura, and Inokuti[42].

aspect in the evaluation of the yield of O^- ions resulting from the electron dissociative attachment,

$$e + CO_2 \rightarrow CO + O^- \tag{22}$$

The cross section for this process is seen in Fig. 7, together with other cross sections. There are two peaks in this cross section as functions of electron energy T, the smaller peak at about 4.3 eV and the larger one at about 8.2 eV.

The electronic-excitation threshold E_1 is effectively at about 6.2 eV, although an optically forbidden electronic excitation occurs at 5.4 eV with a minute cross section. Thus, the 4.3-eV peak of the dissociative attachment belongs to the subexcitation domain, in which the electron moderation is slow. The 8.2-eV peak, in contrast, belongs to the domain of electronic excitation, in which the moderation is fast. Consequently, the 4.3-eV peak is more important than the 8.2-eV peak in the total yield of the O^- ions. Yet the contribution of the 8.2-eV peak is appreciable and amounts to about one-third of the total yield, because the disparity of the stopping cross section above and below E_1 is rather modest, amounting to only about a factor of twenty.

The study on CO_2 thus illustrates the need for a careful analysis in the evaluation of the yield of negative ions and of other products due to subexcitation electrons in general. The study also shows that the distinction between the subexcitation domain and the domain of electronic excitation in molecules is not in general as straightforward and clear-cut as it is in rare gases.

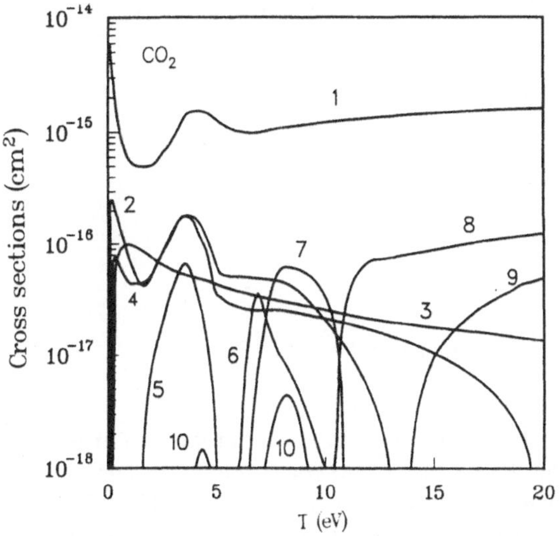

Fig. 7. Cross sections of CO_2 for electrons. Curve 1 represents the momentum transfer cross section $\sigma_m(T)$. Curves 2-5 represent cross sections for four kinds of vibrational excitation, (010), (100), (001), and (200) in the spectroscopic notation. Curve 6 represents the cross section for dissociation. Curves 7 and 8 represent the cross sections for the first two discrete electronic excitations. Curve 9 represents the total ionization cross section. Finally, curve 10 represents the cross section for electron dissociative attachment, eq. (22), multiplied by a factor of ten. Note the two peaks in curve 10 at about 4.3 eV and about 8.2 eV. This figure is reproduced with publisher's permission from Pagnamenta et al.[43].

5.5. Oxygen: The Role of Weak Electronic Excitation

Oxygen, studied by Ishii et al.[44], represents a case in which weak electronic excitations occur with low threshold energies. These excitations are to the a $^1\Delta_g$ state with threshold at 0.977 eV, and to the b $^1\Sigma_g^+$ state with threshold at 1.627 eV. The excited states both occur as multiplets belonging to the same configuration $(\sigma_g 2s)^2(\sigma_u 2s)^2(\pi_u 2p)^4(\pi_g 2p)^2$ as the ground state, $^3\Sigma_g^-$. and therefore have low excitation energies. Generally speaking, the occurrence of such low electronic excitations is expected of any atom or molecule with an open-shell structure in the ground state.

Major cross sections of O_2 for electron collisions are shown in Fig. 8. The vibrational excitation dominates in the contribution to the stopping cross section. In addition, there are conspicuous structures in the vibrational-excitation cross section in the domain $0.2 < T < 1.5$ eV.

Ishii et al. studied the electron degradation spectra $y(T_0, T)$ for various incident electron energies $T_0 \leq 10$ eV. A major finding is that the CSDA fails to reproduce even qualitatively $y(T_0, T)$ obtained as the solution of the Spencer-Fano equation [eq. (7) or eq. (9)]. An example is seen in Fig. 9. The discrepancy of the CSDA from the Spencer-Fano

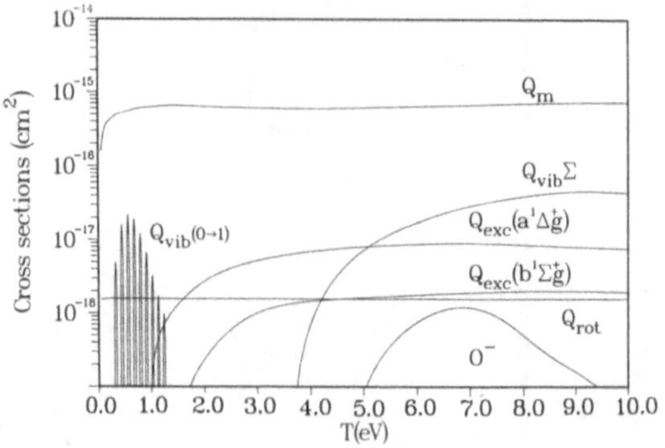

Fig. 8. Cross sections of O_2 for electrons. Curve labeled "Q_m" represents the momentum transfer cross section. Curve labeled "$Q_{vib}\Sigma$" represents the sum of all the vibrational-excitation cross sections. Curve labeled "Q_{vib} (0 → 1)" represents the cross section for the excitation from the vibrational state v = 0 to v = 1. Curve labeled "Q_{exc} (a $^1\Delta_g$)" represents the cross section for the excitation to the a $^1\Delta_g$ state. Curve labeled "Q_{exc} (b $^1\Sigma_g^+$)" represents the cross section for the excitation to the b $^1\Sigma_g^+$ state. Curve labeled "Q_{rot}" represents the total cross section for rotational excitation. Finally, curve labeled "O^-" represents the cross section for dissociative electron attachment. This figure is reproduced with publisher's permission from Ishii et al.[44].

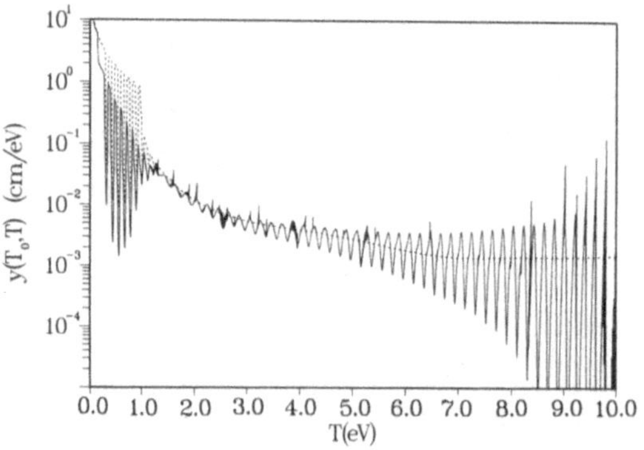

Fig. 9. An example of the degradation spectra in O_2. The spectrum $y(T_0, T)$ is shown as a function of electron energy T for the incident energy T_0 = 10 eV. This figure is reproduced with publisher's permission from Ishii et al.[44].

spectrum is most noticeable in the domain $0.2 < T < 1.5$ eV, as is expected from the derivation of the CSDA [outlined in the paragraph following eq. (17)]. In the same region, the electronic excitations have much smaller cross sections but cause a much larger energy loss per collision than do the vibrational excitations. Thus, the electron moderation is stochastic and cannot be adequately treated within the CSDA.

In the domain 1.5 eV < T < 5 eV, the weak electronic excitations are the principal contributors to the electron degradation. In the domain 5 eV < T < 10 eV, the vibrational excitation dominates. Above about 10 eV, electronic excitation begins to be strong. In other words, the effective domain of subexcitation electrons is below about 10 eV. Thus, O_2 illustrates the need for a flexible use of the term subexcitation electrons depending upon the material under consideration.

5.6. Yields of Negative Ions in Other Gases

Having seen detailed treatments of the negative-ion formation in H_2 and CO_2, the reader is likely to ask, "What about other molecules?" In other words, we wish to determine in which molecules the negative-ion formation may be appreciable. A simple answer to this question was offered by Inokuti, Takayanagi, and Platzman[5]. The basic idea is to see the possibility of the competition of the negative-ion formation with the moderation. For this purpose, it is useful to plot for various molecules the electron energy E_a at which the attachment cross section peaks against E_1, the first strong electronic-excitation threshold.

Figure 10 is such a plot made by the use of recent data, and thus supersedes the figure of Ref. 5. Although the idea of this plot is simple, there are some complications in preparing it because it is not

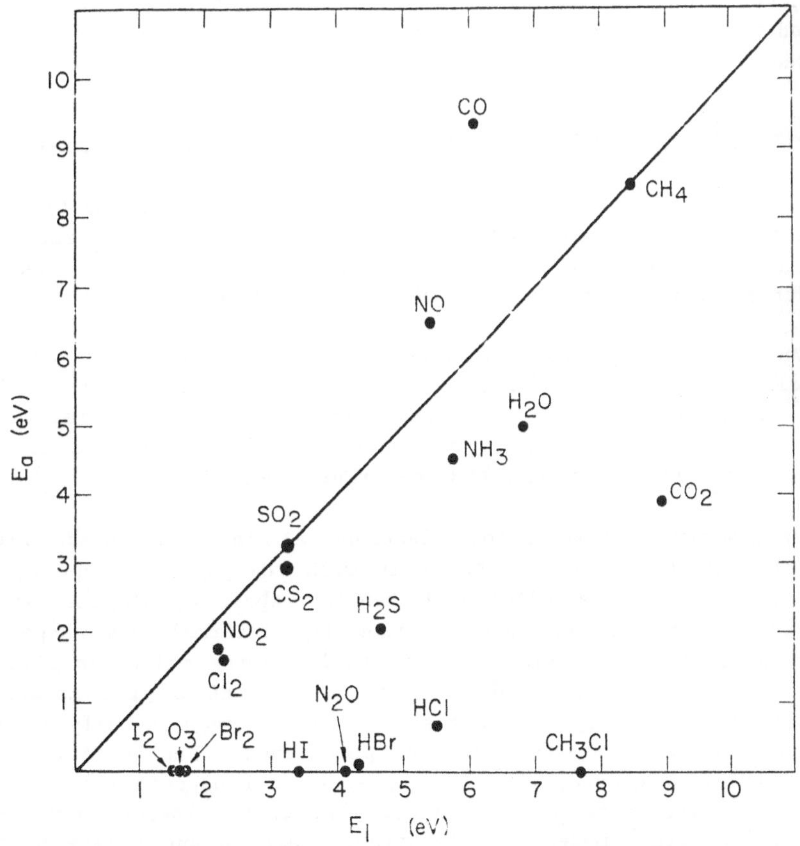

Fig. 10. Plot of the attachment electron energy E_a against the first strong excitation threshold E_1.

always immediately clear what values are appropriate for E_a and E_1. For E_1, we first refer to spectroscopic data given by Herzberg[45,46], but we must also consider the electronic-excitation cross section for electron collisions. This is particularly true when an optically forbidden excitation occurs to a low-lying state, as we saw in O_2. Often relevant electron collision data are inconclusive or unavailable, unlike in O_2. For E_a, the extensive review of the literature by Dillard[47] is extremely helpful. However, thorough consideration of the literature is necessary to determine at which specific energy the electron attachment is fully appreciable. Although I have used my best judgement in preparing Fig. 10, the result should be viewed as tentative.

Notwithstanding the above qualification, Fig. 10 is useful for the general understanding of the negative-ion yield. First, note that CO and NO belong to the domain $E_a > E_1$. This means that the formation of negative ions (O^-) is inappreciable in these gases because it must compete with strong electronic-excitation processes. Second, note that I_2, O_3, Br_2, HI, N_2O, HBr, HCl, and CH_3Cl appear near the horizontal axis, i.e., have $E_a \ll E_1$. For these molecules, there is little need for detailed analysis; practically all of the subexcitation electrons will give rise to negative ions. Therefore, the yield of negative ions will be about the same as the total number of subexcitation electrons, which we can readily obtain from the W value[32]. Finally, molecules such as CH_4, NH_3, H_2O, CO_2, and H_2S appear close to or below the diagonal in Fig. 10. For these molecules the competition of the negative-ion formation with the moderation requires a detailed analysis, as is exemplified by CO_2 (treated in Section 5.4).

6. ADDITIONAL REMARKS: SUBEXCITATION ELECTRONS IN CONDENSED MATTER

So far I have restricted the discussion to gases, for the sake of simple presentation. However, the notion of subexcitation electrons is pertinent to condensed matter, too; indeed, Weiss[1] and Platzman[2] considered liquids such as water and solids such as ionic crystals.

Energy losses of subexcitation electrons in condensed matter have been clarified in lesser detail than those in gases. For molecular solids composed of N_2, CO, H_2O, and hydrocarbons, substantial knowledge exists about the interactions of subexcitation electrons, most notably because of the work of Sanche and co-workers[48-55].

In summary, subexcitation electrons at large distances from any particular molecule interact not only with that molecule but also with many other molecules simultaneously because their de Broglie wavelength is comparable to or even greater than the intermolecular separations. In contrast, at short distances (comparable to the molecular dimension), electrons in molecular solids interact with each of the molecules in virtually the same way as in gases. This is most readily seen in a resonance, which is a short-distance phenomenon. Yet, once a resonance occurs, its consequences are not the same in condensed phases as in a gas. The consequences most often are some energy losses to vibrational excitation or other modes of excitation. This is due to the electronic polarization of the medium caused by the localized electron, as was discussed by Fano, Stephens, and Inokuti[56].

The work by Sanche and co-workers is extremely stimulating to theoreticians. Indeed, Fano and co-workers[56-59] have conducted exploratory studies on various aspects of subexcitation electron interactions in condensed matter. The coming decade will see a fruition of efforts in this direction.

The role of subexcitation electrons in liquid water and aqueous solutions has been pursued for many years by Hart and co-workers[60-63]. A full review of all these topics is deferred to another occasion, in view of the space limitation of the present chapter.

Acknowledgments The present essay owes much to daily discussions with my immediate colleagues, M.A. Dillon, Mineo Kimura, Ken-ichi Kowari, and David Spence, who have also commented on an earlier version of the manuscript. I also thank R. Cooper, D.A. Douthat, U. Fano, E.J. Hart, Y. Itikawa, L. Sanche, M.C. Sauer, Jr., I. Shimamura, and J.A. Stephens for valuable remarks and information that they have generously given me over the years.

REFERENCES

1. J. Weiss, Chemical action of ionizing radiations: Excitation of optical levels by particles of relatively low energy, Nature 174, 78-79 (1954).
2. R.L. Platzman, Subexcitation electrons, Radiat. Res. 2, 1-7 (1955).
3. W.P. Jesse and J. Sadauskis, Ionization by alpha particles in mixtures of gases, Phys. Rev. 100, 1755-1762 (1952).
4. M. Inokuti, Introduction to the Symposium on the Jesse Effect and Related Phenomena, Radiat. Res. 59, 343-349 (1974), and succeeding papers in the same issue.
5. M. Inokuti, K. Takayanagi, and R.L. Platzman, Role of dissociative electron attachment in radiation chemistry, in: Abstracts of Papers, The Third International Congress of Radiation Research, July 1966, Cortina d'Ampezzo, Italy (G. Silini, ed.) p. 115 (1965). See also Argonne National Laboratory Report ANL-7220, p. 11 (1966).
6. M. Inokuti, M. Kimura, and K. Kowari, Energy spectra of subexcitation electrons, Chem. Phys. Lett. 152, 504-506 (1988).
7. M. Inokuti, M.A. Dillon, J.H. Miller, and K. Omidvar, Analytic representation of secondary-electron spectra, J. Chem. Phys. 87, 6967-6972 (1987), and references therein.
8. L.V. Spencer and U. Fano, Energy spectrum resulting from electron slowing down, Phys. Rev. 93, 1172-1181 (1954).
9. D.A. Douthat, Electron degradation spectra in helium, Radiat. Res. 61, 1-20 (1975).
10. D.A. Douthat, Energy deposition and electron energy degradation in molecular hydrogen, J. Phys. B 12, 663-678 (1979).
11. D.A. Douthat, Electron terminal times in irradiated helium and hydrogen, J. Chem. Phys. 79, 4599-4601 (1983).
12. I.G. Kaplan, A.M. Miterev, and V. Ya. Sukhonosov, Comparative study of yields of primary products in tracks of fast electrons in liquid water and in water vapor, Radiat. Phys. Chem. 27, 83-90 (1986).
13. I. Shimamura and K. Takayanagi (eds.), Electron-Molecule Collisions, Plenum Press, New York (1984).
14. P.G. Burke and I. Shimamura, Theory of electron- and photon-molecule collisions, in Molecular Processes in Space (T. Watanabe, I. Shimamura, M. Shimizu, and Y. Itikawa, eds.) pp. 17-39 Plenum Press, New York (1990).

15. H. Ehrhardt, Experiments on low-energy electron-molecule collisions, in Molecular Processes in Space, (T. Watanabe, I. Shimamura, M. Shimizu, and Y. Itikawa, eds.) pp. 41-64 Plenum Press, New York (1990).
16. Y. Itikawa, Momentum-transfer cross sections for electron collisions with atoms and molecules, At. Data Nucl. Data Tables 14, 1-10 (1974); 21, 69-75 (1978).
17. G.J. Schulz, Resonances in electron impact on diatomic molecules, Rev. Mod. Phys. 45, 423-486 (1973).
18. S.I. Drozdov, Scattering of fast neutrons by nonspherical nuclei. III, Zh. Eksp. Teor. Fiz. 30, 786-788 (1955) [Sov. Phys. --JETP 3, 759-761 (1956)].
19. D.M. Chase, Adiabatic approximation for scattering processes, Phys. Rev. 104, 838-842 (1956).
20. I. Shimamura, State-to-state rotational transition cross sections from unresolved energy-loss spectra, Chem. Phys. Lett. 73, 328-333 (1980). See also I. Shimamura, pp. 89-189 of Ref. 13.
21. Y. Itikawa, M. Hayashi, A. Ichimura, K. Onda, K. Sakimoto, K. Takayanagi, M. Nakamura, H. Nishimura, and T. Takayanagi, Cross sections for collisions of electrons and photons with nitrogen molecules, J. Phys. Chem. Ref. Data 15, 985-1010 (1986).
22. K. Takayanagi, Electron-molecule collisions: Data needs versus data available, in: Electron-Molecule Collisions and Photoionization Processes (V. McKoy, H. Suzuki, K. Takayanagi, and S. Trajmar, eds.) pp. 133-143, Verlag Chemie International, Deerfield Beach, Florida (1983).
23. M. Hayashi, Electron collision cross-sections for molecules determined from beam and swarm data, in: Swarm Studies and Inelastic Electron-Molecule Collisions (L.C. Pitchford, B.V. McKoy, A. Chutjian, and S. Trajmar, eds.) pp. 167-187, Springer-Verlag, New York (1985).
24. M. Inokuti, M.A. Dillon, and M. Kimura, Theory of electron degradation and yields of initial molecular species produced by ionizing radiation, Int. J. Quantum Chem. Symp. 21, 251-266 (1987).
25. M.A. Dillon, M. Inokuti, and M. Kimura, Time-dependent aspects of electron degradation. I. Subexcitation electrons in helium or neon admixed with nitrogen, Radiat. Phys. Chem. 32, 43-48 (1988).
26. M. Inokuti, M. Kimura, and M.A. Dillon, Time-dependent aspects of electron degradation. II. General theory, Phys. Rev. A 38, 1217-1224 (1988).
27. M. Inokuti, M. Kimura, and M.A. Dillon, Time-dependent aspects of electron degradation. III. High-energy electrons incident on molecular hydrogen, Radiat. Phys. Chem., in press (1989).
28. M. Kimura, M. Inokuti, K. Kowari, M.A. Dillon, and A. Pagnamenta, Time-dependent aspects of electron degradation. IV. Subexcitation electrons in nitrogen and carbon dioxide, Radiat. Phys. Chem., in press (1989).
29. R, Cooper, L. Denison, and M.C. Sauer, Jr., A pulse radiolysis study of the excitation of nitrogen and aromatic hydrocarbons in rare-gas-additive mixtures: Evidence for subexcitation electron interactions, J. Phys. Chem. 86, 5093-5097 (1982).
30. R. Cooper and M.C. Sauer, Jr., Time-resolved observations on subexcitation electrons in gaseous systems, Radiat. Phys. Chem., in press (1989).
31. R. Cooper and M.C. Sauer, Jr., Picosecond pulse radiolysis of rare gases: Evidence for the time evolution of the subexcitation spectrum, J. Phys. Chem., in press (1989).
32. International Commission on Radiation Units and Measurements, Average Energy Required to Produce an Ion Pair, ICRU Report 37, Washington, D.C. (1974).
33. D.A. Douthat, Initial vibrational and rotational yields from subexcitation electrons in molecular hydrogen, Astrophys. J. 314, 419-421 (1987).

34. J.M. Warman and M.C. Sauer, Jr., Determination of electron thermalization time in irradiated gases, J. Chem. Phys. 52, 6428-6429 (1970).
35. J.M. Warman and M.C. Sauer, Jr., An investigation of electron thermalization in gases using CCl_4 as an electron energy probe, J. Chem. Phys. 62, 1971-1981 (1975).
36. J.M. Warman and M.P. de Haas, The delayed absorption of microwaves due to electron thermalization in nanosecond pulse irradiated N_2, He, and Ar at atmospheric pressure, J. Chem. Phys. 63, 2094-2100 (1975).
37. G.L. Braglia, G.L. Caraffini, and M. Diligenti, A study of the relaxation of electron velocity distributions in gases, Nuovo Cimento B 62, 139-163 (1981).
38. K.D. Knierim, M. Waldman, and E.A. Manson, Moment theory of electron thermalization in gases, J. Chem. Phys. 77, 943-950 (1982).
39. B. Shizgal, Electron thermalization in gases, J. Chem. Phys. 78, 5741-5744 (1983).
40. B. Shizgal and D.R.A. McMahon, Electron distribution functions and thermalization times in inert gas moderators, J. Phys. Chem. 88, 4858-4862 (1984).
41. J.A. Stephens and F. Robicheaux, Resonance effect on the moderation of slow electrons in diatomic gases, Radiat. Res. 110, 19-34 (1987).
42. K. Kowari, M. Kimura, and M. Inokuti, Electron degradation and yields of initial products. II. Subexcitation electrons in molecular nitrogen, J. Chem. Phys. 89, 7229-7237 (1988).
43. A. Pagnamenta, M. Kimura, M. Inokuti, and K. Kowari, Electron degradation and yields of initial products. III. Dissociative attachment in carbon dioxide, J. Chem. Phys. 89, 6220-6225 (1988).
44. M.A. Ishii, M. Kimura, M. Inokuti, and K. Kowari, Electron degradation and yields of initial products. IV. Subexcitation electrons in molecular oxygen, J. Chem. Phys. 90, 3081-3089 (1989).
45. G. Herzberg, Molecular Spectra and Molecular Structure. I. Spectra of Diatomic Molecules, 2nd ed., D. Van Nostrand Co., Inc., Princeton, New Jersey (1950).
46. G. Herzberg, Molecular Spectra and Molecular Structure. III. Electronic Spectra and Electronic Structure of Polyatomic Molecules, D. Van Nostrand Co., Inc., Princeton, New Jersey (1966).
47. J.G. Dillard, Negative ion mass spectrometry, Chem. Rev. 73, 589-643 (1973).
48. M. Michaud and L. Sanche, Interaction of low-energy electrons (1 - 30 eV) with condensed molecules: I. Multiple scattering theory, Phys. Rev. B 30, 6067-6077 (1984).
49. L. Sanche and M. Michaud, Interaction of low-energy electrons (1 - 30 eV) with condensed molecules: II. Vibrational-librational excitation and shape resonances in thin N_2 and CO films, Phys. Rev. B 30, 6078-6092 (1984).
50. L. Sanche, Dissociative attachment in electron scattering from condensed O_2 and CO, Phys. Rev. Lett. 53, 1638-1641 (1984).
51. T. Goulet and J.-P. Jay-Gerin, Theoretical study of the transmission of low-energy (0-10 eV) electrons through thin-film organic molecular solids: Benzene, Radiat. Phys. Chem. 27, 229-239 (1986).
52. R. Azria, L. Parenteau, and L. Sanche, Dissociative attachment from condensed O_2: Violation of the selection rule $\Sigma^- \leftrightarrow \Sigma^+$, Phys. Rev. Lett. 59, 638-640 (1987).
53. M. Michaud and L. Sanche, Total cross sections for slow-electron (1 - 20 eV) scattering in solid H_2O, Phys. Rev. A 36, 4672-4683 (1987).
54. M. Michaud and L. Sanche, Absolute vibrational excitation cross sections for slow-electron (1 - 18 eV) scattering in solid H_2O, Phys. Rev. A 36, 4684-4699 (1987).
55. G. Bader, J. Chiasson, L.G. Caron, M. Michaud, G. Perluzzo, and L. Sanche, Absolute scattering probabilities for subexcitation electrons in condensed H_2O, Radiat. Res. 114, 467-479 (1988).

56. U. Fano, J.A. Stephens, and M. Inokuti, Absence of resonances in the elastic scattering of electrons in molecular solids, J. Chem. Phys. 85, 6239-6240 (1986).
57. U. Fano and J.A. Stephens, Slow electrons in condensed matter, Phys. Rev. B 34, 438-441 (1986).
58. U. Fano, Studies of slow electron action on condensed media, Radiat. Phys. Chem. 32, 95-97 (1988).
59. J.A. Stephens and U. Fano, Slow electrons in condensed matter. The large polaron, Phys. Rev. A 38, 3372-3376 (1988).
60. W.G. Brown and E.J. Hart, Effect of pH on oxygen (^3P) atom formation in γ-ray irradiated aqueous solutions, J. Phys. Chem. 82, 2539-2542 (1978).
61. E.J. Hart and W.G. Brown, O(^3P) atom formation in γ- and 184.9-nm-irradiated aqueous perchlorate solutions, J. Phys. Chem. 84, 2237-2240 (1980).
62. E.J. Hart and W.G. Brown, O(^3P) atom formation in γ-ray irradiated aqueous bromate solutions, Radiat. Phys. Chem. 15, 163-167 (1980).
63. E.J. Hart, W.G. Brown, and E. Bjergbakke, The mechanism of O ^3P atom formation in γ-ray irradiated and UV photolyzed aqueous solutions, Radiat. Phys. Chem. 23, 181-186 (1984).

ENERGY TRANSFER PROCESSES IN COLLISIONS INVOLVING IONIC FORCES

F. A. Gianturco

Department of Chemistry
University of Rome
Città Universitaria,00185 Rome
Italy

1. INTRODUCTION

The development of theoretical methods to describe both chemical reactions and energy transfer processes that occur in the gas phase and involve neutral partners, continues to be a very active field. The theories employed are largely based on quantum mechanics and are applied to a wide variety of systems[1-3] which are becoming increasingly more complicated and with a greater number of constituent atoms. Recently, however, the general interest has also increased on the experimental and theoretical study of inelastic processes in which at least one of the partners is an ion and therefore the insurgence of particularly strong long-range forces becomes a dominant feature of the overall energy-deposition, energy-exchange and energy-scrambling processes. One reason for this increase of interest is that these reactions play an important role in many diverse areas of research like atmospheric chemistry, combustion processes and interstellar molecular processes. Another, equally important, reason has been the development of a wide range of experimental results which involve proton beam interacting with fairly simple molecules and that have revealed marked propensity rules when discussing the excitation of rotational and/or vibrational modes or the selective excitation of specific vibrational modes in polyatomic species.

Moreover, the processes that we shall be discussing occur in the low-energy regime and this is an area of study in which Professor Takayanagi has made several important contributions and that he has reviewed on several occasions[4]. Thus, it seems only fitting to discuss the recent progress made in a subject which has been obviously so dear to him.

Because of the presence of charged particles, one of the possible occurrences is that during the collisional event a charge-transfer process takes place and therefore more than one potential energy surface (PES) needs to be invoked to describe the phenomenon. Moreover, the strongly orientational forces which are acting during the dynamical event are highly selective in depositing energy in one internal mode rather than in another and therefore the electronic coupling between the surfaces further complicates the picture when it is 'switched on' during encounters which are already causing inelastic events with rather large probability.

In the present review we will therefore focus our attention on one class of dynamical processes involving at least one charged particle : the inelastic interaction of a beam of protons with simple molecular targets, diatomics and polyatomics. In spite of the apparent simplicity of the process, in fact, we shall see that several novel aspects are brought out by the experiments and that both structural factors and dynamical factors play an important role in shedding some light on the microscopic causes of the various encounters.

In the following section we will therefore briefly describe the experimental findings. Since Section 4 will discuss the various features of the computed PES and their bearing on the scattering effects, Section 3 will provide beforehand the background of the dynamical treatments and their bearings on possible mechanisms involved in explaining the experiments.

2. THE EXPERIMENTAL FINDINGS

2.1. Inelastic Processes with Diatomics

In principle, the most general method for studying state-to-state inelastic collisions is the crossed-beam, energy-loss method. The ability to measure differential cross sections (DCS), thereby selectively probing different regions of the PES, is the most attractive feature of such a method. However, from a practical standpoint, it is rather difficult to implement for energy losses involving only rotational and/or vibrational energy content. In many systems of interest, in fact, the vibrationally inelastic cross sections are very small at thermal energies, while at higher collision energies it is difficult to achieve the energy resolution needed to separate inelastic from elastic scattering. In many systems it is also rather difficult to distinguish between rotational and vibrational contributions to the inelastic scattering.

The simplest of the diatomic targets is the H_2 molecule and therefore the $H^+ + H_2$ system has been studied first and its PES has been obtained long ago at a reasonable level of accuracy. One drawback, of course, is that the system is reactive and that charge-transfer channels become open, as we shall see below, at energies greater than 1.8 eV. Fortunately, the rearrangement reaction only becomes important for E_{coll} below about 2 eV and, at higher energies, the charge-transfer cross section is much smaller than the inelastic cross section. It therefore follows that at collision energies greater than 3 eV the inelastic processes dominate the scattering and the collisions can be considered as occurring on a single surface corresponding to the ground electronic state of the system.

The vibrationally inelastic processes were first studied in the early seventies[5-7], when people measured absolute integral cross sections for vibrational excitation (up to v = 4) in H^+, D^+ collisions with H_2 at energies between 100-1500 eV and also vibrationally state-resolved differential cross sections for collisions with H_2, HD and D_2 at much lower relative energies (4-20 eV), the latter being the ones of more direct interest for the present review.

The results at 10 eV showed that the vibrational transition probabilities from v=0 to v=1-4 increased monotonically with scattering angles up to the region of the classical rainbow angle. The elastic DCS showed a well-resolved rainbow and a steep increase at small scattering angles, while the inelastic angular distributions exhibited rainbow maxima shifted slightly to larger angles and a flat distribution between 5° and 20° in the c.m.-system. Thus, their increase is mainly due to the rapid fall-off of the elastic component. At angles beyond the classical rainbow both types of cross sections fall off very steeply so that most of the inelastic scattering which occurs in this system is probably accounted for by the angular range covered by those measurements. The experiments also found that the average energy transfer is approximately independent of the isotopic composition of the molecule[7], thus it appears that the main effect of the passing proton is to temporarily withdraw electron density from the H_2 bond, resulting in bond extension and simultaneous vibrational excitation.

A further set of experiments with greatly improved angular resolution were carried out later by the Kaiserslautern group[8-10], which reported detailed measurements at several collision energies (E_{cm} = 4.67, 6, 10, 15.3, 20 eV). At the lower energies pure rotational excitation is the dominant process, while at energies above about 7 or 8 eV the vibrational excitation process takes over. In both cases the measurements were able to show quantum undulations in the DCS before the rainbow and a slower fall-off of the cross sections at larger scattering angles.

Since the processes were all compared with either semiclassical or quantum[12] calculations performed on a single PES, the main 'bond dilution' mechanism and an impulsive, Poisson distribution of relative probabilities between final vibrational states seem to be sufficient to produce a rather good agreement between computed and measured angular distribution[13-14]. Further details of pure rotationally inelastic processes and of their relation with the orientational forces which are present in this system will be discussed in the following section on the dynamics.

The energy-loss technique described above has been applied to other diatomics heavier than H_2, the systems studied include N_2, CO, NO, O_2, HCl and HF, but most of the work has been focussed on the isoelectronic molecules N_2 and CO.

The earlier studies[15] examined small-angle scattering of protons from N_2, CO, HCl and HF at collision energies between 10-30 eV. Large energy losses were observed in the scattering off HF and HCl molecules. Due to the large vibrational spacings in these molecules, it was possible to attribute most of the energy loss to the rotational excitation of the diatomic targets.

In contrast, no energy loss was observed with N_2 and only a small energy loss was detected in the case of CO. Further studies with higher resolution showed indeed[16-18] that the H^+-molecule collisions excite more vibrational energy in CO than in N_2. In the Kaiserslautern experiments[16] proton energy losses were measured at various angles out to and just beyond the rainbow angle which, at E_{cm} = 30 eV, is approximately at 10°. These data were employed to derive vibrationally state-resolved DCS and vibrational transition probabilities as functions of scattering angle. Both v=1 and v=2 excitations of CO and NO were observed but only v=1 could be detected for the N_2 target. In the case of the CO system both transition probabilities were found to increase monotonically with the scattering angle, a trend which mainly reflects the more rapid fall-off of the vibrationally elastic DCS and not a real increase with scattering angle of the inelastic angular distributions. In fact, there is very little elastic or inelastic scattering at angles larger than the rainbow angle. On the whole, these systems exhibit a smaller extent of vibrational energy transfer than in the case of the H_2 target. In the latter case, in fact the average energy transfer $\langle \Delta E \rangle$ at E_{cm} = 20 eV was found to be 0.44 eV or ~ 2% of the available energy. In contrast, the H^+-CO scattering 30 eV showed[16,17] a total of $\langle \Delta E \rangle$ of 0.07 eV or 0.25% of the available energy, a value which was about 2 times larger than that for N_2 and slightly smaller than for NO. Other experimental observations[18] indicate that: (i) the vibrational transition probabilities for N_2, CO and NO increase monotonically with scattering angle over the angular range studied; (ii) only v=1 is excited in N_2 whereas both v=1 and 2 are excited in CO and NO; (iii) the amount of vibrational energy transfer increases in the order N_2, CO, NO with $\langle \Delta E \rangle_{vib}$ being 2-3 times higher for CO than for N_2 at most angles. These results agree quantitatively with the earlier ones and also manage to extract by difference an estimate of the amount of average rotational energy transfer in all the above systems; they found $\langle \Delta E \rangle_{rot}$ to be comparable to or larger than $\langle \Delta E \rangle_{vib}$ at all scattering angles. The rotational excitation that occurs during the collision events was found to increase in the order NO, CO, N_2. Thus, it appears that there is no correlation between the average rotational excitation and the permanent dipole moment of the diatomic target of these systems. The various aspects of the orientational forces that come to play a role during the inelastic dynamics will be more specifically discussed in the following sections, where the need to combine knowledge of structural properties with specific features of the dynamical time scales will be clearly shown.

2.2. Inelastic Collisions with Polyatomics

A number of very intriguing studies of vibrationally inelastic scattering of atomic ions from polyatomic target molecules have been also reported in recent years. The employed projectiles have been protons, deuterons and alkali ions, while the target molecules include the triatomics CO_2 and N_2O, the spherical top molecules CH_4, CF_4 and SF_6, the asymmetric top H_2O and a number of small fluorohydrocarbons. To begin with, a surprising degree of mode-selectivity has been observed in some of the above systems and this effect has been the highest with H^+ and D^+ as projectiles. Second, both the features and the magnitude of the vibrational excitation can be strong functions of the collision energy and scattering angle. Third, many of the experimental results can be rationalized by using a simple model in which the vibrationally inelastic transitions are viewed as "spectroscopic" transitions induced by the electronic field of the passing ion, whereby the molecular properties mainly determine the amount of inelasticity while the overall features of the PES explain the relative probability of channeling energy into either rotations or vibrations.

In the case of the CO_2 target, the first results of scattering with protons were reported by Linder and collaborators[19] at three different c.m. collision energies (14.7, 28.8 and 48.9 eV) and at c.m. scattering angles between 0° and the rainbow region. The rainbow angles are approximately 18°, 9° and 5.5° at the above energies and the small differences between laboratory and c.m. scattering angles in this system are less than the experimental resolution, and thus can be ignored.

Strong variations in the excitation pattern were observed in the above experiments as function of collision energy and scattering angle. The CO_2 vibrational levels will be classified in the usual way, by calling v_1, v_2 and v_3 the number of quanta in the symmetric stretching mode (0.166 eV each), the doubly degenerate bending mode (0.083 eV each) and the asymmetric stretching mode (0.291 eV each), respectively. Thus, in the small-angle region near $\vartheta = 0°$, the (010) bending mode excitation appears to dominate the inelastic energy transfer. Some (001) asymmetric stretch excitation is also observed in the forward scattering region and this becomes relatively more important at higher collision energies. As the scattering angles increase, however, the pattern of excitation also changes dramatically; the (001) and (100) transition probabilities increase monotonically with scattering angle, while the (010) transition probability turns out to be nearly independent of angle. At the higher collision energies, on the other hand, asymmetric stretch excitation accounts for nearly all of the large-angle inelastic scattering, especially in the rainbow region. The general behaviour of the individual excitation processes as the energy increases and as one moves away from the forward direction can be related both to structural features of the PES and to time factors from the dynamics of the process. They will be further discussed in the following sections.

The interactions responsible for H^+-CO_2 inelastic scattering at larger angles, particularly in the rainbow region, are clearly more involved than just charge-dipole interaction. The most interesting experimental result in this angular region is the dominance of asymmetric stretch excitation over bending excitation[19] and some of the possible causes for such a behaviour can be gleaned both from an exam of the relevant PES and from specific dynamical calculations, as we shall see below. The very similar experimental behavior of the N_2O molecule during inelastic collisional events has also been a very intriguing finding which is once more related with the structure of the relevant PES for the system at hand[20].

A clearer picture, however, seems to emerge from the extensive body of work on the scattering from targets like CH_4, CF_4 and SF_6. Most of the results on these systems are summarized in a 1982 paper from the Göttingen group[21]. The experiments were performed at c.m. collision energies in the range 4-10 eV and most of the energy loss spectra were detected at angles between 5° and 15°. Since the spherically averaged potentials for the H^+-CH_4, CF_4 and SF_6 systems have fairly deep wells (around 3.6-3.8 eV) then the corresponding rainbow angles are rather large, i.e., $\vartheta_r \sim 25°-30°$ at $E_{cm} = 10$ eV. This means that all the measured data pertain to scattering angles much less than ϑ_r, where one expects that the inelastic scattering should be mainly determined by the long-range part of the PES.

In all three molecules, only triply degenerate ν_3 and ν_4 modes are infrared(IR)-active. In each case also the ν_3 mode has the highest frequency and the largest dipole moment derivative. By far the best resolved spectra were given by CF_4, where ν_3 excitations clearly dominate. At $E_{cm} = 9.7$ eV and $\vartheta \sim 10°$ the average energy transfer for excitation of the ν_3 mode was found to be about 2.6% of the available energy and to increase monotonically with scattering angle over the range studied. The SF_6 spectra turn out to be less well resolved because of the lower ν_3 frequency but the behavior of the average energy transfer of the latter system is very similar to that of the CF_4 case.

In the case of H^+-CH_4, on the other hand, the data gathered are less clear and the situation is still not completely understood. In earlier, low-resolution studies[22] a large average excitation energy was observed at $E_{cm} = 20$ eV, while later experiments at $E_{cm} = 9.2$ eV and $\vartheta_{cm} = 10°$ observed[23] a much lower energy transfer and did not find that a single vibrational mode stands out prominently, as in the cases mentioned before, as being the dominant mode of excitation. Further studies with even more detailed resolution reported inelastic collisions of this system[24] over a broader energy range and found additional evidence of the unexpected excitation of IR inactive modes (e.g., ν_2) of tetrahedral methane. Thus, it seems reasonable to invoke in this case, as shall be reported in the next paragraph, the existence of temporary charge-transfer mechanisms and of dynamical vibronic coupling.

In a more general way, further experimental studies carried out by the Göttingen group[25] have shown that the phenomenon of mode-selective vibrational excitation in collisions with simple ions is not unique to highly symmetrical molecules such as CO_2, CH_4 and SF_6 but can be detected also in fluorohydrocarbon systems like CH_3F, CH_2F_2...CF_4 and C_2H_4, C_2H_3F ...C_2F_4 and C_2F_6. All of these molecules present at least one strong IR-active mode, in most cases a C-F stretch, with a mode energy of 110-160 meV (900-1300 cm^{-1}). For each of those systems the energy-loss spectra were measured at several angles between 5° and 20° at a c.m. collision energy of 98 eV. The results for CH_4, C_2H_4 and for slightly fluorinated molecules do not exibit a clear structure attributable to a specific, dominant mode while, in contrast, the energy loss distributions for the highly fluorinated targets show three or more regularly spaced maxima with spacings of one or more CF stretching modes. Their relative magnitudes also correlate well with absolute IR intensities in these systems.

In conclusion, the energy-loss spectra from proton inelastic scattering seem to provide a new, well articulate area of research where interesting correlations could be made between structural properties of the systems at hand and the dynamics of the collisional events. Furthermore, the existence of strong ionic forces has opened up the inclusion of another class of phenomena which appear to take place at the same time as the inelastic events: the charge-transfer, near-thermal collisions.

2.3. Charge-Transfer Collisions

The investigation of the dynamics of proton-molecule vibrational interactions at low collision energies, $10\ eV \leq E_{cm} \leq 30\ eV$, has recently received a further, considerable impetus from the experimental data obtained for a number of different systems, and over a wide range of scattering angles, on vibrationally resolved charge-transfer spectra. Besides the interesting information that such data yield on the charge-transfer process itself, they also complement and elucidate the earlier investigations on roto-vibrational inelastic scattering events discussed before. They contribute to the newly revived field of nonadiabatic charge-transfer vibronic coupling phenomena[26], where very little information based on scattering events has been available till very recently[27].

In an electronically nonadiabatic collision at least two potential energy surfaces are involved as the partners approach each other and interact. In that instance the coupling region will be passed twice, first along the incoming branch of the trajectory and second along the outgoing one. This generally leads to at least four different collision pathways depending on whether the system stays on the diabatic or adiabatic energy surface on the ingoing and/or outgoing way.

One of the simplifying aspects of such couplings at the relative energies of the experiments considered here is that the collision energies pertain to a range of values where the target molecules can be considered to be stationary during the time needed for traversing a single, well-localized charge-transfer coupling region ($t \sim 10^{-15}$ s), while the molecular vibrations ($t_{vib} \sim 10^{-14}$ s) take place during the time that is likely to elapse between the crossing of one coupling region and the next. This implies the reliability of various dynamical approximations, as we will discuss below, and the near 'vertical' electron transitions between the two participating PES and, consequently a dominant role of the Franck-Condon factors in determining the relative importance of the final vibrational states. Deviations from such a simple overlap factor, however, are expected whenever the forces at play act for a long enough time to distort the vibrational coordinate in the molecule in question even before the actual curve crossing has taken place[28].

The experimental findings on systems like O_2[27], H_2[29], CO_2[20], H_2O[30] and others have been very valuable in clarifying the competitive role of resonant conditions for the amount of energy exchanged and the Franck-Condon (FC) factors. They have been considered important element for a long time, but the dynamics that has been revealed by the energy-loss collision experiments has been very useful in creating a much deeper understanding of the microscopic role of the forces at play in this broad class of ion-molecule collisions.

3. THE SCATTERING MODELS

As is well known from elementary considerations the collisions of simple molecules (sometimes polar ones) with ions are characterized by long-range interactions which strongly depend on molecular orientations. When the collision velocities are not too low, the probability of, say, rotational transitions caused by such forces can be appreciable even in a relatively distant encounter. This orientation-dependent interaction effects in a complicated way the relative motion of the collision pair and the standard Langevin cross section for ion-molecule reactions need to be modified.

In the specific case of protons being the projectiles there are a few additional features related both to their lack of structure and to the strong 'chemical' forces in which they can be involved.

First of all, because of their light masses, it was found that for spherical molecules rotational excitation was completely negligible. Even for molecules with large permanent dipole moment, with an exception discussed below for HF, only a small amount of rotational excitation was observed. Compared to all other ions with the same energy, the protons lead to the smallest kinematical smearing due to target motion during collisions. Moreover, being devoid of electrons they can come much closer to the molecule than ions with electronic shells and therefore the related field strengths can be unusually large ($\geq 10^8$ V/cm) and provide for an extremely strong coupling with molecular electrons, albeit for rather short time intervals, typically of the order of 10^{-14}-10^{-15} s. Since such an excitation time is always shorter than typical vibrational periods (of about 2×10^{-14} s) one can qualitatively say that the oscillator is excited coherently and that simultaneous transitions with significant probabilities can occur into very high overtone states which lead to unusually large energy transfers.

This type of behavior is strongly reminiscent of laser experiments where IR multiphoton excitations occur in molecular systems. In the latter case, however, because of the longer laser pulse duration ($\sim10^{-9}$ s) compared to typical nuclear motions, sequences of incoherent steps can take place and can lead to intramolecular relaxation processes which are different from those observed in the ionic collisions we are discussing here.

The corresponding dynamical treatments have therefore taken into consideration all the points discussed above and have employed different computational schemes depending on the relative collision energies that were discussed.

3.1. The Forced Oscillator

At the intermediate and high E_{cm} values (> 5 eV), where prevalent vibrational excitations were detected, the main approach was that of the forced harmonic oscillator model[31,32] which has been applied in earlier years for predicting vibrational relaxation rates in several systems[33].

In classical terms, if one defines ξ as the displacement of the oscillator from its equilibrium position, then the corresponding equation of motion is

$$\ddot{\xi} + \omega^2 \xi = \frac{1}{\mu} F(t) \tag{1}$$

where ω is the frequency and μ the reduced mass of the oscillator. The $F(t)$ is a time-dependent force from which one can calculate the classical expression for the energy being transferred to the oscillator:

$$\Delta E_{vib} = \frac{1}{2\mu} \left| \int_{-\infty}^{\infty} F(t) e^{-i\omega t} dt \right|^2 \tag{2}$$

which can also be viewed as the modulus squared of the Fourier component of the acting force field at the resonant frequency.

Corresponding quantum calculations for the same model produce the expectation value of the energy transfer as given by[34]

$$\langle \Delta E_{vib} \rangle = \sum_{n_f=0}^{\infty} P_{0\rightarrow n_f} \, n_f \, \hbar\omega \tag{3}$$

where n_f is the final state of the vibrator and $P_{0\rightarrow n_f}$ is the quantum probability for the excitation $|0\rangle \rightarrow |n_f\rangle$. The latter is given by a Poisson distribution

$$P_{0\rightarrow n_f} = \frac{1}{n_f!} \varepsilon_k^{n_f} \exp(-\varepsilon_k) \tag{4}$$

where $\varepsilon_k = \Delta E_k/\hbar\omega$ and ΔE_k is the average energy transfer equal to the first moment of the energy transfer distribution. This model has been extended quite easily to cases where two or more modes of the molecule can absorb energy and the various transition probabilities are obtained from a product of independent Poisson distributions

$$P_{(0,0\ldots,0)\rightarrow(0,\ldots n_k,n_{k+1},\ldots 0)} = \prod_{i=n_k,n_{k+1}} \frac{\varepsilon_i^{n_i}}{n_i!} \exp(-\varepsilon_i) \tag{5}$$

the product now runs over all modes that are excited during the process and n_k, n_{k+1} designate the overtone states of the kth and the (k+1)th vibrational modes.

The relevant proton-molecule potential can now be expanded in a Taylor series for each normal coordinate Q_k into which a classical trajectory $\underline{R}(t)$ is inserted

$$V[\underline{R}(t),Q_k] = V(t,0) + \sum_k \frac{\partial V}{\partial Q_k} Q_k + \frac{1}{2} \sum_{k,\ell} \frac{\partial^2 V}{\partial Q_k \partial Q_\ell} Q_k Q_\ell + \ldots$$

$$= V_0(t) + \sum_k F_k \, Q_k + \frac{1}{2} \sum_{k,\ell} G_{k,\ell} \, Q_k \, Q_\ell + \ldots \tag{6}$$

where the first term on the r.h.s. is a spherically symmetric part which only contributes to elastic deflection of the ion. On the other hand, if rotational excitations and/or electronic excitations are to be included, this term should contain terms leading to pure rotational excitation and/or to Franck-Condon electronic excitation. These two aspects will also be further discussed below.

Usually the dynamics is carried out by further assuming that the ion deflection angle is mainly determined from the spherical part of the induction potential and that both the repulsive region of the interaction and its dispersion attractive interaction can be disregarded in the region of b values which dominate the trajectories[25]. This essentially means that, if one writes $V_0(t)$ as

$$V_0[\underline{R}(t)] = \frac{B}{R^8} - \frac{C}{R^6} - \frac{1}{4\pi\varepsilon_0}\frac{e^2\alpha_0}{2R^4} \tag{7}$$

only the last contribution on the r.h.s. of eq. (7) is deemed to control the relevant trajectories.

Furthermore, during the collision at the energy considered it is often assumed that the relative velocity remains unchanged, due to the usually small relative energy transfers (i.e., $\Delta E/E \sim 0\text{-}3$ %), and that the ion trajectory can be approximated by a straight line. As a consequence, one can neglect for the moment all but the first-order infrared coupling terms appearing in eq. (6) and write down the coupling potential for each relevant mode as

$$V(t,Q_k) = -\sum_k \frac{\partial M}{\partial Q_k}\, \varepsilon_\perp(t)\; Q_k \tag{8}$$

where $\partial M/\partial Q_k$ is the dipole moment derivative with respect to the normal coordinate Q_k and $\varepsilon_\perp(t)$ is the time-dependent electric field acting on the molecule along the direction of R_m, the radial distance of closest approach during a classical trajectory from the $V_0(t)$ of eq. (7). One should keep in mind, however, that another coupling exists with $\varepsilon_{//}(t)$, i.e., the field component parallel to the actual trajectory at R_m, but the latter turns out to be markedly smaller from estimates of the acting field strengths in the present systems[32].

The result of eq. (8) is therefore the acting force during the forced motion of eq. (1). The corresponding classical expression for the energy transfer to that particular mode is given by

$$\Delta E_k = \pi\left(\frac{\partial M}{\partial Q_k}\right)^2 |\varepsilon_\perp(\omega_k)|^2 \tag{9}$$

where

$$\varepsilon_\perp(\omega_k) = \frac{1}{\sqrt{2\pi}}\int_{-\infty}^{\infty}\varepsilon_\perp(t)\,\exp(i\omega_k t)dt \tag{10}$$

is naturally the Fourier component of the electric field at the resonance frequency ω_k. Thus the picture is simply given by a normal mode that projects out of the broad frequency distribution produced by the passing proton a particular field component which is matching the oscillating frequency of that mode. The strongest coupling, and the largest energy transfer is thus expected to go to those modes with the largest value of $\varepsilon_\perp$ and/or $\partial M/\partial Q_k$.

For a straight-line trajectory and assuming no field dispersion by charge transfer, one can calculate rather easily the electric field strength at the center of the target molecule[35] as a function of time. At small scattering angles the above type of trajectory is certainly acceptable and the final energy transfer is obtained as

$$\Delta E_k = 2\left(\frac{\partial M}{\partial Q_k}\right)^2\left(\frac{e\,\omega_k}{4\pi\varepsilon_0 g^2}\right)^2[K_0^2(W_k) + K_1(W_k)] \tag{11}$$

here g is the straight-line velocity, $W_k = \omega_k b/g$ and the impact parameter b is estimated, at the given collision energy, through the deflection function from the spherical induction potential of eq. (7). K_0 and K_1 are modified Bessel functions which account for the two components of the acting field during the ion trajectory.

From the above equation one sees that the parameter W_k is proportional to the ratio between the collision time τ_{coll} and the time required for a vibration to occur, τ_{vib}. Thus, for small g values W_k approaches infinity corresponding to the adiabatic limit with negligible energy transfer. At the other limit of very short collisions, i.e., for $\tau_{coll} \ll 2\pi/\omega_k$, then the exponential term of eq. (9) is approximately equal to unity and the energy transfer can be written as

$$\Delta E_k \sim \frac{1}{2}\left(\frac{1}{4\pi\varepsilon_0}\right)^2\left|\frac{\partial M}{\partial Q_k}\right|^2\frac{e^2}{R_m^4}\tau^2 \tag{12}$$

where R_m is the classical turning point and $\tau = 2b/g$.

As mentioned before, the attractive induction potential produces the corresponding classical deflection function from rather simple considerations[36].

$$\Theta(E_{cm}, b) = \frac{1}{4\pi\varepsilon_0}\,\frac{3\pi}{8}\,\frac{e^2\alpha_0}{E_{cm}\,b^4} \tag{13}$$

hence the parameter W_k is given by

$$W_k = \left[\omega_k^2\,\frac{e\mu}{4E_{cm}}\left(\frac{1}{4\pi\varepsilon_0}\,\frac{3\pi\alpha_0}{2E_{cm}\Theta}\right)^{1/2}\right]^{1/2} \tag{14}$$

The final energy transfer for each mode is thus related to the transition dipole moment and to a dynamical function $f(E_{cm},\Theta,\omega_k)$

$$\Delta E_k \cong \left(\frac{\partial M}{\partial Q_k}\right)^2 f(E_{cm},\Theta,\omega_k) \tag{15}$$

it is interesting to note that, if the forced oscillator model is used with nonlinear contributions to eq. (6), then the probability distribution becomes non-Poisson[37]. The differences, however, between such distribution and a Poisson distribution remain rather small. Thus, the energy distribution among modes and overtones is very well described in the excitation experiments by the relationship of eq. (4). One interesting test of its validity is to write

$$\frac{d}{dn_f}\left[\ln(P_{0\to n_f}\, n_f!)\right] = \ln \varepsilon_k \tag{16}$$

whereby the slope of the straight line representing the energy dependence is a measure of the actual energy transfer ε_k. Thus, for $\varepsilon_k < 1$, corresponding to an average energy transfer which is less than the energy in a single quantum, we expect a negative slope. On the other hand, the situations $\varepsilon_k = 1$ and $\varepsilon_k > 1$ should change accordingly the observed experimental slope.

The experiments on polyatomic systems discussed before were all analyzed following the classical model of above and the presence of combined Poisson distributions was detected in most of the cases examined. If several modes were excited, it was found that the dominant mode, the kth mode, still appeared as a Poisson distribution but with all transition probabilities reduced by a constant factor. If that mode is degenerate, the relationship still holds but ΔE_k represents now the sum of the classical energy transfers into each of the degenerate components.

The classical approach discussed above still requires of course the solution of the classical equations of motion for all the trajectories sampled by the chosen interaction. Their use in interpreting experimental findings has been limited, however, to the use of eq. (15) for specific choices of impact parameters as generated by the spherical potential deflection function. The general result has been that, as the scattering angle changes, so do the Poisson distributions associated to a particular mode. Thus the mode-selective excitation should be related also to the specific behavior of the dynamical function $f(E_{cm},\Theta,\omega_k)$ that appears in eq. (15).

For a more detailed evaluation of the proton-molecule dynamics one needs obviously both the knowledge of the relevant PES and a more realistic, i.e., quantum mechanical, approach to the scattering process.

The treatments that have been applied to these systems reflect naturally the relative importance of the specific features of the ion-molecule interaction vis à vis the collision energy involved in the scattering. Thus, we will briefly discuss first the main models employed at higher collision energies, i.e., for situations where vibrational channels are open, and then will follow with the special models suggested for low-energy processes of astrophysical interest.

3.2. The Coupled Equations

The general starting point for solving the dynamical problem is the expansion of the total wave function onto isolated states of the target molecule with coefficients determining the behavior of the function for the relative motion[38]. In the simplest case of a diatomic target one can then write a standard uncoupled representation as

$$\Psi(\underset{\sim}{R},\underset{\sim}{r}) = \sum_{\ell,n,j} R^{-1} u^{\ell}_{nj}(R)\, Y_{\ell m_\ell}(\hat{R})\, Y_{jm_j}(\hat{r})\, \chi_{nj}(r) \tag{17}$$

where the asymptotic quantum numbers $|l,n,j\rangle$ are used to label both the orthonormal functions employed in the expansion and the corresponding unknown coefficients u's.

The usual, total angular momentum representation in the space-fixed (SF) frame of reference allows one to write down the quantum mechanical equations for the coefficients as coupled differential equations (CDE) for the collision coordinate

$$\underline{W}^J \underline{U}^J = 0 \tag{18}$$

where

$$\underline{W}^J = \underline{L} - \underline{V}^J(R) \tag{19a}$$

$$\{\underline{L}\}_{ij} = \{d^2/dR^2 - \ell_i(\ell_i+1)R^{-2} + \kappa_i^2\}\delta_{ij} \tag{19b}$$

$$\{\underline{V}^J(R)\}_{ij} = \frac{2\mu}{\hbar^2} \sum_\lambda V_\lambda^{ij}(R) \langle \mathcal{J}_i^J | P_\lambda | \mathcal{J}_j^J \rangle \tag{19c}$$

$$V_\lambda^{ij}(R) = \langle \chi_i | V_\lambda(R,r) | \chi_j \rangle \tag{19d}$$

and

$$\{\underline{U}^J(R)\}_{ij} = u_{ij}(R) \tag{19e}$$

here the strongly anisotropic ion-molecule potential has been expanded over an orthonormal basis of Legendre polynomials

$$V(\underline{R}, \underline{r}) = \sum_\lambda V_\lambda(R,r) P_\lambda(\hat{\underline{r}}\cdot\hat{\underline{R}}) \tag{20}$$

and usually one writes the internal coordinate dependence within each multipolar coefficient as[39]

$$V_\lambda(R,r) = A_\lambda(R) \sum_k C_{k\lambda}\, \xi^k(r) \tag{21}$$

where

$$\xi = \frac{r-r_{eq}}{r_{eq}} \tag{22}$$

with r_{eq} being the equilibrium geometry of the target diatomics.

Each scattering channel index $i = |n_i, j_i, l_i\rangle$ is labeled by the angular momenta values coupled within each J contribution and by the contributing vibrational state n_i. Each coupling matrix element contains therefore the combined effect of anisotropic couplings (rotational state couplings) and of vibrational state couplings. The size of the $\underline{V}^J$ matrix is therefore controlled by the product $(v_{max} \times r_{max})$, where v_{max} corresponds to the maximum number of vibrational states included in eq. (17) and r_{max} has the same meaning for the rotational states[38].

Because of the light mass of the proton, an efficient excitation of vibrational modes in simple targets is expected to happen from very short collision times. For E_{cm} values from 5.0 eV to 40 eV such times vary from

10^{-15} s to 5×10^{-14} s, which correspond to a time scale shorter than a typical period of vibrational motion ($\sim4\times10^{-14}$ s) and much less than a typical period of rotational motion ($\sim10^{-12}$ s). Moreover, the typical values of average energy transfers discussed before correspond to a rather small fraction of the total energy available, thereby leaving the proton with roughly the same average velocity before and after the encounters[35].

The qualitative considerations above can be introduced into the coupled (CC) equations (18) in order to yield specific decoupling effects between all the existing scattering channels.

Thus, within the range of action of the ion-molecule PES one could essentially disregard the dynamical coupling between the rotor states and the incoming projectile and select a constant value of partial wave describing the motion of the latter (or, equivalently, consider as being constant during the collision the projection of the rotational angular momentum along the SF frame of reference[40]). Moreover, the negligible values of energy transfers allow one to substitute into eq. (19b) a constant wave vector value, k_i, to the diagonal submatirx of values which pertain to the full rotational manifold within each vibrational level. The complete set of CC equations can now be diagonalized with respect to $\hat{j}$ and $\hat{l}$ by using an angle-dependent transformation over the internal orientation $\gamma=\arccos(\hat{R}\cdot\hat{r})$, so that only the dynamical coupling between vibrational states survives during the actual solution of the coupled equations. The rotational coupling is in fact obtained as a converged quadrature over γ-dependent scattering amplitudes[38,41].

This approximate treatment of the quantum-mechanical expansion (17) goes under the name of VCC-RIOS (vibrational close coupling - rotational infinite order sudden) approximation and has been applied both to linear systems and nonlinear polyatomics[42,43]. The final set of coupled equations can now be written as

$$\left\{ \frac{d^2}{dR^2} - \frac{\bar{\ell}(\bar{\ell}+1)}{R^2} + k_{n\bar{j}}^2 \right\} f_n^{\bar{\ell}\bar{j}}(R;\gamma) = \frac{2\mu}{\hbar^2} \sum_{n'} V_{nn'}^{\bar{j}}(R;\gamma) f_{n'}^{\bar{\ell}\bar{j}}(R;\gamma) \tag{23}$$

where $\bar{l}$ is now an arbitrary value of the angular momentum l (CS approximation) and the new, angle-dependent coefficients of the radial motion are coupled via the potential matrix elements already given in eqs. (19c) and (19d), albeit in different form

$$V_{nn'}^{\bar{j}}(R;\gamma) = \langle \chi_n^{\bar{j}} | V(R,r,\gamma) | \chi_{n'}^{\bar{j}} \rangle_r \tag{24}$$

Solving eq. (23) under the appropriate boundary conditions for the outgoing continuum functions f's yields the corresponding multichannel S-matrix elements for each vibrational transition ($n \to n'$), the $S_{nn'}^{\bar{\ell}}(E_{cm},\gamma)$. The effect of rotationally inelastic processes is then obtained from geometrical averaging over the angle

$$T_{m_j}^{\bar{\ell}}(nj\to n'j') = 2\pi \int_0^{\pi} d\gamma \sin\gamma \, Y_{j'}^{m_j *}(\gamma,0) \{ \delta_{nn'} - S_{nn'}^{\bar{\ell}}(\gamma) \} Y_j^{m_j}(\gamma,0) \tag{25}$$

The above elements of the T-matrix ($\underline{T} = \underline{1} - \underline{S}$) are written down for a choice of $\bar{l}$ that keeps the transition diagonal in m_j[44].

A very useful result from the above approach is that the full manifold of rotationally inelastic cross sections, integral and differential, can be obtained from a factorisation relation[40,45], whereby any DCS or ICS from an arbitrary state j reduces to a linear, 3j-symbol weighted superposition of rotationally and vibrationally inelastic cross sections from the rotovibrational ground state of the system under study

$$\sigma(nj\rightarrow n'j') = \frac{k_{n0}^2}{k_{n'j}^2} (2j'+1) \sum_{\bar{j}} C^2(j\bar{j}j'|000)\sigma(n0\rightarrow n'\bar{j}) \tag{26a}$$

$$\frac{d\sigma}{d\Omega}(nj\rightarrow n'j') = \frac{k_{n0}^2}{k_{n'j}^2}(2j'+1) \sum_{\bar{j}} C^2(j\bar{j}j'|000) \frac{d\sigma}{d\Omega}(n0\rightarrow n'\bar{j}) \tag{26b}$$

Even if such factorization relations are expected to be only valid in an approximate fashion and probably for only a limited range of collision conditions, they turned out to apply reasonably well to the experimental situations of proton collisions with simple linear targets[46], and are obviously of great practical value in reducing the number of individual, inelastic cross sections which need to be actually computed and treated as a source of independent information for a better understanding of such inelastic, energy transfer processes.

The advantage of the VCC-RIOS approach to the dynamical problem is given by the rather direct connection that is often possible to make between the outcomes of the collision process, features of the observed experiments and detailed properties of the relevant PES, i.e., the strength of its anisotropy, the extension in space of its long-range part, the orientational dependence of its vibrational coupling, etc.

In a few cases this has been indeed carried out over the years by starting with the computation of the full anisotropic potential energy surfaces[47-49], the evaluation of the vibrational coupling between the proton and the target molecule[39,50], and finally using the detailed knowledge of the interaction to yield the average energy transfer data and the angular distributions after collision[50-52].

The picture that emerges from the above studies, which have been mostly oriented at understanding the reasons for the dominance (or lack) of vibrational excitations versus rotational excitation processes, is one which puts together structural considerations with 'time scale' dynamical factors, i.e., which relates the specific orientational forces of, say, an HF molecule interacting with a proton with the collision times or the relative velocity of the various internal motions[46].

An example of such results is shown in Fig. 1, where the individual ICS for pure rotational excitation with a $\langle\Delta E_{rot}\rangle$ of ≥ 1.5 eV are shown as functions of collision energy.

One clearly sees there that, as the energy transfer increases, the overall probability decreases. This decrease also occurs as the percentage of energy being transferred becomes smaller with the increase of collision energy. Thus, the prediction is of inelastic collisions where most of the energy transfers go into rotational degrees of freedom until

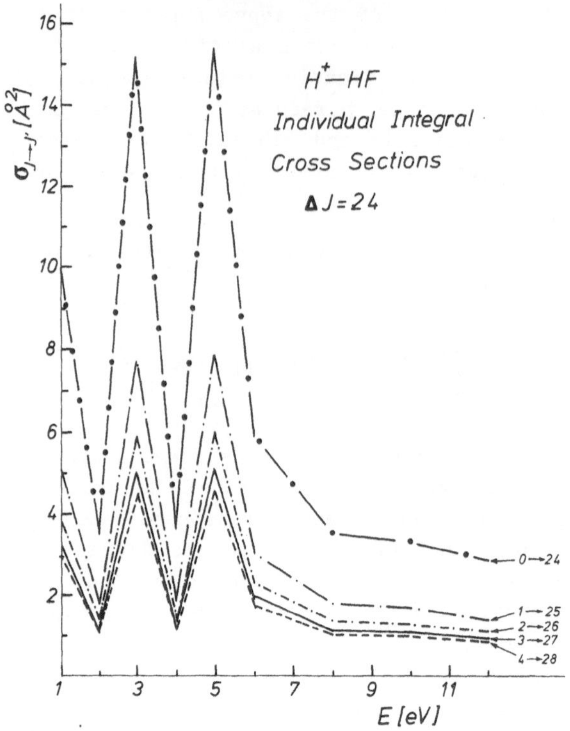

Fig. 1. Computed IOSA integral, partial cross sections for H^+ + HF collisions as functions of energy and for different values of $\langle \Delta E_{rot} \rangle$. See Ref. 51 for the detailes of calculations.

the E_{cm} becomes at least 15-20 eV, as is the case shown by the experiments[46].

3.3. Adiabatic Expansions

As discussed before, the functional expansion of eq. (17) essentially employs diabatic states of the molecular systems, i.e., uses an asymptotic representation of the target molecule even under the perturbing effects of the interaction with the ion. Another possibility is of course given by selecting a more efficient representation of the molecular states under the influence of the ion, i.e. to choose some type of adiabatic expansion which could hopefully converge more rapidly even in the presence of interaction regions with strong potential coupling. One also expects that, as the collision energy decreases, only the long-range part of the PES plays an important role and therefore the spacings between adiabats remain larger than the couplings among them[53].

One of the approaches suggested has been applied to polar molecules and to low-energy ion-molecule collisions involving only rotational excitations[54-57] and has been called the perturbed rotational state (PRS) approach, since the molecule is thought as having sufficient time to have its rotations deformed by the long-range dipole interaction.

The starting point of the theory is the calculation of the PRS wave functions χ by solving the eigenvalue problem

$$\{\mathcal{H}_{rot}(\hat{\underline{r}}) + \frac{M \cdot e}{R^2} \cos \vartheta\}\chi = \varepsilon(R)\chi \tag{27}$$

where $\mathcal{H}_{rot}$ is the molecular rotational Hamiltonian, M the permanent molecular dipole moment and $\vartheta = \arccos \hat{\underline{R}} \cdot \hat{\underline{r}}_{eq}$. The solution of eq. (27) can be obtained in a coordinate system where the polar axis is chosen to be along $\underline{R}$.

One can then expand χ in terms of spherical harmonics with respect to the Euler angles which define the molecular orientation in a rotating coordinate system as

$$\chi = \sum_k X_{km} Y_{km}(\vartheta', \phi') \tag{28}$$

By introducing a set of reduced quantities

$$\varepsilon/B = u \ , \quad R/(D \cdot e/B)^{1/2} = x \tag{29}$$

one can write eq. (27) as a universal equation with the new dimensionless variable x for the coefficients of (28)

$$[k(k+1) - u] X_{km} + \frac{1}{x^2} \sum_k \langle km|\cos \vartheta'|k'm\rangle X_{k'm} = 0 \tag{30}$$

At this point, one can turn to the equation of motion for the total wave function and obtain the transition probabilities between different rotational states by solving classically the relative motion along the R-coordinate[54] and determine the relevant time-dependent coefficients for the various final states. The simplest approach of a straight-line trajectory and of a constant velocity during the motion turned out to produce very good agreement with more elaborate CC calculations[54] and for several choices of the parameters involved[57]. A further modification that treated the motion more realistically and solved the classical equations of motion via the actual long-range potential also produced good agreement with earlier, more elaborate calculations which needed a markedly longer computational time.

The idea of treating low-energy collisions involving light ionic partners via an efficient set of adiabatic curves has been applied recently to processes involving negative charges. Although the phenomena under study involved mostly electrons as projectiles, it is instructive to describe them here because of their obvious importance in the case of ionic partners like H^+ or H^-.

In case one chooses a reference frame fixed with the molecular system (MF) and the interaction potential is known in that frame, i.e., the matrix of coupling potential (19c) is known over the whole range of the radial variable R, then one might use a more appropriate angular basis that takes into account the direct distorsion by the centrifugal potential within the MF frame of the local potential

$$\sum_{\ell'} \{ V^{\lambda}_{\ell\ell'}(R) + \frac{\ell(\ell+1)}{R^2} \delta_{\ell\ell'} \} \mathcal{T}^{\lambda}_{\ell' p}(R) = \varepsilon^{\lambda}_{p}(R) \mathcal{T}^{\lambda}_{\ell p}(R) \tag{31}$$

here λ defines the particular projection of l along the main molecular axis, l and l' define the channel partial waves in the MF. The adiabatic basis which diagonalizes locally the new effective potential is now contained in the new, orthogonal matrix $\mathcal{T}^{\lambda}$ and the corresponding eignvalues are labeled by the adiabatic channel index p(R). The new adiabatic angular eigenstates are now dependent on the relative motion coordinate

$$Z^{\lambda}_{p(R)}(\hat{\underline{R}};R) = \sum_{\ell} Y_{\ell\lambda}(\hat{\underline{R}})\ \mathcal{T}^{\lambda}_{\ell p(R)}(R) \tag{32}$$

and only a few of them should be needed in expansion (17):

$$R^{-1}u^{\ell}_{nj}(R)\ Y_{\ell m_{\ell}}(\hat{\underline{R}}) \Rightarrow R^{-1}\ u^{p(R)}_{n\lambda}(R)\ Z^{\lambda}_{p(R)}(\hat{\underline{R}};R) \tag{33}$$

where the new angular adiabatic basis is employed.

Similar adiabatic expansions with model potentials have been suggested before for polar molecules[58,59] and for processes involving the formation of temporary anions[6] and seem to indicate a marked improvement of convergence in expansion[17] and a better insight into the process under study.

An interesting example of some of such curves for a negative charge interacting with the HCl molecule ($^1\Sigma$ state) is shown in Fig. 2. One clearly sees that at low collision energies ($E_{cm} \leq 6$ eV) only a few curves are involved and that the coupled equations are likely to converge at a reasonable level with only the l = 1 and l = 2 states. Moreover, the latter adiabat is the only one likely to support pseudo-bound states of the predissociating anionic system, behaving like d-wave shape resonances in conventional electron-molecule scattering[61,62]. Thus, even before

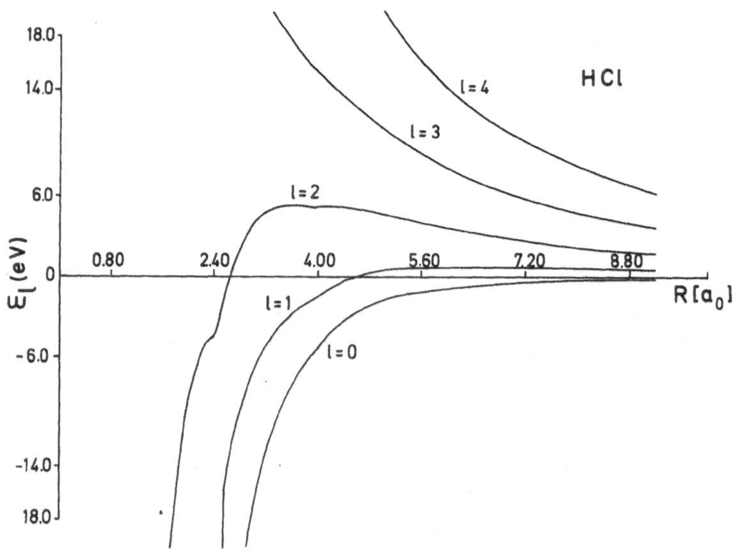

Fig. 2. Adiabatic potential curves for HCl ($^1\Sigma$) interacting with a negative charge as functions of relative distance. The index l corresponds to the contributing partial wave in the projectile expansion. Details of calculations from Ref. 62.

performing actual dynamical calculations one could make use of the adiabatic analysis to get some further understanding of the forces at play in this region of collision energies.

3.4. Nonadiabatic Models

In the cases were charge-exchange processes take place during encounters, as discussed in the previous section, then a network of potential curves usually needs to be employed. This situation naturally complicates the treatment of eq. (19) and therefore it was not implemented in a rigorous way for a rather long time, as people preferred to devise model treatments that essentially reduce the surface crossing to simpler curve crossings.

In the early seventies Tully and Preston[63] employed an idea of Bjerre and Nikitin[64] that suggested to couple the Landau-Zener formula[65] with the classical trajectory method in order to extend the quasiclassical approach to multisurface processes. That particular study produced energy-dependent integral cross sections for the $H^+ + D_2$ system which turned out to be in reasonably good agreement with the available experiments[65]. This success has directed the main applications of this simpler trajectory surface hopping method (TSHM) to integral quantities like ICS or product translational energy distributions.

The idea of replacing atom-diatom surfaces by curves was introduced for modeling charge-exchange processes by Bauer, Fisher and Gilmore (BFG)[66] who used it first for studying the quenching reaction of $Na(3^2P)$ by N_2. To find an explanation for the large cross sections observed experimentally for the process and for other similar processes, BFG devised a network of potential curves related to three different potential energy surfaces: the initial one, which correlated with the excited Na atom and the starting vibrational state of N_2, the final correlated with the ground state of Na and a different vibrational state of N_2 and an intermediate ionic PES which was assumed to cross the earlier two and correlated with Na^+ and N_2^-. The various curves can now form a grid of crossing points, to each of which is attributed a diabatic and an adiabatic transition probability computed via Landau-Zener formulas:

$$P_d = \exp(-\varphi) \tag{34}$$

$$P_a = 1 - \exp(-\varphi) \tag{35}$$

where

$$\varphi = \frac{2\pi u^2}{\Delta F \hbar v} \tag{36}$$

Here u describes the diabatic coupling term, v is the relative velocity and ΔF represents the difference in the gradients of the two intersecting curves

$$\Delta F = | F_1 - F_2 | \tag{37a}$$

with

$$F_i = \frac{\partial V_{ii}}{\partial R}, \quad i = 1,2 \tag{37b}$$

Within such a simple model one can then devise a quantitative evaluation of the quenching process as function of the charge-exchange probability for each of the relevant vibrational states of the N_2 molecule.

The more recent treatments have been directed to the study of the new set of observations discussed in the previous section, i.e., to charge-exchange encounters like

$$H^+ + M(v_i,j_i) \rightleftharpoons H + M^+(v_f,j_f) \tag{38}$$

where different experimental outcomes, as already said, are expected in going from highly endothermic systems like $H^+ + H_2$ ($\Delta E_{charge\ transfer}$ = 1.83 eV) to near resonant systems like $H^+ - CO_2$ (ΔE_{ct} = 0.2 eV) and further down to exothermic systems like $H^+ - O_2$ (ΔE_{ct} = -1.53 eV)[67].

A further, simplified approach often used when interpreting the experimental data is that given by the Franck-Condon (FC) model. Within that context one assumes that the transition from one surface to the other is sudden, without any intermediate transition as that postulated in the BFG model. The system is therefore described as moving from one given vibrational state in one surface to the final vibrational state of the other. The transition probability therefore is obtained from a Landau-Zener (LZ) branching ratio between surfaces, multiplied by an appropriate FC overlap factor

$$P^{\alpha}_{1,j} = P^{\alpha} \, |S_{1,j}|^2 \tag{39}$$

where 1 represents the initial vibronic state on surface 1 and j the intersecting state on the second surface. P^{α} is the usual LZ probability function.

In general the BFG and FC models produce different results which are supposed to become identical when either the coupling term or the overlap integral becomes small enough. Obviously there should be a parametric factor according to which either one or the other approach produces the best agreement with experiments.

Thus, one can say that the FC model should become more relevant the higher the collision energy while one should expect that the BFG model, because it is based on a step-by-step transition process, should be more realistic for low collision energies.

One should keep in mind, however, that such an energy-related parameter is not sharp enough since the relevance of the models is also related to the way in which the energy is distributed among the various degrees of freedom after the CT collision, as well as on the relative position of the full seam at the crossing point. For a simple model case a parameter γ was recently proposed[68] which is able to collect a good part of the needed information. If one, in fact, considers the equation for the crossing seam

$$G(R,r) = V_1(R,r) - V_2(R,r) = 0 \tag{40}$$

one can show that a parameter γ takes the form

$$\gamma = \frac{v_R}{v_r}\frac{\Delta G_R}{\Delta G_r} \tag{41}$$

where

$$\Delta G_s = \frac{\partial G}{\partial s}, \qquad s = r,R \tag{42}$$

and v_R, v_r are the translational and vibrational velocities, respectively. Next, it was established[68] that when $\gamma \gg 1$ the FC model becomes exact as the relative velocity dominates over the local vibrational motion, but when $\gamma \ll 1$ the BFG model should be preferred since the coupling between vibronic states occurs several times during the whole encounter and makes more realistic a step-by-step mechanism.

The systems which have been studied by using the above models for a qualitative explanation of the data involve the $H^+ - O_2$ case[27,69], the $H^+ - H_2$ system[29,70] and the $H^+ - H_2O$ system[30], while only very recently some work has been started on the direct solution of the VCC-RIOS approach, in three dimension, with the inclusion of off-diagonal diabatic coupling in an adiabatic representation[71-73].

4. THE CALCULATION OF ORIENTATIONAL FORCES

As we have mentioned several times in the previous sections, to carry out dynamical calculations requires the previous knowledge of the full PES, or of an array of PES's, for the system at hand. This necessarily means that the radial extension, angular dependence and overall strength of the forces at play need to be obtained in all the main regions of interaction: (i) the long-range region where polarization forces often dominate, (ii) the intermediate region where 'chemical' forces begin to act and new bonds are formed in ionic systems and (iii) the short-range region where strongly repulsive, highly orientational forces are created by the screened Coulomb effects of the nuclear cusps at the nuclear positions.

Over the last few years several attempts have been made at evaluating such forces for atom (ion)-diatomics situations over a broad enough range of relative distances that could allow one to carry out realistic scattering calculations. In the present section we will try to show how the knowledge of the PES could often help to get a qualitative picture of the dynamics even before the scattering calculations are performed.

In the case of protons as projectile, and whenever one can rely on a single PES for describing the process of interest, SCF calculations have been rather successful in describing orientational forces in cases like $CO - H^+$[47], $CO_2 - H^+$[49], $HF - H^+$[51] and the two lowest surfaces (adiabatic) for $O_2 - H^+$[14], that is in cases where the strongly orientational forces of chemical origin could be realistically described by a single determinant expanded over a large basis set (usually of the TZ+P type) of Gaussian orbitals.

Another possible tool for analyzing the more interesting situations where several curves are likely to play a role and where one expects to observe (or has actually observed) inelastic effects related to charge-transfer processes is given by the use of the diatomics-in-molecules (DIM) approach. In this case one needs to provide as input data several properties concerning both atomic and diatomic subspecies of the supermolecule, and the latter properties are obtained both from ab initio calculations and from experiments[75]. The final result, however, is usually fairly reliable for the description of that intermediate region of relative distances where the mixing of the various chosen configurations is most important and where the nonadiabatic curve crossing events are most likely to occur.

In the case of the H^+ - HF system, for instance, the DIM surface calculations[76] were carried out for both the asymptotic situations that lead to H^+ + HF ($^1\Sigma^+$) or to $H(^2S_g)$ + HF^+ ($^2\Pi$) fragments, using a wide range of relative distances and internal geometries. They were also composed with ab initio results and turned out to be in very good agreement with the latter calculations.

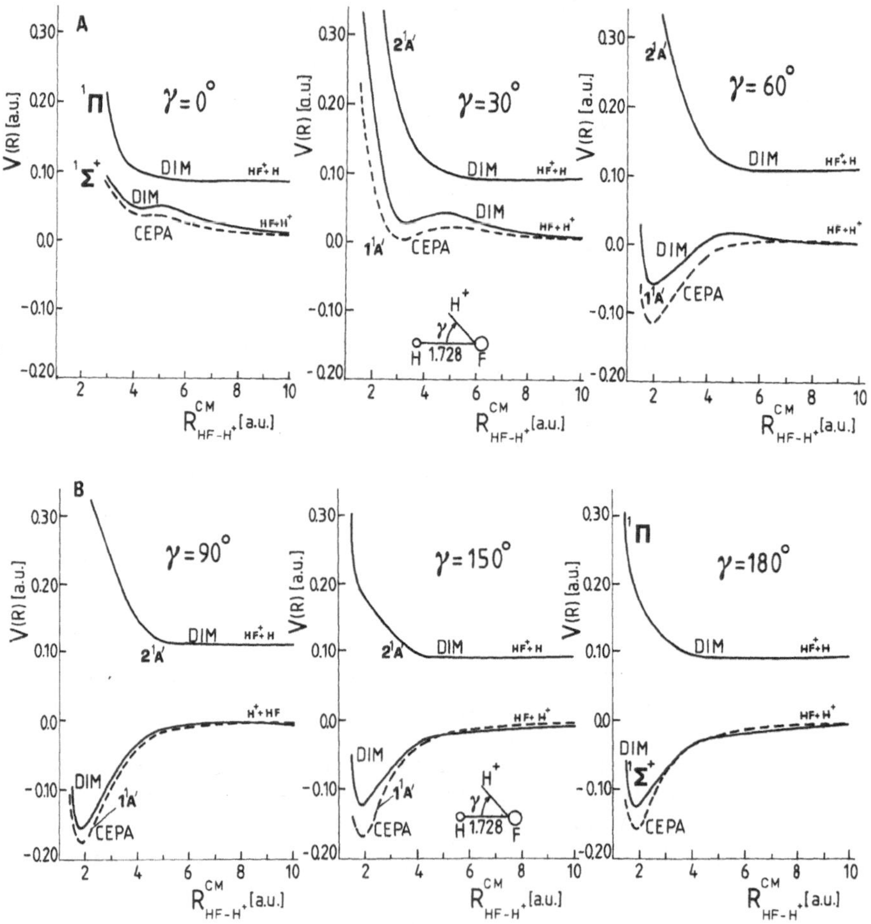

Fig. 3. (a) DIM potential curves for H^+ + HF and H + HF^+ at different orientations, from the collinear $(H\text{-}H\text{-}F)^+$ to γ=60°. The details of calculations are given in Ref. 76. (b) Same as in (a) but for different orientations leading to the other collinear structure of the $(H\text{-}F\text{-}H)^+$ configurations. Details also in Ref. 76.

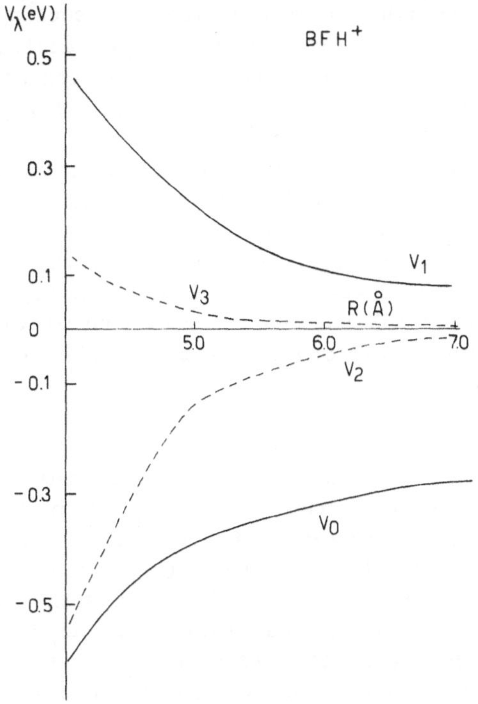

Fig. 4. Computed multipolar coefficients for H^+ interaction with BF at its equilibrium geometry. The calculations were carried out at the SCF level. See Ref. 81 for details of calculations.

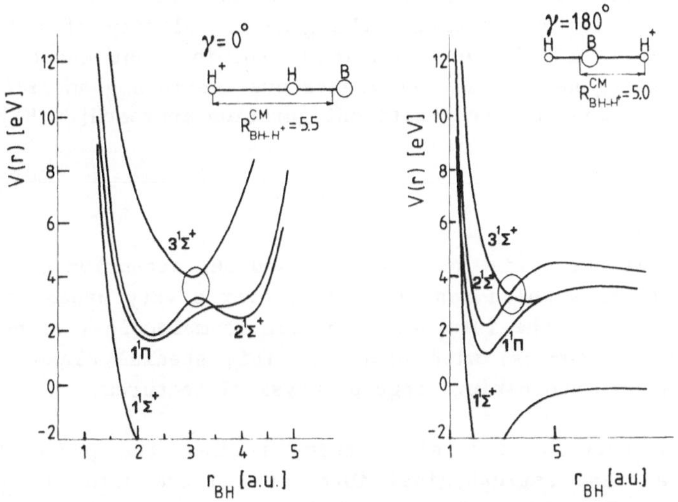

Fig. 5. DIM calculations of two collinear configurations of the $(HBH)^+$ system from H^+ + BH collisions and as functions of BH bond length. See details in Ref. 78.

The analysis of the various features shown by these two PES indicates clearly the strong orientational nature of the proton-molecule interaction and the different likelihoods of nonadiabatic couplings depending on the 'hitting angle' of the impinging proton. The results reported in Figs. 3(a) and 3(b) are typical of the present system and also show good agreement with earlier, CEPA results from an ab initio approach[77].

Another clear example of the orientational nature of ion-molecule forces is given by the computed multipolar coefficients of the PES for H^+-BF interaction when the equilibrium geometry is chosen for the target molecule. From Fig. 4 one clearly sees the attractive nature of the V_0 and V_2 coefficients, dominated in the long-range region by the polarization forces, and the repulsive character of contributions with odd index, that relate to charge-dipole and charge-quadrupole induction forces.

Of course one is aware of the fact that the analysis of the full seam is also needed to estimate nonadiabatic coupling effects, and an example of such a dependence is shown in Fig. 5 for the $(HBH)^+$ system as an example[78]. The distance from the center of mass has been fixed at values where no avoided crossing were observed in the calculations. The collisions on the H-side of the molecule ($\gamma=0°$) indicate a good possibility of rotational coupling with the $1^1\Pi$ state with formation of $BH^+(A^2\Pi)$ over a wide range of internuclear distances and in its lower vibronic states. On the other hand, the encounters on the B-end of the molecule show a strong avoided crossing at low collision energies but only with a BH molecule that is already vibrationally excited. Thus, the target preparation in the crossing beam is deciding in which direction the energy deposition will go as the nonadiabatic coupling occurs the most effectively for vibrationally excited target molecules.

More recent cases of complicated systems like BF and O_2 interacting with protons have also confirmed the general validity of a DIM approach for multisurface calculations[79], while showing that quantitative estimates of crossing seams and energy gaps over various geometries certainly require a more sophisticated multiconfiguration approach[80,81].

5. CONCLUSIONS

In the present review we have focussed our attention on a variety of inelastic processes which tend to occur, over a very broad range of collision energies, in the interaction of simple molecules, either diatomics or few-atom polyatomics, with protons. This special class of processes present a rather interesting range of physical features:

(i) In the very-low collision energy regimes the processes involved have a great astrophysical interest and are often responsible for resonant formation of charged clusters or for rotational break up of protonated species.

(ii) Due to the presence of highly anisotropic long-range forces the dynamics could be treated rather successfully using simplified force field and computationally undemanding physical models.

(iii) The existence of charge-transfer (CT) processes which accompany the inelastic collisions, and often condition the outcome of them, is an additional physical complication that is reflected in the obviously more demanding preparation of dynamical models. On the other hand, because of the rather simple nature of the projectiles, such CT events are still amenable to realistic treatments with resonable computational requirements, provided that the experiments allow the introduction of a few, very important, simplifying features[18,69].

(iv) The overall analysis of the experimental findings opens up in this case a new and extended area of molecular physics where the interplay between structural properties and dynamical time scales for the various degrees of freedom becomes even more close and significant.

Acknowledgments. The financial support of the Italian National Research Council (CNR) and of the Italian Ministry of Education (MPI) for the present research is gratefully acknowledged. It is a great pleasure to dedicate this review to Professor K. Takayanagi, to his indefatigable activity in the area of atomic and molecular collisions and to the many pleasant and illuminating conversations on various aspects of this subject that I had the honour to have with him over the years.

REFERENCES

1. F.A. Gianturco, The Transfer of Molecular Energy by Collisions, Springer Verlag, Heidelbelg (1979).
2. R.B. Walker and J.C. Light, Ann. Rev. Phys. Chem. 31, 401-44 (1967).
3. M. Baer, Adv. Chem. Phys. 49, 191-213 (1982).
4. e.g.: K. Takayanagi, Comm. Atom. Mol. Phys. 9, 143-52 (1980).
5. F.A. Herrero and J.P. Doering, Phys. Rev. A5, 702-11 (1972).
6. H. Udseth, C.F. Giese, and W.R. Gentry, Phys. Rev. A8, 2483-97 (1973).
7. C.F. Giese and W.R. Gentry, Phys. Rev. A10, 2156-68 (1974).
8. H. Schmidt, V. Hermann, and F. Linder, Chem. Phys. Lett. 41, 365-71 (1976).
9. H. Schmidt, V. Hermann, and F. Linder, J. Chem. Phys. 69, 2734-41 (1978).
10. V. Hermann, H. Schmidt, and F. Linder, J. Phys. B11, 493-504 (1978).
11. R. Schinke, H. Kruger, V. Hermann, H. Schmidt, and F. Linder, J. Chem. Phys. 67, 1187-99 (1977).
12. R. Schinke and P. McGuire, Chem. Phys. 31, 391-402 (1978).
13. R. Schinke, M. Dupuis, and W.A. Lester, Jr., J. Chem. Phys. 72, 3909-22 (1980).
14. R. Schinke, J. Chem. Phys. 72, 3916-29 (1980).
15. H. Udseth, C.F. Giese, and W.R. Gentry, J. Chem. Phys. 60, 3051-63 (1974).
16. J. Krutein and F. Linder, J. Chem. Phys. 71, 599-612 (1979).
17. J. Krutein and F. Linder, Chem. Phys. Lett. 51, 597-611 (1977).
18. F.A. Gianturco, U. Gierz, and J.P. Toennies, J. Phys. B14, 667-75 (1981).
19. J. Krutein and F. Linder, J. Phys. B10, 1363-78 (1977).
20. G. Niedner, M. Noll, and J.P. Toennies, J. Chem. Phys. 87, 2067-83 (1987).
21. T. Ellenbrock, U. Gierz, M. Noll, and J.P. Toennies, J. Phys. Chem. 86, 1153-63 (1982).
22. W.R. Gentry, H. Udseth, and C.F. Giese, Chem. Phys. Lett. 36, 671 (1975).
23. T. Ellenbrock, U. Gierz, and J.P. Toennies, Chem. Phys. Lett. 70, 459 (1980).
24. Y.-Nan Chiu, B. Friedrich, W. Maring, G. Niedner, M. Noll, and J.P. Toennies, J. Chem. Phys. 88, 6814-30 (1988).
25. U. Gierz, M. Noll, and J.P. Toennies, J. Chem. Phys. 82, 217-29 (1985).

26. e.g., see: A.W. Kleyn, J. Los, and E.A. Gislason, Phys. Rep. 90, 1-76 (1982).
27. M. Noll and J.P. Toennies, J. Chem. Phys. 85, 3313-28 (1986).
28. E.A. Gislason and E.M. Goldfield, Phys. Rev. A25, 2002-16 (1982).
29. G. Niedner, M. Noll. J.P. Toennies, and Ch. Schlier, J. Chem. Phys. 87, 2685-98 (1987).
30. B. Friedrich, G. Niedner, M. Noll, and J.P. Toennies, J. Chem. Phys. 87, 5256-67 (1987).
31. E.H. Kerner, Can. J. Phys. 36, 371-92 (1958).
32. T. Ellenbrock and J.P. Toennies, Chem. Phys. 71, 309-343 (1982).
33. K.H. Shin, Chem. Phys. Lett. 5, 137-46 (1970).
34. M.S. Bartlett and J.E. Moyal, Proc. Camb. Phil. Soc. 45, 545-61 (1949).
35. U. Gierz, M. Noll, and J.P. Toennies, J. Chem. Phys. 83, 2259-79 (1985).
36. R.J. Cross, J. Chem. Phys. 46, 609-21 (1976).
37. H.D. Meyer, Chem. Phys. 61, 365-78 (1981).
38. e.g., see: F.A. Gianturco, Atomic and Molecular Collision Theory, Plenum, New York (1981).
39. F.A. Gianturco, E. Semprini, F. Stefani, U.T. Lamanna, and G. Petrella, J. Phys. Chem. 92, 925-931 (1988).
40. e.g., see: R. Goldflam, S. Green, and D.J. Kouri, J. Chem. Phys., 67, 4149-72 (1977).
41. D. Secrest, J. Chem. Phys. 62, 710-26 (1975).
42. D.C. Clary, J. Chem. Phys. 75, 209-19 (1981).
43. D.C. Clary, J. Chem. Phys. 75, 2899-907 (1981).
44. R. Schinke and P. McGuire, J. Chem. Phys. 71, 4201-23 (1979).
45. V. Khare, J. Chem. Phys. 68, 4631-48 (1978).
46. B. Friedrich, F.A. Gianturco, G. Niedner, M. Noll, and J.P. Toennies, J. Phys. B20, 3725-35 (1987).
47. F.A. Gianturco, U.T. Lamanna, and D. Ignazzi, Chem. Phys. 48, 387-99 (1980).
48. F.A. Gianturco, U.T. Lamanna, and M. Attimonelli, Chem. Phys. Lett. 71, 49-57 (1980).
49. F.A. Gianturco, E. Semprini, and F. Stefani, Chem. Phys. Lett. 126, 81-7 (1986).
50. F.A. Gianturco, U.T. Lamanna, and G. Petrella, Nuovo Cim. 6, 75-92 (1985).
51. F.A. Gianturco, A. Palma, E. Semprini, F. Stefani, H.P. Diehl, and V. Staemmler, Chem. Phys. 107, 293-309 (1986).
52. F.A. Gianturco, U.T. Lamanna and M. Attimonelli, Chem. Phys. 48, 399-412 (1980).
53. e.g., see: F.T. Smith and D. Mukherjee, Phys. Rev. A17, 954-67 (1978).
54. K. Takayanagi, J. Phys. Soc. Jpn. 45, 976-89 (1978).
55. K. Sakimoto, J. Phys. Soc. Jpn. 48, 1683-90 (1980); 51, 2657-65 (1982).
56. K. Sakimoto and K. Takayanagi, J. Phys. Soc. Jpn. 48, 2076-91 (1980).
57. K. Takayanagi, Comm. Atom. Mol. Phys. 9, 143-52 (1980).
58. C.W. Clark, J. Phys. B13, L27-L30 (1980).
59. C.W. Clark and J. Siegel, J. Phys. B13, L31-L37 (1980).
60. D.C. Clary, J. Phys. Chem. 92, 3173-81 (1988).
61. F.A. Gianturco, Physica Scripta T23, 141-145 (1988).
62. F. Battaglia and F.A. Gianturco, unpublished results.
63. J.C. Tully and R.K. Preston, J. Chem. Phys. 55, 562-78 (1971).
64. A. Bjerre and E.E. Nikitin, Chem. Phys. Lett. 1, 179-87 (1967).
65. L.D. London, Phys. Z. Sowjetunion 2, 46-59 (1932).
66. E. Bauer, E.R. Fisher, and F.R. Gilmore, J. Chem. Phys. 51, 4173-85 (1969).
67. M. Noll and J.P. Toennies, in: Collision Theory for Atoms and Molecules (F.A. Gianturco, ed.), Plenum, New York, in press (1989).
68. M.S. Child and M. Baer, J. Chem. Phys. 74, 2832-49 (1981).

69. F.A. Gianturco, A. Palma, and F. Schneider, in preparation.
70. M. Baer, G. Niedner, and J.P. Toennies, J. Chem. Phys. 88, 1461-63 (1988).
71. M. Baer, G. Drolshagen, and J.P. Toennies, J. Chem. Phys. 73, 1690-704 (1980).
72. M. Baer, Chem. Phys. Lett. 35, 112-21 (1975).
73. M. Baer, G. Niedner, and J.P. Toennies, unpublished results.
74. F.A. Gianturco and V. Staemmler, Int. J. Quantum Chem. 28, 553-74 (1985).
75. P.J. Kuntz, in: Atom-Molecule Collision Theory (R.B. Bernstein, ed.), Plenum, New York (1979).
76. F.A. Gianturco and F. Schneider, Chem. Phys. Lett. 129, 481-85 (1986).
77. V. Staemmler and W. Kutznelnigg, J. Chem. Phys. 62, 1225-37 (1975).
78. F.A. Gianturco and F. Schneider, Chem. Phys. 111, 113-120 (1987).
79. F. Schneider, L. Zülicke, F. Di Giacomo, F.A. Gianturco, I. Paidarova, and R. Polàk, Chem. Phys. 128, 311-320 (1988).
80. F.A. Gianturco and F. Schneider, J. Phys. B22. 49-63 (1989).
81. F.A. Gianturco and F. Schneider, J. Phys. B21, 329-37 (1988).

EXPERIMENTS ON ION-MOLECULE COLLISIONS

Y. Kaneko

Department of Physics
Tokyo Metropolitan University
Setagaya-ku, Tokyo 158
Japan

1. INTRODUCTION

This chapter deals with experiments on ion molecule collisions at low energies. Ion molecule reactions first attracted attention in space research when it was pointed out that reactions like

$$O^+ + N_2 \rightarrow NO^+ + N \qquad (1)$$

and

$$O^+ + O_2 \rightarrow O_2^+ + O \qquad (2)$$

might play an important role in controlling the electron densities in the ionosphere[1,2]. Because dissociative recombination of electrons with molecular ions has a much larger cross section than radiative recombination with atomic ions has, the above reactions to convert atomic ions to molecular ions are supposed to be the rate limiting processes of the electron recombination in the upper ionosphere. A number of both theoretical and experimental studies have been devoted to ion molecule reactions in connection with ionospheric research of the Earth, and also of other planets. The knowledge and techniques established in this ionospheric research have been successfully applied to the study of synthesis of molecules in interstellar space. The importance of ternary collisions and association processes has been increasingly recognized. Nowadays, ion molecule reactions are believed to be the most important processes in synthesizing such a complex molecular ion as CH_3HCN^+ from very simple atomic ions in interstellar clouds in an awfully long period of time[2].

A long time before geophysicists paid attention to the role of ion molecule reactions in the ionosphere, mass spectroscopists were interested in ion molecule reactions concerning the characteristics of ion sources. The so-called cracking pattern of mass spectra was a main concern, but their interest grew into basic reaction chemistry and many applications including ionospheric interests. Nowadays, ion molecule reactions are a basic idea behind the chemical ionization source which is a most promising technique in mass spectroscopy.

On the other hand, it has been known since the beginning of this century that highly stripped iron ions exist in solar corona. Until recently the study of ion molecule collisions involving highly charged ions was a formidable task for laboratory experiments. Astronomical observation and theoretical work were the only available sources of information. Recently, new-type sources of highly stripped ions were developed in cooperation with controlled-thermonuclear-fusion research, and collision experiments using them are progressing rapidly[3].

There have been a number of original and review papers published on the subject of low energy ion molecule collision experiments. As mentioned above, these experiments and the investigators originate in certain roots; namely, mass spectroscopy, gaseous electronics, astrochemistry, physical chemistry, radiation science and orthodox atomic collision physics. There have been three volumes published in the NATO Advanced Study Institute series[4-5]. Though it is a little old (1975) the first volume edited by Ausloos is very comprehensive and instructive. Physicists and chemists working on the common subjects in different fields were gathered to form a community and wrote this book. Though also old, two-volume books edited by Franklin[7] and even older, a monograph by McDaniel et al.[8] are also instructive. Other three volumes entitled "Gas Phase Ion Chemistry" were edited by Bowers[9,10]. The first and second volumes are also comprehensive, and the third is especially devoted to processes involving light. Recently, a collection of invited papers on ion molecule collisions has been published, which is dedicated to Ferguson[11].

There are many data compilations as well. The one edited by Anicich and Huntress, Jr.[12], is especially for the chemistry of planetary atmosphere, cometary comae, and interstellar clouds. Very recently a new data compilation including very large molecules has been edited by Ikezoe et al.[13]. Both compilations are fairly comprehensive, and include data already in some previous compilations.

Since the present subject is diversely spread and so many review papers are already available, it seems meaningless to attempt to write another general review in a short article, and it exceeds at least the present author's ability. Therefore this chapter concentrates on only a few topics with which the author has been concerned for a long time. They are the orbiting effect, the reaction window, and state-to-state experiments. Through these topics, it is attempted to illustrate some features of low energy ion molecule collisions. Some of the examples will be ion atom collisions rather than ion molecule collisions. Some of the examples may not be directly related to space research nor be low energy experiments. They will be useful, however, for understanding the features of the low energy ion molecule collisions. References are rather arbitrarily chosen and are not comprehensive at all.

Professor K. Takayanagi has made a great contribution to the study of atomic and molecular processes in space. In addition to the famous work on rotational excitation of molecules by electron impact, he developed a theory on ion scattering by polar molecules. The details are given by Dalgarno in Chapter 1 of this book, and are briefly described in a later section of this chapter. In addition to this, he has made every endeavor to promote atomic collision research in Japan since after world war II. In order to introduce Japanese activity in this field to the world, Professor Takayanagi started to publish a series of "Progress Reports on the Study of Atomic Collisions and Related Topics in Japan" in 1971[14]. The publication of these was succeeded by the Society for Atomic Collision Research, the first chairman of which was Professor Takayanagi. The society, together with the Japan Science Council, organized the XIth International Conference on the Physics of Electronic and Atomic Collisions (XI-ICPEAC, Kyoto, 1979). The Kyoto ICPEAC stimulated very much Japanese activity in this field. The present high activity of atomic collision research in Japan is deeply indebted to Professor Takayanagi's efforts. It is a pleasure for the author who is a long time friend of his to dedicate this article to Professor Takayanagi in commemoration of the sixtieth anniversary (1986) of his birth and his retirement (1990) from the Institute of Space and Astronautical Science (ISAS). The author is grateful to the editors for giving this opportunity to him.

2. ORBITING EFFECT

2.1. Orbiting Effect by Polarization Interaction

2.1.1. Theoretical background

Orbiting effect is the unique feature of the low energy ion molecule collisions. When the relative velocity is low enough, ions and neutrals are attracted to each other by the induced polarization force or some other long-range forces, so that geometrical collision cross sections will increase with the decrease of the relative velocity. Needless to say, this has a very important meaning in applications to space research. In practice, however, it is not necessarily easy to estimate the orbiting effect from the experimental data. The observed cross section is not the geometrical collision cross section itself but the one including reaction probability which is a complicated function of the radial velocity. It is our concern how the orbiting effect would appear in the experimental cross sections.

In the standard procedure, the binary collision is represented by the scattering of a virtual particle of reduced mass μ with initial relative velocity v by a central force. For the polarization force, the interaction potential is given by

$$V(R) = -\frac{q^2\alpha}{2R^4} \tag{3}$$

where R is the internuclear distance, q the ionic charge, and α the polarizability of the neutral target. The effective potential taking the centrifugal force into account is given by

$$V_{eff}(R) = -\frac{q^2\alpha}{2R^4} + \frac{Eb^2}{R^2} \tag{4}$$

where $E = (1/2)\mu v^2$ is the initial energy, and b the impact parameter. The effective potential $V_{eff}(R)$ is shown in Fig. 1 for some values of b while E is kept constant.

In order that the virtual particle can hit the scattering center (R = 0), the height of the centrifugal barrier must be equal to or lower than E unless the tunneling effect is considered. Putting the centrifugal barrier equal to E when $b = b_o$ and $R = \rho_o$, we have the following equations:

$$\left(\frac{dV_{eff}(R)}{dR}\right)_{R=\rho_o} = \frac{2q^2\alpha}{\rho_o^5} - \frac{2Eb_o^2}{\rho_o^3} \tag{5}$$

$$-\frac{q^2\alpha}{2\rho_o^4} + \frac{Eb_o^2}{\rho_o^2} = E \tag{6}$$

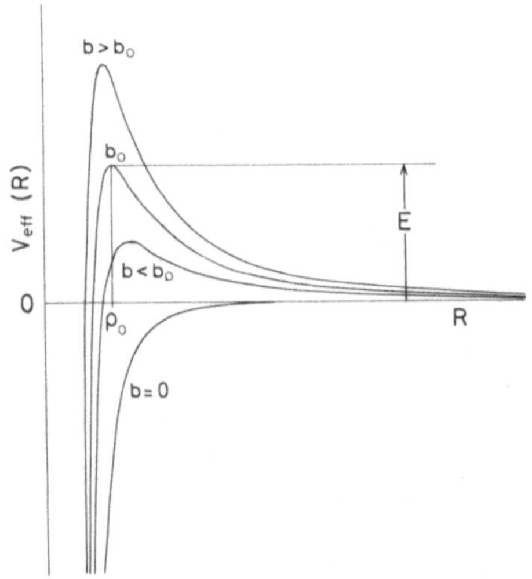

Fig. 1. Effective potential $V_{eff}(R)$ for various impact parameters b as the incident energy E is kept constant. When the height of the centrifugal barrier is equal to E, the virtual particle is captured into a circular orbit of radius ρ_o.

From eqs. (5) and (6) we obtain ρ_o and b_o as

$$\rho_o = \left(\frac{q^2\alpha}{2E}\right)^{1/4} \tag{7}$$

and

$$b_o = \left(\frac{2q^2\alpha}{E}\right)^{1/4} = \sqrt{2}\,\rho_o \tag{8}$$

The meaning of ρ_o and b_o is as follows: when the virtual particle approaches the scattering center with impact parameter b_o, it is trapped into a circular orbit whose radius is ρ_o and can not get to the scattering center. If the impact parameter is smaller than b_o, the particle can get into the inside of the circular orbit and get to the scattering center along spiral orbits. If the impact parameter is larger than b_o, the particle can never get to the scattering center. Thus, b_o is called the critical impact parameter, and ρ_o is called the orbiting radius.

The cross section for the virtual particle to hit the scattering center is given by

$$\pi b_o^2 = \pi q\sqrt{\frac{2\alpha}{E}} = \frac{\pi q}{v}\sqrt{\frac{\alpha}{\mu}} = \sigma_L \tag{9}$$

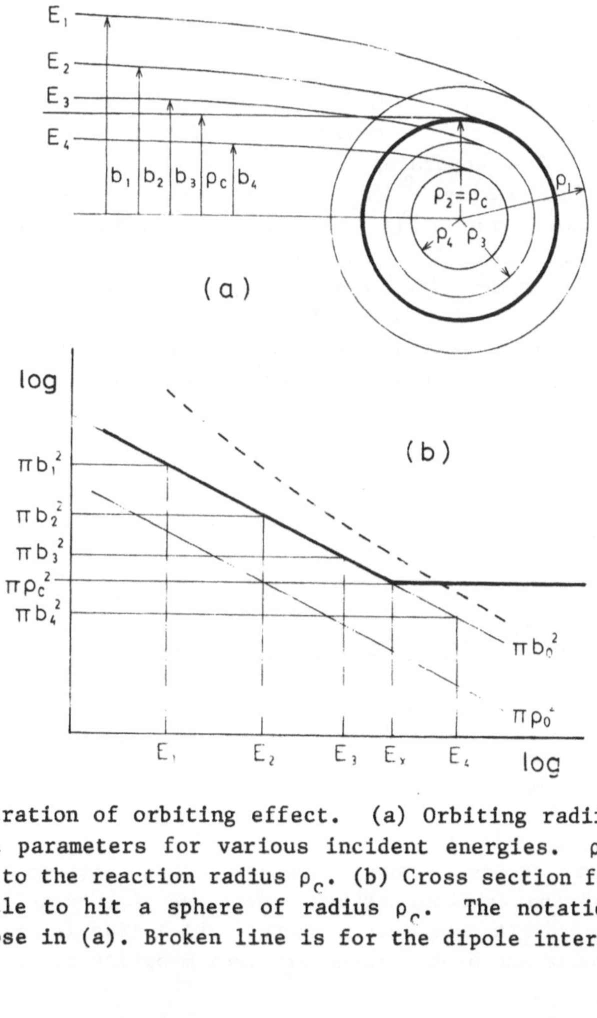

Fig. 2. Illustration of orbiting effect. (a) Orbiting radii and critical impact parameters for various incident energies. ρ_2 is taken as equal to the reaction radius ρ_c. (b) Cross section for the virtual particle to hit a sphere of radius ρ_c. The notations correspond to those in (a). Broken line is for the dipole interaction.

and it is called the Langevin cross section. Numerical values of the Langevin cross section are calculated by

$$\sigma_L = 16.9\, q' \sqrt{\frac{\alpha}{E}} \tag{10}$$

where σ_L is given in A^2, α in A^3, E in eV, and q' is the charge state of the ion.

Now, we assume that a boundary sphere inside which the reaction occurs with an appreciable probability exists. The radius of the sphere, ρ_c, is called the reaction radius. Then, if $\pi\rho_c{}^2$ is larger than σ_L, the geometrical cross section for the particle to hit the reaction sphere is $\pi\rho_c{}^2$ while it is σ_L if $\pi\rho_c{}^2$ is smaller than σ_L. Let the energy for this transition to occur be E_x. The situation is illustrated in Fig. 2.

2.1.2. Experimental evidence

The best experimental evidence of what is mentioned in the last paragraph in Section 2.1.1 is found in the cross sections of symmetric resonance charge transfer processes. Figure 3 shows cross sections of the following processes:

$$Kr^+ + Kr \rightarrow Kr + Kr^+ \tag{11}$$

$$Kr^{++} + Kr \rightarrow Kr + Kr^{++} \tag{12}$$

These cross sections were first measured by the drift tube technique at room temperature[15], then remeasured at 82 K with a liquid-N_2 cooled drift tube[16]. The apparatus used is shown schematically in Fig. 4. Isotopic ions, for example, $^{84}Kr^+$ or $^{84}Kr^{++}$ were injected as primary ions into a drift tube. The drift tube is filled with He buffer gas (~0.3 Torr) admixed with a small amount of natural Kr gas (~0.005 %). The injected ions are thermalized quickly by collisions with He atoms, and then drift toward the end of the drift tube under the action of a uniform electric field. In this condition, it can be assumed that the average ion energy is controlled by collisions solely with He atoms because the Kr content is small. On the other hand, injected ions react only with Kr

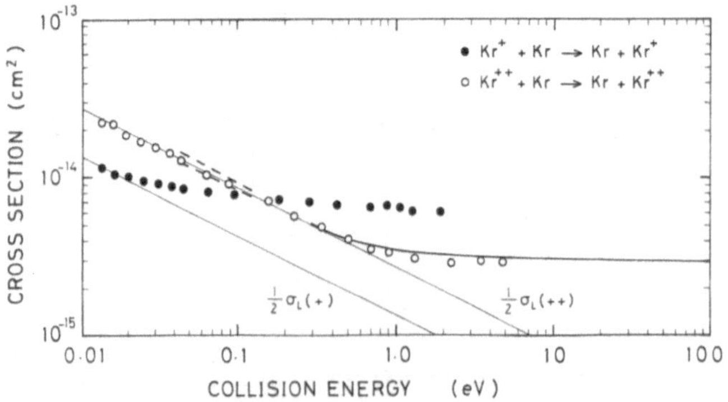

Fig. 3. Symmetric resonance charge transfer cross sections as an illustration of the orbiting effect. Open and full circles are the results of drift tube measurements. Thick line is by the beam guide technique, and broken lines are from mobility data.

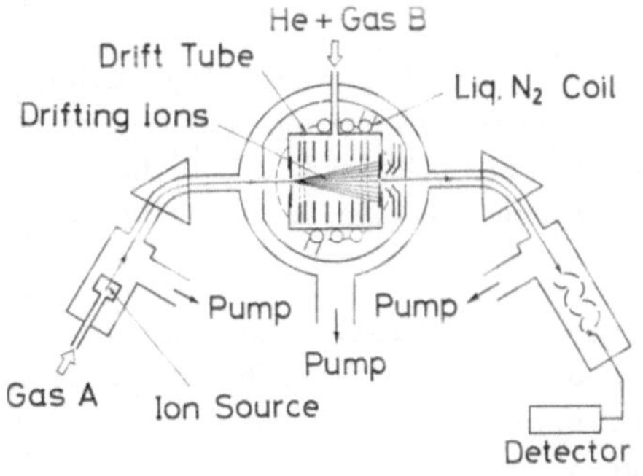

Fig. 4. Schematic diagram of the drift tube apparatus used for the measurement obtaining the results shown in Fig. 3.

atoms and He atoms do not participate in the reaction, because the ionization energy of He is higher than that of Kr or Kr^+.

The mean energy of drifting ions is given by Wannier's formula[17]

$$<E_i> = \frac{1}{2}(M_i + M_b)v_d^2 + \frac{3}{2}kT \tag{13}$$

where M_i and M_b are the masses of the ion and the buffer gas molecule (He), respectively, and T is the temperature of the buffer gas. By means of Wannier's formula, the mean square relative velocity and the mean collision energy between ions and reactant molecules in the center-of-mass system is calculated by[18]

$$<E> = \frac{1}{2}\mu <v_r^2> \tag{14}$$

where $$< v_r^2 > = \frac{2<E>}{M_i} + \frac{3kT}{M_r} \tag{15}$$

and M_r is the mass of the reactant molecule, and μ the reduced mass. The drift velocities v_d were measured separately for Kr^+ and Kr^{++} as a function of the ratio of the electric field strength to the gas pressure in the drift tube. By changing the ratio, <E> can be controlled from the gas temperature to several eV.

Reaction rates κ were obtained directly from the deviation of the isotope abundance ratio of the Kr ions leaving the drift tube from the natural abundance. Cross sections were determined simply by

$$\sigma(<E>) = \frac{\kappa(T_{eff})}{<v_r^2>^{1/2}} \tag{16}$$

where $$T_{eff} = \frac{2<E>}{3K} \tag{17}$$

The validity of Wannier's formula and the velocity distribution of drifting ions had been a controversial subject for a long time. It was finally concluded in the late 1970's, however, as long as the buffer gas is He, Wannier's formula is correct within less than 10% allowance and the velocity distribution is almost Maxwellian with T_{eff}[19]. In Fig. 3, <E> is simply taken as the collision energy.

The data shown together in Fig. 3 are those obtained by the beam guide technique[20]. The apparatus used is shown schematically in Fig. 5. The beam guide technique is another promising technique in low energy ion molecule collisions, but there is no space to give the details here. The readers are referred to Refs. 20 and 21. The results, given here for illustration only, show how low an energy the beam guide technique can attain.

The data extracted from mobility experiments are also shown in Fig. 3[22]. Two components of mobility have been reported, which are attributed to metastable states of Kr^{++} ions though they are not identified[23]. All of the experimental results are in good agreement with each other within the experimental accuracy.

As the energy decreases, the double charge transfer cross section increases slowly at first, then turns off to a steep increase along the line of $(1/2)\sigma_L$ below 1 eV, where σ_L is the Langevin cross section given by eq. (10). On the other hand, the single charge transfer cross section increases slowly in the whole energy range studied except that there is an indication of of an increase along the line of $(1/2)\sigma_L$ at the lowest energy end.

It is well known that symmetric resonance charge transfer occurs through u-g oscillation[24]. Roughly speaking, cross sections are given by $(1/2)\pi\eta^2$ where a factor 1/2 is the probability of having been charge exchanged after leaving the u-g oscillation region η. The range η which may be called the reaction radius in this case is only a weak function of energy[24]. Therefore, the cross section is given by $(1/2)\pi\eta^2$ when $(1/2)\pi\eta^2 > (1/2)\sigma_L$, while it is given by $(1/2)\sigma_L$ when $(1/2)\pi\eta^2 < (1/2)\sigma_L$. Since σ_L is larger and η is smaller for doubly charged ions than for singly charged ions, the change from $(1/2)\pi\eta^2$ to $(1/2)\sigma_L$ occurs at higher energies for double charge transfer than single charge transfer. Thus the cross section for double charge transfer will exceed that for single charge transfer at low energies.

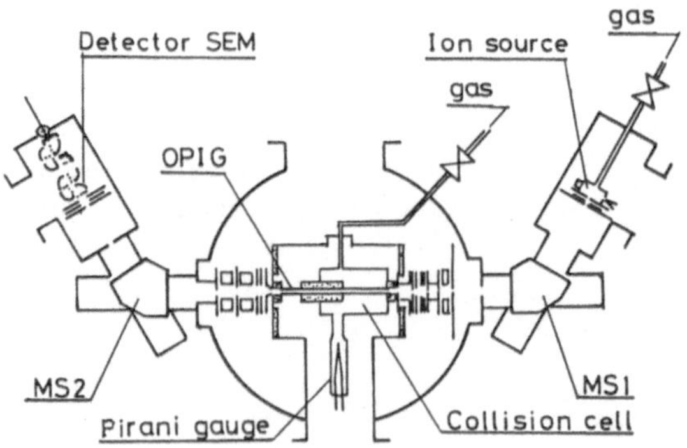

Fig. 5. Schematic diagram of the apparatus with an octopole ion beam guide (OPIG). The primary ion beam is decelerated before entering OPIG, and the reactant gas is introduced into the envelope of OPIG. Cross section measurements can be made even below 1 eV as shown in Fig. 3.

These results may be one of the very few unambiguous pieces of evidence that a transition is taking place between the so-called rectilinear collision and the orbiting collision because the reaction probability of resonance charge transfer can be assumed as constant (1/2) throughout this energy region. Similar results have been obtained for symmetric resonance triple charge transfer of Kr^{3+}, but in this case the situation is a little complicated by competition with the one-electron capture process[25] which can be neglected in the double charge transfer process[20].

As far as the cross section is proportional to $E^{1/2}$, the rate constant is independent of E or temperature. In fact, it is known that many of the ion molecule reactions have a temperature-independent rate constant in the thermal region. It has been discovered also, however, that many reactions have a cross section minimum around a few tenth eV since the drift tube technique could bridge the gap between thermal energies and the lowest energy limit of beam experiments. Before this discovery, thermal data were often connected with high energy data by a straight line. An example is shown with reaction (1) in Fig. 6. As mentioned in the introduction, reaction (1) is a process important in the ionosphere. One can imagine that the linear extrapolation of beam experiments could bring serious confusion into ionospheric research. The confusion was amplified by the fact that the low-lying metastable state of $O^+(^2D)$ has a much larger cross section than the ground state $O^+(^4S)$ just at the energy where the cross section minimum appears[26].

In general, the cross section minimum is explained as follows: Though the reaction radius changes little, the cross section decreases with the decrease of collision energy because the reaction probability decreases, but the cross section tends to increase when the orbiting starts. A problem is that the gradient $d\sigma/dE$ on the low energy side of the minimum is often steeper than $E^{-1/2}$ expected for σ_L, and an E^{-1} or even steeper dependence is quite common. Ferguson et al. argued from a statistical point of view that the reaction channels with large exothermicity will be less favored as the lifetime of the collision complex becomes shorter[27]. Another explanation is that at very low energies, a

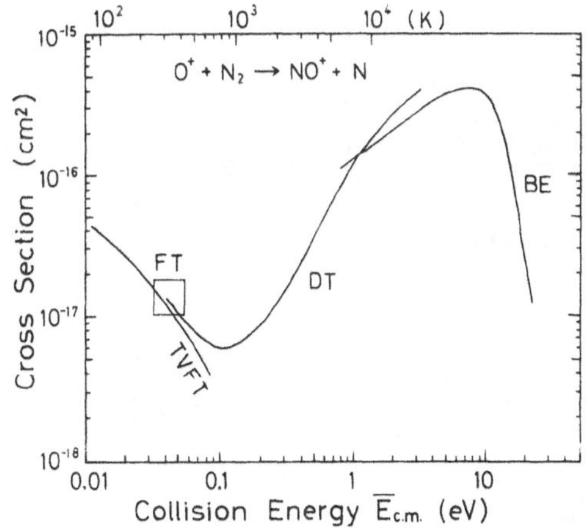

Fig. 6. Cross sections obtained by various techniques for $O^+ + N_2 \rightarrow NO^+ + N$. FT: flow tube. VTFT: variable temperature flow tube. DT: drift tube. BE: beam experiment.

part of the kinetic energy of the incident ion will be lost to excite the rotational levels of the target molecules, so that the ion can not escape from the orbiting circle until the exothermic reaction occurs to decompose the complex. The latter argument implies that ion molecule reactions ultimately have a cross section close to the hitting cross section, so that the cross section goes up anyway after passing the minimum. Whether it is true or not is still an interesting problem, and of course, very important in space research. It is worthwhile to note that cross section minima tend to appear when the absolute values of cross sections are considerably smaller than the Langevin cross section at thermal energies. We shall come back to this problem later.

2.2. Orbiting Effect by Dipole and Quadrupole Interactions

When neutral molecules have a permanent dipole moment, the hitting cross sections are expected to be larger than those for the polarization force only. It is again difficult, however, to estimate accurately the orbiting effect from the experimental cross sections since it is not easy even without the permanent dipole as mentioned before.

When the target molecule has a dipole moment μ_D, the charge dipole interaction is given by

$$V(R) = -\frac{q\mu_D}{R^2}\cos\Theta \tag{18}$$

where, Θ is the angle between the molecular axis and the intermolecular axis. Assuming that the angle Θ is kept constant during the collision, so that the total interaction potential can still be considered as a central force, we have

$$V(R) = -\frac{q^2\alpha}{2R^4} - \frac{q\mu_D}{R^2}\cos\Theta \tag{19}$$

and

$$V_{eff}(R) = -\frac{q^2\alpha}{2R^4} - \frac{q\mu_D}{R^2}\cos\Theta + \frac{Eb^2}{R^2} \tag{20}$$

In a manner similar to that used for the polarization interaction, we obtain

$$b_o = \left\{\left(\frac{2q^2\alpha}{E}\right)^{1/2} + \frac{q\mu_D}{E}\cos\Theta\right\}^{1/2} \tag{21}$$

$$\rho_o = \left(\frac{q^2\alpha}{2E}\right)^{1/4} \tag{22}$$

and

$$\pi b_o^2 = \pi\left\{\left(\frac{2q^2\alpha}{E}\right)^{1/2} + \frac{q\mu_D}{E}\cos\Theta\right\}$$

$$= \sigma_L + \frac{\pi q\mu_D}{E}\cos\Theta \tag{23}$$

It is noted that the hitting cross section πb_o^2 is given as a summation of the Langevin cross section and the second term in eq. (23). It is also noted that the orbiting radius ρ_0 is the same as that for the case of the polarization interaction only. The hitting cross section is shown schematically by a broken line in Fig. 2(b).

There are two theories for evaluating the orientation angle Θ. Locked-dipole (LD) theory[28] assumes that the dipole is locked along the intermolecular axis during the collision so that $\Theta = 0$ and $\cos\Theta = 1$. The LD theory gives maximum values of cross sections, which have been found to be too large compared with experiments. Average dipole orientation (ADO) theory[29] assumes an average angle $\bar{\Theta}$ which is a function of collision velocity and is kept constant during the collision.

In contrast to these classical theories, Takayanagi and Sakimoto have proposed perturbed rotational state (PRS) theory[30,31] in which the rotational motion of a molecule is treated quantum mechanically while the orbital motion is treated classically. The rotational motion is assumed to be adiabatically perturbed when the ion approaches it slowly. The theory was named PRS on the analogy of the well-known PSS theory. Reduced hitting cross sections were calculated for rotational states of molecules. Obviously the hitting cross section is the largest when the rotational state is $j = 0$. Then, the temperature-averaged rate constants will increase with the decrease of temperature, because the population of the $j = 0$ state will increase.

Clary et al.[32] calculated the rate constants for some ion polar molecule collisions by the so-called adiabatic (AD) theory which is essentially the same as the PRS theory and obtained good agreement with experiments around 200 K. They expected a steep increase below that temperature. This has very important implications in interstellar chemistry. Unfortunately, laboratory experiments at such low temperatures are not easy to perform because polar molecules will condense to form polymers. Recently, a new technique called CRESU which uses stationary uniform supersonic flows has been developed[33]. It was reported that the rate constants of reactions with polar molecules like H_2O could be measured below 70 K[34]. This technique seems very promising.

Takayanagi applied the PRS theory to the quadrupole interaction and calculated reduced hitting cross sections[35]. They are larger than the Langevin cross section by a factor of two or so but the energy dependence does not differ very much from that of the Langevin cross section. So far few comparisons have been made between the theory and experiments.

3. REACTION WINDOW

3.1. Potential Curve Crossing

In the previous section, the reaction radius ρ_c was introduced without clear definition. The definition of ρ_c is rather conceptual because it depends on the reaction mechanism itself. When a reaction proceeds on an adiabatic potential surface, namely the collision system forms a collision complex with a certain lifetime and then proceeds to unimolecular decomposition to complete the reaction, the size of the collision complex may be a measure of ρ_c. On the other hand, as in the usual case of charge transfer, when a reaction takes place through a nonadiabatic transition between two potential curves at the potential crossing R_c, the definition of ρ_c is clearer. In this case R_c can be considered as the synonym of ρ_c although the transition could occur not only at R_c but also within a finite range around R_c.

In order to have an appreciable probability of reaction, the potential crossing must be within a certain region of intermolecular distance. This ragion is often called the reaction window[36]. The reaction window is a very useful concept in predicting features of the reaction. If the colliding system has potential curve crossings within the reaction window, the reaction will take place with an appreciable cross section. If the crossing is outside the window the reaction will be negligible.

The physics of the reaction window may be explained by the standard Landau-Zener theory. Let the potential curves cross each other as shown in Fig. 7. If they have the same symmetry, the adiabatic potential curves will be those shown by thick lines due to the non-crossing rule. In dynamical processes, however, there is a probability of nonadiabatic transition between the adiabatic potential curves. The probability of remaining on the adiabatic curve (thin line) when the system passes through the crossing is given by the Landau-Zener theory as

$$p = \exp\left[-\frac{4\pi^2 {H_{12}}^2}{h(\frac{dR}{dt})_{R_c}\Delta F}\right] \tag{24}$$

where, h is the Planck constant, H_{12} the transition matrix element which is one half of the adiabatic splitting at the curve crossing, and ΔF the difference in slope of adiabatic potential curves at R_c[37]. The radial velocity $(dR/dt)_{R_c}$ is given by

$$(dR/dt)_{R_c} = \sqrt{\frac{2E}{\mu}}\sqrt{1 - \frac{b^2}{{R_c}^2}\frac{V(R_c)}{E}} \tag{25}$$

where the polarization force is taken into account. The probability of the collision system exiting along the final channel is

$$P = 2p(1-p) \tag{26}$$

and the cross section is given by

$$\sigma = 2\pi \int_0^X Pbdb \tag{27}$$

where b is the impact parameter, and

$$x = \frac{1}{R_c}\left({\rho_c}^4 + R_c^4 \right)^{1/2} \tag{28}$$

The transition matrix element H_{12} is usually not known unless the potential curves are well determined. This is the reason why quantitative comparison of the Landau-Zener theory with experiments was delayed by half a century since the theory was established in the 1930's. In the case of charge transfer between multicharged ions and neutral atoms, the potential

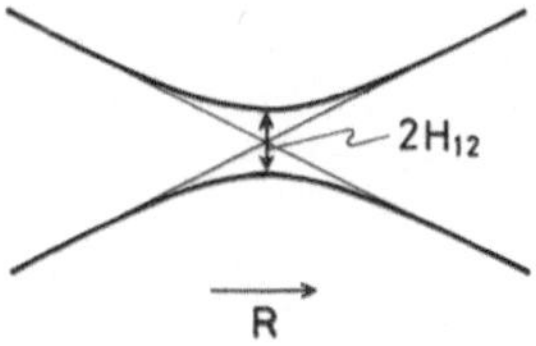

Fig. 7. Illustration of the potential energy curve crossing.

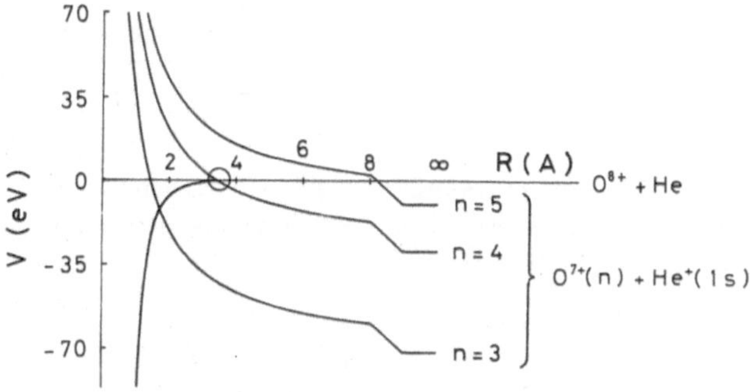

Fig. 8. Potential energy curves for the O^{8+} + He and O^{7+} + He^{+} systems. Only the polarization and Coulomb forces are taken into account.

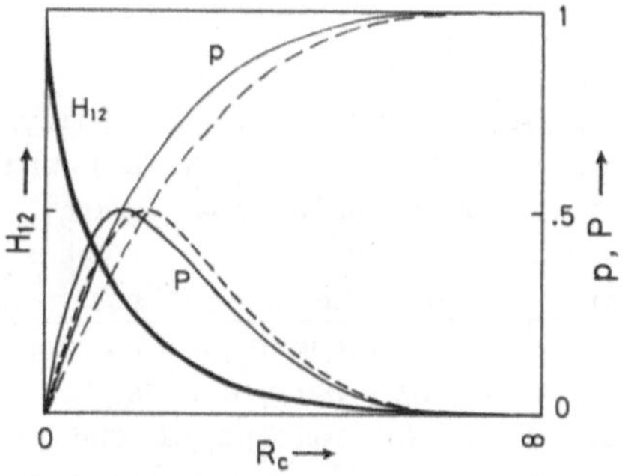

Fig. 9. Illustration of H_{12}, p and P as functions of R_c. Both the vertical and horizontal axes are to an arbitrary scale. Dotted lines are for a lower $(dR/dt)_R$.

curves can be accurately determined because the initial potential curve can be represented by only the polarization force and the final one by Coulomb repulsion. Figure 8 shows the potential curves of the O^{8+} + He system as an example. The crossing radius is known definitely, and the interaction energy H_{12} can be calculated on much simpler assumptions than other cases. Since H_{12} is a decreasing function of the internuclear distance R, p is an increasing function of R. The probability P is a parabolic function of p. Therefore, P must have a maximum at a certain distance R where p = 0.5, and P must be 0 at R = 0 and ∞. These relations are shown schematically in Fig. 9. The region where P has an appreciable

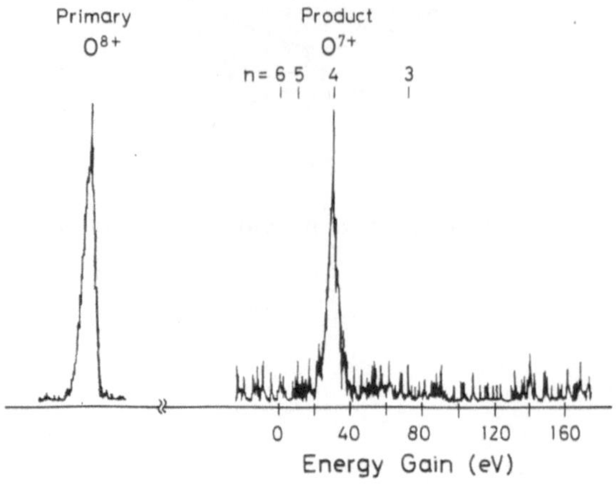

Fig. 10. Translational energy spectrum of O^+ produced from the process $O^{8+} + He \rightarrow O^{7+}(n) + He^+ + \Delta E$. Electrons are captured selectively into the n = 4 state of O^{7+}.

value, say P > 0.1%, is considered to be a reaction window. If the potential crossing is inside this region the reaction will occur, but if it is outside this region the reaction will be negligible.

Figure 10 shows a translational energy spectrum of O^{7+} produced in one-electron capture from the He target by O^{8+},

$$O^{8+} + He \rightarrow O^{7+}(n) + He^+ + Q \tag{29}$$

where n is the principal quantum number[3,38]. Fully stripped oxygen ions O^{8+} were produced in an ion source named NICE-1 which is of EBIS type. The exothermicity Q is just the translational energy change ΔE measured at 0° angle (see Section 4.2). By knowing ΔE from the spectrum, one can show that electrons are captured into the n = 4 state of O^{7+}. Neither the n = 3 nor the n = 5 state of O^{7+} captures an electron. Such a high selectivity of the capturing state has been always found for highly charged ions, and can be explained by the concept of the reaction window. That is, the n = 3, 4 and 5 states of $O^{7+}(n) + He^+(1S)$ cross with the initial state O^{8+} + He at R_c = 1.7 A, 3.5 A and 9.7 A, respectively, as seen in Fig. 8, and only R_c for the n = 4 state is considered to be within the reaction window.

It must be emphasized that a reaction window is not universal but depends on H_{12}, $(dR/dt)_{Rc}$, etc[39]. It is worthwhile to examine the effect of $(dR/dt)_{Rc}$ on the reaction window using Fig. 9. When $(dR/dt)_{Rc}$ is decreased in eq. (24), p will be decreased. Therefore the R_M, at which p = 0.5 and P is the maximum, will shift to larger R. As the results show, the reaction window is widened to larger R. When $(dR/dt)_{Rc}$ is increased the reaction window will be narrowed to smaller R by the reverse reason.

Figure 11 shows an experimentally obtained reaction window[40] for charge transfer reaction

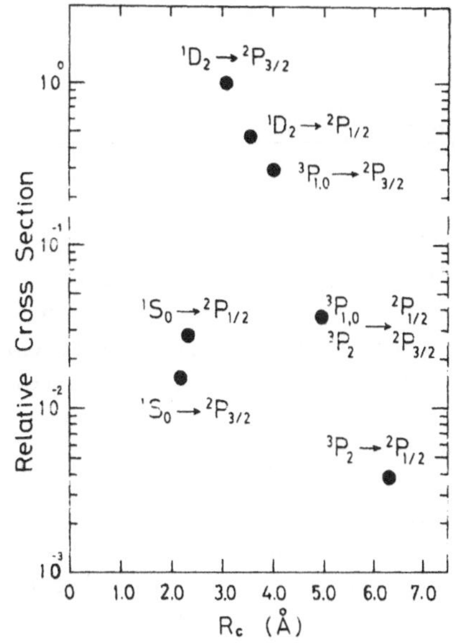

Fig. 11 Relative state-to-state cross sections and R_c of the process $Ne^{++}(^3P_{0,1,2}, {}^1D_2, {}^1S_0) + Xe \rightarrow Ne^+(^2P_{3/2,1/2}) + Xe^+(^2P_{3/2,1/2}) + \Delta E$. Reaction window may be considered as a range 2 - 6 A with a maximum at 3 A.

$$Ne^{++}(^3P_{0,1,2}, {}^1D_2, {}^1S_0) + Xe$$

$$\rightarrow Ne^+(^2P_{3/2,1/2}) + Xe^+(^2P_{3/2,1/2}) + \Delta E \tag{30}$$

The energy gain ΔE can be accurately measured by translational energy spectroscopy as mentioned later. The relative cross section for each state-selected channel is plotted against the crossing radius which is calculated from ΔE. The figure shows that a sharp cross section peak exists around 3 A, and the reaction window may be considered as being from 2 A to 5 A. Smith et al.[36] found a reaction window which lies in the range 2-6 A with a maximum around 4 A for various charge transfer reactions including reaction (30). The slight difference between the two results may be explained by the fact that Smith et al.'s data were taken at thermal energy while the data shown in Fig. 11 were taken at 1334 eV. Although the concept of a reaction window is very useful, it should not be considered in isolation because there are factors which must be taken into account, for example, the symmetry property[41] and the number of subleveles.

3.2. Reaction Window and Orbiting Effect

As mentioned in Section 2.1.1, orbiting effect will appear when the hitting cross section is larger than $\pi\rho_c{}^2$. Since ρ_c must be within the reaction window, the critical impact energy E_x below which orbiting effect will appear can be easily estimated. From eq. (10),

$$\pi\rho_c^{\,2} = \sigma_L(E_x) = 16.9\ q'\sqrt{\frac{\alpha}{E_x}}$$

$$E_x = \frac{(16.9)^2 q'^2 \alpha}{\pi^2 \rho_c^{\,4}} = \frac{28.8\ q'^2 \alpha}{\rho_c^{\,4}} \tag{31}$$

For the usual charge transfer process, ρ_c is around 3 A as mentioned above, and α is 1 ~ 3 A^3 so that E_x is around 0.3 ~ 1 eV. In this energy region it is well known for cross section minima to appear. If the maximum of the reaction probability, which never exceeds unity, exists at a much higher energy than E_x, the absolute magnitude of the cross section minimum at E_x will be much smaller than the Langevin cross section. This is the case mentioned in the last paragraph of Section 2.1.2. On the other hand, for example in the case that the reaction proceeds along the adiabatic potential surface, the reaction radius ρ_c is often small and the reaction probability is high even at low energies, so the cross section will be close to the Langevin cross section up to a few eV.

4. STATE-TO-STATE EXPERIMENTS

4.1. General Description

It is needless to say that an experiment in which internal states of reactant and product particles are well defined is the most desirable to do. In the recent decade, such a state-to-state experiment has been challenged by many investigators in ion molecule collision study.

There are various techniques which can be used for this purpose, and each technique has its own merits and demerits. Light emission is often used for the determination of product states. Though it has an obvious merit in good energy resolution of the products' states, it requires a high density of products because the light is emitted in a wide solid angle, in addition to this usual detectors have a low quantum yield. It has other demerits in that the range of wavelength is limited by the optical system used, and metastable states and ground states which do not emit light can not be observed.

The laser technique is undoubtedly the most promising one. State selection of reactants by laser irradiation and state identification by laser-induced fluorescence may be one of the ideal ways of performing state-to-state experiments. However, it has not yet been too successful, because the wavelength range is limited by the lasers used and detailed knowledge of the level structure, which is often not fully available, is required to analyze the results.

For state selection of primary ions, new techniques have been developed recently. The TESICO (threshold electron and secondary ion coincidence) technique has great potential compared with simple photo-ionization. This technique was reviewed by Koyano[42]. On the other hand, charge transfer of multicharged ions is strongly state-selective as mentioned above, so that the charge-changed ions could be used as a state-selected ion beam[43].

In the case of ion molecule collisions, in contrast to the case of neutral-neutral collisions, translational energy spectroscopy (TES) is a very effective technique because the velocities of ions is easy to measure. Since the pioneering work by Moore and Doering[44] this technique has been used by many investigators. This method has a unique feature that the internal states of both reactants and products are simultaneously determined. A demerit of this method is in the energy resolution which is usually lower than optical spectroscopy. In some cases this demerit can be overcome by using state-selected ions for primary ion beams[43]. The following sections are devoted to the high-resolution translational energy spectroscopy, which the author's group has been concerned with.

4.2. High-Resolution Translational Energy Spectroscopy

High-resolution TES was discussed in detail by Kobayashi[45,46]. The relation between the excitation energy (endothermicity) Q and the translational energy E_i of the ion scattered at a laboratory angle θ can be exactly determined from the energy and momentum conservation laws. Assuming that the target particle was at rest before the collision, we obtain

$$Q = 2\gamma(E_0 - E_i)^{1/2} \cos\theta + (1-\gamma)E_0 - (1+\gamma)E_i \qquad (32)$$

where $\gamma = M_1/M_2$ is the mass ratio of the incident to target particle and E_0 is the incident energy. When the scattering angle θ is small and the kinetic energy loss $\Delta E = (E_0 - E_i) \ll E_0$,

$$Q = \Delta E - E_0\theta^2 \qquad (33)$$

When the scattering angle $\theta = 0$,

$$Q = \Delta E \qquad (34)$$

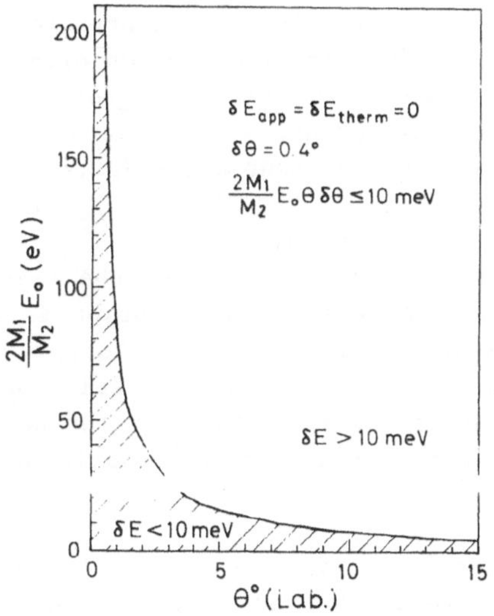

Fig. 12. Necessary condition for obtaining an energy resolution better than 10 meV. Only the recoiling effect is considered, and both the instrumental and thermal broadenings are neglected (see text)

Once Q is known, the internal states of the particles involved in the reaction can be identified with the help of the knowledge of the energy levels. Figure 10 is an example of the TES from which the final state was well identified. Figure 12 is another example that relative cross sections of state-to-state charge transfer reactions can be obtained by TES.

The characteristic feature of translational energy spectroscopy is that the internal states of all the particles involved in the reaction are simultaneously identified as far as the resolution permits. This is an advantage over the other technique by which only one particle can be state-identified at a time. This feature, however, can be also a demerit. When the resolution is not good, the identification of spectra will be ambiguous because many combinations of internal states could correspond to the observed peak.

Necessary conditions for obtaining a good resolution in TES was examined in detail by Kobayashi[45]. The resolution is limited by three factors; instrumental resolution, thermal motion of target particles and recoil of target molecules. The first and second factors can be minimized by technical improvements. The third factor, however, is essentially due to the physics of the process concerned. When the scattering angle is zero which means that the recoil energy is negligibly small, the resolution can be improved to as good as a few meV. Figure 12 shows a necessary condition for achieving a resolution better than 10 meV (FWHM) with a detector whose angular resolution is 0.4° (FWHM)[45]. For large angle scattering, say over 10°, $2M_1E_0/M_2$ must be < 10 eV in order to have resolution better than 10 meV.

This is an essential difference from electron spectroscopy in which the recoil energy can be neglected. Though this might be considered as a serious demerit of ion translational energy spectroscopy, the usefulness of TES should not be underestimated. At first, instrumental improvement is much easier than for electron spectroscopy. Because ions scatter into a much smaller angle than electrons, the detection is easier and the beam intensity can be weaker than in electron spectroscopy. This is a big help because the surface contamination effect which often causes trouble can be neglected. No care is needed for the geomagnetic field because the ions are much heavier than electrons. Furthermore, zero angle scattering implies that the process is a distant collision, namely the process should be the dominant one.

Figure 13 shows a schematic diagram of the high-resolution apparatus used in our laboratory. A detailed description is given in Ref. 47. It consists of an ion source, an energy selector, a collision chamber, an energy analyzer and a detector. The energy selector and the analyzer are of electrostatic hemispherical type and have the same geometry. The mean radius is 75 mm, and both the entrance and exit apertures are 0.5 mm in diameter.

Since no care is needed for geomagnetism in ion TES, both the selector and the analyzer were made as large as possible to increase the geometrical resolution which is determined by the ratio of the radius and the aperture size. The selector assembly is mounted on a vertical plane which can be rotated around the collision center. In order to reduce

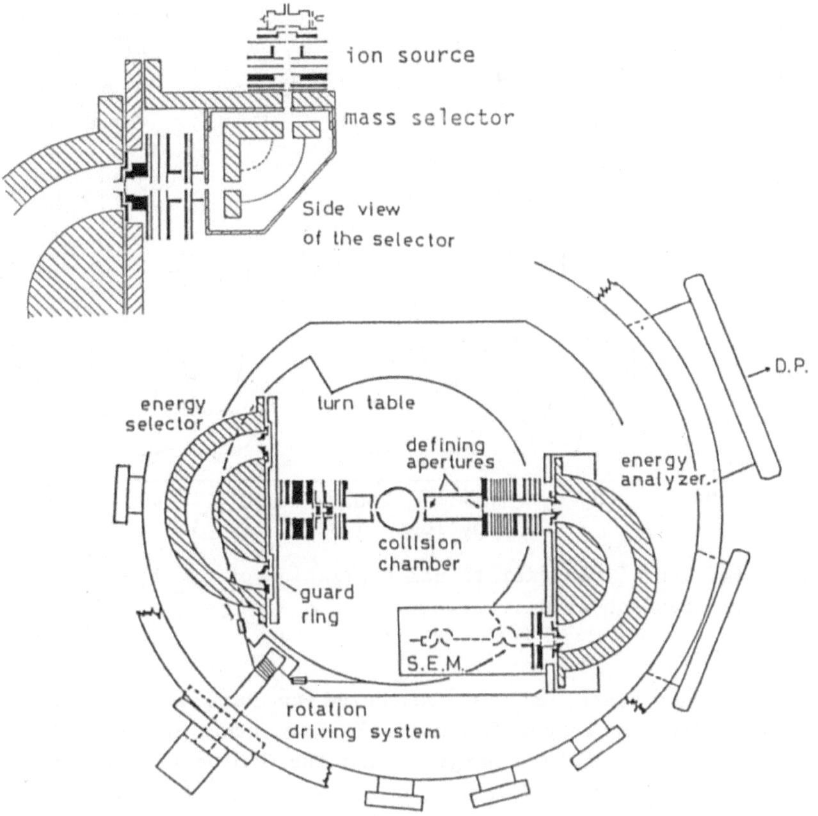

Fig. 13. Schematic diagram of the high-resolution ion translational energy spectrometer with a versatile ion source (top figure).

their weight, the selector and the analyzer were made of aluminum alloy. Aluminum alloy is said to be a bad material for these purposes, because charge up on the oxydized surface would cause trouble. Our machine, however, has been working for ten years without any such trouble probably because the ion current is small compared with that in electron spectroscopy as mentioned above. As the resolution is limited by the transmission energies in the selector and the analyzer, they are kept as low as possible. The best resolution so far achieved is 7 meV (FWHM) with Li^+ ions produced by a surface ionization source, which is nearly equal to the expected value. Since the ion source was replaced by a versatile one of electron impact type with a mass analyzer made of a ferrite magnet, the best resolution is 20 meV (FWHM). It is rather surprizing if one notes that the translational energy itself is 100-1000 eV which corresponds to VUV or soft X-ray region.

4.2.1. Vibrational and rotational excitation

Figure 14 shows an example of high-resolution TES for vibrational excitation of molecules by Li^+ impact[47]. Lithium ions Li^+ emitted from heated β-eucryptite were used as the projectiles. Since Li^+ has a He-like

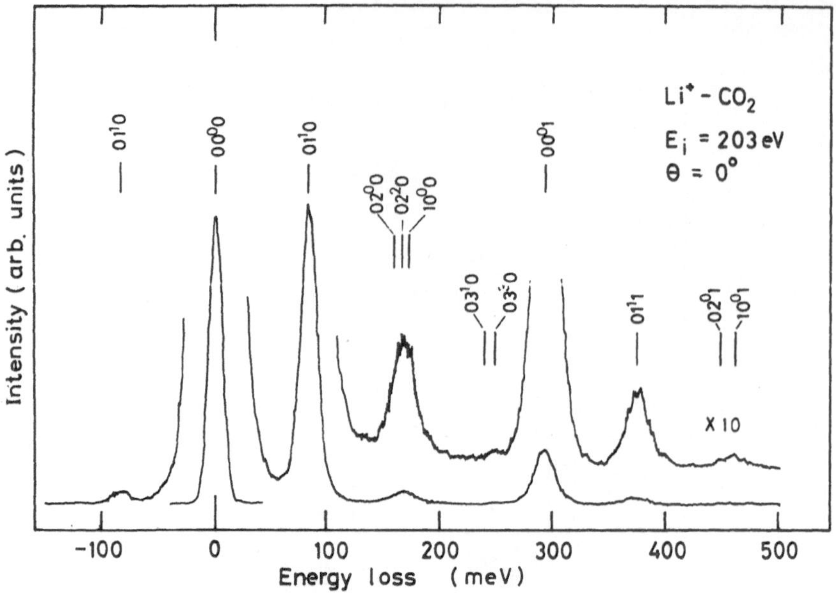

Fig. 14. TES of the vibrational excitation of CO_2 by Li^+ impact by Itoh et al.[47].

closed shell structure, it is considered to have no electron exchange interaction with the target CO_2 molecule. In other words, Li^+ acts just like a structureless point charge. Then the dominant interaction is the electrostatic interaction between the charge of Li^+ and the dipole moment of CO_2. Thus, only the infrared-active modes of vibration, namely the bending (010) and asymmetric stretching (001) modes and their harmonics, are excited. No infrared-inactive mode, namely the symmetric stretching mode (100), is excited. A peak corresponding to the energy gain due to deactivation of the bending mode is seen. This is the so-called superelastic collision. These excitation and deexcitation are strongly forward scattering, and the angular profile of each peak almost coincides with that of the primary peak as long as the energy is high enough. Therefore, the cross section can be determined directly from the relative peak height. This is another feature of ion translational energy spectroscopy.

It has been found that infrared-active modes, for example vibrational excitation of heteronuclear molecules CO and NO are stronger than Raman-active modes, for example those of the homonuclear molecules N_2 and O_2[48,49]. This vibrational excitation, however, decreases rapidly with the decrease of impact energy.

In contrast to Li^+ impact, H^+ can excite vibrational levels of many diatomic molecules even in the few eV region, because H^+ has exchange interactions with molecules though H^+ itself is a structureless particle. In Fig. 15 the TES obtained by Schmidt et al.[50] is shown for H^+ colliding with H_2 in a 10 eV range.

In the low energy region, the time-of-flight (TOF) technique gains an advantage over electrostatic analysis in energy resolution. Rudolph and

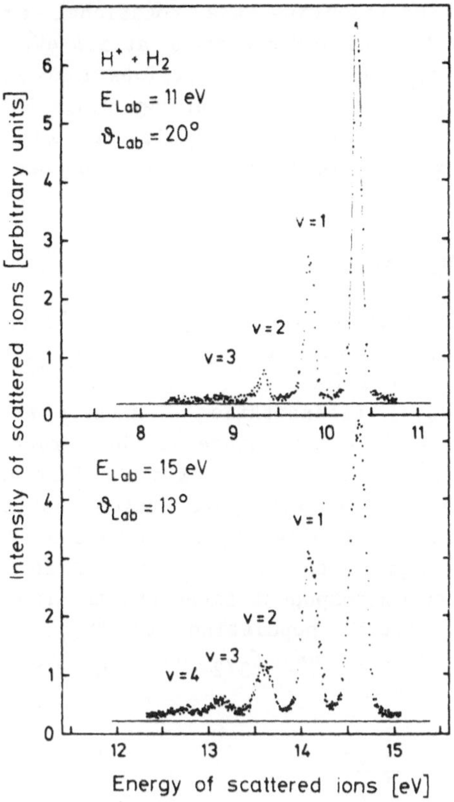

Fig. 15. TES of the vibrational excitation of H_2 by H^+ impact by Schmidt et al.[50].

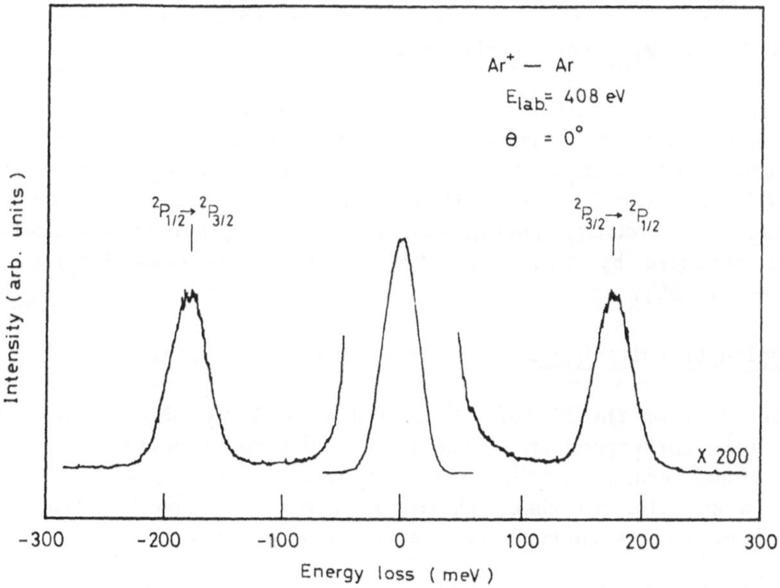

Fig. 16. TES of the Ar^+ colliding with Ar by Itoh et al.[53].

Toennies[51] succeeded in resolving the rotational excitation of H_2 molecules by H^+ impact using the TOF technique at 5.5 eV. Itoh et al.[52] measured cross sections of rotational transitions (J=0→2, 1→3, 2→4, 3→5, 2→0, 3→1, 1→5) of H_2 by Li^+ impact in the energy range 50-400 eV with the same apparatus as used for vibrational excitation of CO_2. These are examples of the very few experiments which have succeeded in separating rotational levels.

4.2.2. Transitions of ionic states

Figure 16 shows TES for Ar^+ colliding with Ar[53]. There are two peaks in symmetrical positions with reference to the primary peak. They are the fine-structure transitions $Ar^+(^2P_{3/2}) \rightarrow Ar^+(^2P_{1/2})$ (excitation) and $Ar^+(^2P_{1/2}) \rightarrow Ar^+(^2P_{3/2})$ (deexcitation). Each pair has almost the same height. This feature does not change if the target gas is changed, and is explained as follows. The transition probabilities for excitation and deexcitation between the fine-structure states must be the same, but one half of the $^2P_{3/2}$ states are eligible to go to the $^2P_{1/2}$ states while all of the $^2P_{1/2}$ states can go to the $^2P_{3/2}$ states in view of the conservation of angular momentum component along the intermolecular axis. On the other hand, the fractional populations of $^2P_{3/2}$ and $^2P_{1/2}$ in the primary Ar^+ beam are different, say $f_{3/2}$ and $f_{1/2}$. So the peak height ratio must be

$$\frac{I(1/2 \rightarrow 3/2)}{2I(3/2 \rightarrow 1/2)} = \frac{f_{1/2}}{f_{3/2}} \tag{35}$$

If the ratio $I(1/2 \rightarrow 3/2)/I(3/2 \rightarrow 1/2)$ is exactly unity, it means that the fractional populations of fine-structure states coincide with the statistical weights, and this is almost so for the experimental results. A careful study showed, however, that the ratio $f_{1/2}/f_{3/2}$ is 0.498 + 0.002 for Ar^+ while it is 0.479 + 0.002 for Kr^+ when the electron energy is above 50 eV[54]. The lower values obtained for heavier ions are attributed to the fact that more autoionizing states that turn into the $^2P_{3/2}$ state exist below $^2P_{1/2}$ for heavier ions.

This is probably the first direct experimental determination of fine-structure populations in rare gas ion beams. This technique can be applied to the determination of other metastable state populations in ion beams, which are usually difficult to detect in ion molecule collision experiments. For example, populations of low-lying metastable states in Kr^{++} beams produced by single electron impact have been determined by Kobayashi et al.[55].

4.2.3. Coupled transitions of ionic states and target states

Figure 17 shows the TES of Kr^+ colliding with N_2, NO, CO and O_2[56]. There are peaks corresponding to the fine-structure transitions of Kr^+ and pure vibrational transitions of target molecules in common with all the spectra. In addition to that, there are peaks corresponding to coupled transitions of fine-structure excitation (or deexcitation) of Kr^+ and vibrational or electronic excitation of the target molecules in the case of CO and O_2 targets. Contrary to these, no such coupled transition is seen in the case of N_2 and NO targets. What is the difference between the two groups? A hint is that Kr^+-CO and Kr^+-O_2 systems have a large cross section for charge transfer, while Kr^+-N_2 and Kr^+-NO systems have a negligibly small one. This suggests that these coupled transitons may

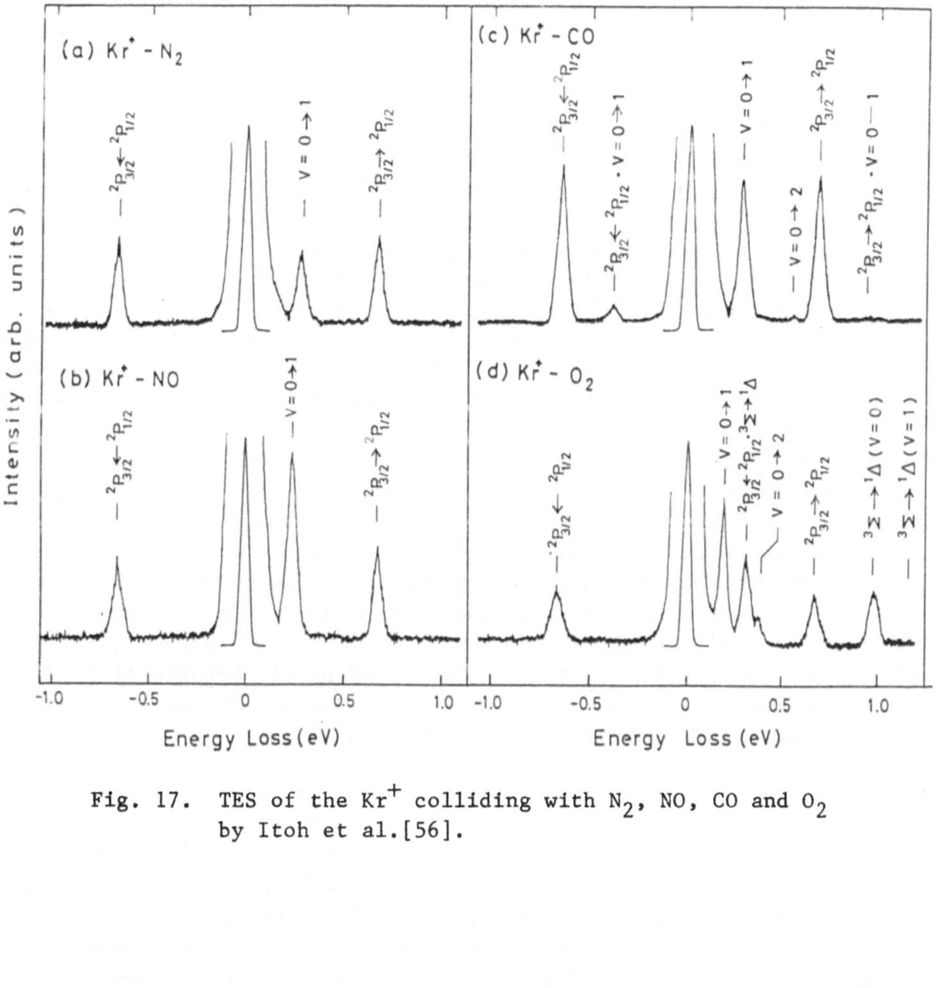

Fig. 17. TES of the Kr^+ colliding with N_2, NO, CO and O_2 by Itoh et al.[56].

occur via charge transfer channels. Nakamura et al.[57] also showed that in $Ar^+ + H_2$ and $Ar^+ + N_2$ systems, both of which have a large cross section for charge transfer, coupled fine-structure and vibrational transitions occur strongly. Parlant and Gislason[58] calculated the cross sections of these transitions as well as pure fine-structure transitions in $Ar^+ + N_2$ collision assuming that they occur via charge transfer channels, and obtained results agreeing very well with Nakamura's experiments.

4.2.4. Charge transfer

In the case of charge transfer of singly charged ions, the fast product is neutral. Therefore, the time-of-flight technique must be used instead of electrostatic energy analysis for translational energy spectroscopy. It has been found that neutral atoms like He and Ar with kinetic energies as low as a few tens of eV can liberate electrons from metal surfaces though the yield is very low ($\sim 10^{-4}$)[59]. Matsuo et al.[60,61] used this phenomenon to measure the translational energy spectra of product neutrals of the reactions

$$Ar^+ + B \rightarrow Ar + B^+ + \Delta E \tag{36}$$

When Ar gas is used as the target, ΔE must be zero. This was used as the standard of the time of flight, and the shift from it was measured when the target gas was changed to B. Though the resolution was not as good, they could obtain ΔE spectra with H_2[60], N_2 and O_2[61] targets. From these ΔE spectra, they could show that the dominant processes of charge transfer between Ar^+ and N_2 are $Ar^+(^2P_{3/2})$ producing N_2^+ in the v = 1 state and $Ar^+(^2P_{1/2})$ producing the v = 2 state. That is, charge transfer between Ar^+ and N_2 is endothermic even though the reaction looks exothermic by a few tenths of an eV if all the particles involved are in their ground state. This result coincides with early drift tube experiments[62] in which a decreasing cross section with the decrease of energy was found even below 0.1 eV. It was an exceptional behavior among exothermic charge transfer reactions. Recently, however, the rate constant of this reaction has been found to increase below 300 K by low temperature SIFT experiments[63]. A similar trend was observed by the author's group as well[64]. This charge transfer reaction is certainly the one which has been most intensively studied by many investigators partly because it looks easy to handle by experimentalists, and partly because it is difficult to understand. There are variety of elaborate experiments[65-68] and they are compactly summarized in Ref. 58.

The vibrational state distribution of product ions in charge transfer has been studied recently. Figures 18 and 19 show examples of high-resolution TES in which the vibrational distribution of the product molecules is clearly resolved. Noll and Toennies[69] measured vibrational distributions of H_2^+ produced by the charge transfer reaction $H^+ + H_2 \rightarrow H + H_2^+$. They used a TOF technique for neutral H atoms in a similar way to that which Matsuo et al. used. In this case the secondary electron yield was about one order of magnitude higher than for neutral Ar atoms because H is much lighter than Ar. Another example is the TES for Ne^+ produced in the one-electron capture process $Ne^{2+} + N_2 \rightarrow Ne^+ + N_2^+$[70]. The sharp peaks are identified as

$$Ne^{2+}(^1S_0) + N_2 \rightarrow Ne^+(2s2p^6\ ^2S_{1/2}) + N_2^+(X\ ^2\Sigma_g^+, v'=0,1) \tag{37}$$

and

$$\rightarrow Ne^+(2s2p^6\ ^2S_{1/2}) + N_2^+(A\ ^2\Pi_u, v'=0\text{-}4) \tag{38}$$

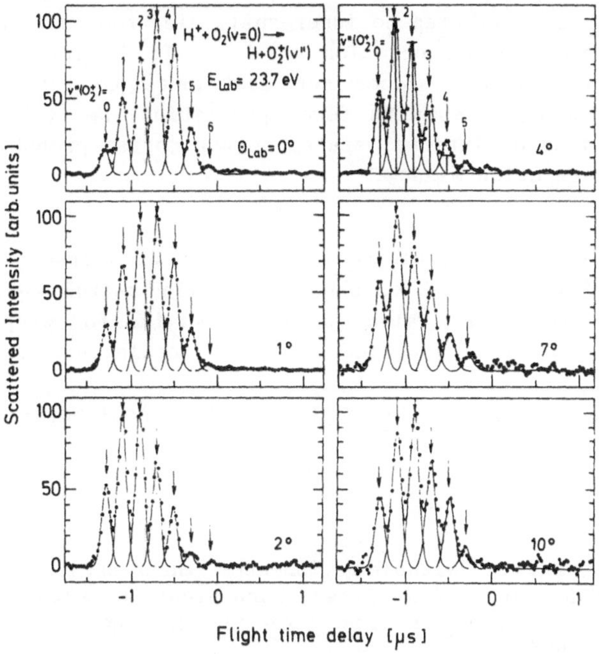

Fig. 18. TOF spectrum of the H produced in the charge transfer process $H^+ + H_2 \rightarrow H + H_2^+$ by Noll and Toennies[69].

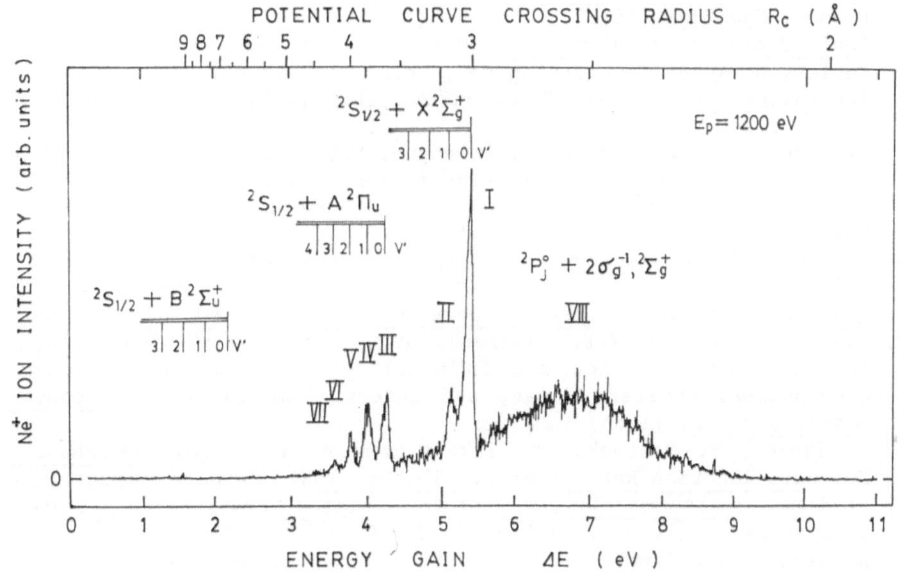

Fig. 19. TES of the Ne^+ produced in the charge transfer process $Ne^{2+} + N_2 \rightarrow Ne^+ + N_2^+$. Upper scale is for R_c determined from ΔE. For the peak assignment, see text.

The broad peak is attributed to inner-shell electron capture by $Ne^{2+}(^1D_2)$, which is followed by dissociation of N_2^+. It is interesting that the ground state of $Ne^{2+}(^3P_J)$ does not contribute to the charge transfer process but higher states with fewer population, $Ne^{2+}(^1S_0$ and $^1D_2)$, do. Other combinations of internal states do not appear probably because they are out of the reaction window.

Acknowledgments The author is grateful to Y. Itikawa, I. Koyano, K. Sakimoto, N. Kobayashi and K. Okuno for their useful advice on writing this article. Thanks are also due to T. Kojima for preparation of the drawings. The author is indebted to all his colleagues in his laboratory at Tokyo Metropolitan University, and in the NICE group at the Institute of Plasma Physics, Nagoya University, for their collaboration in the whole course of study.

REFERENCES

1. D.R. Bates and M. Nicholet, Ion-atom interchange, J. Atmos. Terr. Phys. 18, 65 (1960); D.R. Bates, Recombination, Electronic and Atomic Collisions, Invited Papers, XV-ICPEAC, pp.3-12, North-Holland, Amsterdam (1988).
2. N.G. Adams and D. Smith, Recent advances in studies of reaction rates relevant to interstellar chemistry, Astrochemistry (M.S. Vardya and S.P. Tarafdar, eds.), pp.1-18, D. Reidel Publishing Co., Dordrecht (1987).
3. The Collected Papers of NICE Project/IPP, Nagoya (H. Tawara, ed.), TPPJ-AM-43, Nagoya University (1985).
4. Interactions between Ions and Molecules (P. Ausloos, ed.), NATO ASI Series, Plenum, New York (1975).
5. Kinetics of Ion-Molecule Reactions (P. Ausloos, ed.), NATO ASI Series, Plenum, New York (1979).
6. Ionic Processes in Gas Phase (M.A. Almoster Ferreira, ed.), NATO ASI Series, D. Reidel Publishing Co., Dordrecht (1984).
7. Ion-Molecule Reactions, Vols. 1, 2 (J.L. Franklin, ed.), Plenum, New York (1972).
8. E.W. McDaniel, V. Cermak, A. Dalgarno, E.E. Ferguson, and L. Friedman, Ion-Molecule Reactions, Wiley-Interscience, New York (1970).
9. Gas Phase Ion Chemistry, Vols. 1, 2 (M.T. Bowers, ed.), Academic, New York (1979).
10. Gas Phase Ion Chemistry, Vol. 3 (M.T. Bowers, ed.), Academic, New York (1984).
11. Ion Chemistry, Int. J. Mass. Spectrom. Ion. Phys. 80, 81 (1987).
12. V.G. Anicichi and W.T. Huntress, Jr., A survay of bimolecular ion-molecule reactions for use in modeling the chemistry of planetary atmospheres, cometary comae, and interstellar clouds, Astrophys. J. Suppl. Ser. 62, (1986).
13. Y. Ikezoe, S. Matsuoka, M. Takebe, and A. Viggiano, Gas Phase Ion-Molecule Reaction Rate Constants Through 1987, Maruzen, Tokyo (1987).
14. Studies of Atomic Collisions and Related Problems in Japan (K. Takayanagi, ed.), No.1 (1971), No.2 (1974), No.3 (1977).
15. K. Okuno, T. Koizumi, and Y. Kaneko, Symmetric resonance double charge transfer in Kr^{++} + Kr and Xe^{++} + Xe systems, Phys. Rev. Lett. 40, 1708-1710 (1978).
16. T. Koizumi, K. Okuno, and Y. Kaneko, Drift tube study of symmetric resonance single- and double-charge transfer in Kr^+ + Kr, Xe^+ + Xe, Kr^{++} + Kr, Xe^{++} + Xe systems, J. Phys. Soc. Jpn. 51, 2650-2656 (1982).
17. G. Wannier, Motion of gaseous ions in strong electric fields, Bell System Tech. J. 32, 170-254 (1953).

18. Y. Kaneko, L.R. Megil, and J.B. Hasted, Study of inelastic collisions by drifting ions, J. Chem. Phys. 45, 3741-3751 (1966).
19. H.R. Skullerud, Monte-Carlo investigation of the motion of gaseous ions in electrostatic fields, J. Phys. B: At. Mol. Phys. 6, 728-742 (1973).
20. K. Okuno, Charge transfer of Ar^{2+} and Kr^{2+} in their own gases studied by the beam guide technique, J. Phys. Soc. Jpn. 55, 1504-1515 (1986).
21. D. Gerlich, Low energy ion reactions measured with guided beam, Electronic and Atomic Collisions, Invited Papers, XIV-ICPEAC, pp.541-553, North-Holland, Amsterdam (1986).
22. Y. Kaneko, Ionic mobilities and charge transfer collisions, Electron and Ion Swarms (L.G. Christophorou, ed.), pp.179-188, Pergamon, New York (1981).
23. R. Johnsen and M.A. Biondi, Mobilities of doubly charged rare gas ions in their parent gases, Phys. Rev. A18, 989-995 (1978).
24. D. Rapp and W.E. Francis, Charge exchange between gaseous ions and atoms, J. Chem. Phys. 37, 2631-2645 (1972).
25. K. Okuno and Y. Kaneko, Low energy collision experiments using the beam guide technique. —Charge transfer cross sections of Ar^{3+} and Kr^{3+} in their own gases—, J. Mass Spectroscopy 34, 351-365 (1986).
26. Y. Kaneko and N. Koabayashi, Low energy ion neutral reactions. IV. Formation of NO^+ and N_2^+ in $O^+ + N_2$ collisions, J. Phys. Soc. Japn. 36, 1649-1654 (1974).
27. E.E. Ferguson, D.K. Bohme, F.C. Fehsenfeld, and D.B. Dunkin, Temperature dependence of slow ion-atom interchange reactions, J. Chem. Phys. 50, 5039-5040 (1969).
28. T.F. Moran and W.H. Hamill, Cross sections of ion-permanent-dipole reactions by mass spectrometry, J. Chem. Phys. 39, 1413-1422 (1963).
29. T. Su and M.T. Bowers, Theory of ion-polar molecule collisions. Comparison with experimental charge transfer reactions of rare gas ions to geometric isomers of difluorobenzene and dichloroethylene, J. Chem. Phys. 58, 3027-3037 (1973).
30. K. Takayanagi, Low-energy ion-polar molecule collision —The perturbed rotational state approach , J. Phys. Soc. Jpn. 45, 976-985 (1978).
31. K. Sakimoto and K. Takayanagi, Influence of the dipole interactions on the low-energy ion-molecule reactions, J. Phys. Soc. Jpn. 48, 2076-2083 (1980).
32. N.G. Adams, D. Smith, and D.C. Clary, Rate coefficients for the reactions of ions with polar molecules at interstellar temperatures, Astrophys. J. 296, L31-L34 (1985).
33. B.R. Rowe, G. Dupeyrat, J.B. Marquette, and P. Gaucherel, Study of the reactions $N_2^+ + 2N_2 \rightarrow N_4^+ + N_2$ and $O_2^+ + 2O_2 \rightarrow O_4^+ + O_2$ from 20 to 160 K by the CRESU technique, J. Chem. Phys. 80, 4915-4921 (1984).
34. J.B. Marquette, B.R. Rowe, G. Dupeyrat, and G. Poissant, Ion-molecule reaction studies below 80 K by the CRESU technique, Astrochemistry (M.S. Vardya and S.P. Tarafdar, eds.), pp.19-23, D. Reidel Publishing Co., Dordrecht (1987).
35. K. Takayanagi, Low-velocity ion-molecule collisions with quadrupole interaction, J. Phys. Soc. Japn. 51, 3337-3344 (1982).
36. D. Smith, N.G. Adams, E. Alge, H. Villinger, and W. Lindinger, Reactions of Ne^{2+}, Ar^{2+}, Kr^{2+}, and Xe^{2+} with the rare gases at low energies, J. Phys. B: At. Mol. Phys. 13, 2787-2799 (1980).
37. See, for example, A.M. Chang and D.E. Pritchard, Effects of interatomic attraction on total cross sections in curve crossing collisions, J. Chem. Phys. 70, 4524-4533 (1979).
38. S. Ohtani, Y. Kaneko, M. Kimura, N. Kobayashi, T. Iwai, A. Matsumoto, K. Okuno, S. Takagi, H. Tawara, and S. Tsurubuchi, Observation of electron capture into selective state by fully stripped ions from He atoms, J. Phys. B: At. Mol. Phys. 15, L533-L535 (1982).
39. M. Kimura, T. Iwai, Y. Kaneko, N. Kobayashi, A. Matsumoto, S. Ohtani, K. Okuno, S. Takagi, H. Tawara, and S. Tsurubuchi, Landau-Zener model

calculations of one-electron capture from He atoms by highly stripped ions at low energies, J. Phys. Soc. Jpn. 53, 2224-2232 (1984).
40. T. Nakamura, N. Kobayashi, and Y. Kaneko, Ion energy-loss spectroscopy of Kr^{2+} - He and - Ne collision, II. One electron capture processes, J. Phys. Soc. Jpn. 54, 1743-1749 (1985).
41. J.J. Kaufman, Theoretical considerations of potential energy surfaces for ion-molecule reactions, Interactions between Ions and Molecules (P. Ausloos, ed.), NATO AIS Series, pp.185-213, Plenum, New York (1975).
42. I. Koyano, K. Tanaka, and T. Kato, State selected ion-molecule reactions by TESICO, Electronic and Atomic Collisions, Invited Papers, XII-ICPEAC, pp.355-367, North-Holland, Amsterdam (1982).
43. B.A. Huber, H.-J. Kahlert, and W.W. Wiesemann, Study of electron capture reactions by means of double translational spectroscopy, J. Phys. B: At. Mol. Phys. 17, 2883-2895 (1984).
44. J.H. Moore, and J.P. Doering, Ion-impact excitation of pure vibrational transitions in diatomic molecules, Phys. Rev. Lett. 23, 564-567 (1969).
45. N. Kobayashi, Ultrahigh-resolution ion energy-loss spectroscopy, Electronic and Atomic Collisions, Invited Papers, XII-ICPEAC, pp.355-367, North-Holland, Amsterdam (1982).
46. N. Koabayashi, High resolution energy spectroscopy on ion-molecule collisions, Electronic and Atomic Collisions, Invited Papers, XV-ICPEAC, pp.333-343, North-Holland, Amsterdam (1988).
47. Y. Itoh, N. Kobayashi, and Y. Kaneko, High resolution ion energy-loss spectroscopy of Li^+ - CO_2 collisions, J. Phys. Soc. Jpn. 51, 2977-2981 (1982).
48. N. Kobayashi, Y. Itoh, and Y. Kaneko, Vibrational excitation of CO, CO_2 and N_2O by Li^+ impact in the energy range from 70 eV to 1500 eV, J. Phys. Soc. Jpn. 45, 617-625 (1978).
49. N. Kobayashi, Y. Itoh, and Y. Kaneko, Vibrational excitation of H_2, D_2, N_2 and O_2 by Li^+ impact in the energy range from 100 eV to 1200 eV, J. Phys. Soc. Jpn. 46, 208-214 (1979).
50. H. Schmidt, V. Hermann, and F. Linder, Spectroscopy of low-energy H^+ + H_2 collisions; rotational and vibrational excitation of H_2, Chem. Phys. Lett. 41, 365-369 (1976).
51. K. Rudolph and J.P. Toennies, Time-of-flight studies of rotational quantum excitation of H_2 in collision with protons at E_{CM} = 3.7 eV, J. Chem. Phys. 65, 4483-4491 (1976).
52. Y. Itoh, N. Kobayashi, and Y. Kaneko, Rotational excitation of H_2 molecules by Li^+ impact in the energy range from 50-400 eV, J. Phys. B: At. Mol. Phys. 14, 679-691 (1981).
53. Y. Itoh, N. Kobayashi, and Y. Kaneko, Fine-structure transitions in $Ar^+(^2P_j)$ + $Ar(^1S_0)$ collisions in the energy range 60 eV - 1500 eV, J. Phys. Soc. Jpn. 50, 3541-3542 (1981).
54. N. Kobayashi, T. Nakamura, and Y. Kaneko, Fine-structure transitions in collisions of $Kr^+(^2P_j)$ with rare gases, J. Phys. Soc. Jpn. 52, 1581-1584 (1983).
55. N. Kobayashi, T. Nakamura, and Y. Kaneko, Ion energy-loss spectroscopy of Kr^{2+} - He and -Ne collisions. I. Transitions among $^3P_2$, $^3P_1$, $^3P_0$, $^1D_2$ and $^1S_0$ of Kr^{2+}, J. Phys. Soc. Jpn. 52, 2684-2691 (1983).
56. Y. Itoh, T. Nakamura, N. Kobayashi, and Y. Kaneko, Ion energy-loss spectroscopy for the collisions of $Kr^+(^2P_j)$ with N_2, CO, NO, and O_2, J. Phys. Soc. Jpn. 52, 1091-1094 (1983).
57. T. Nakamura, N. Kobayashi, and Y. Kaneko, Ion energy-loss spectroscopy for the collision of $Ar^+(^2P_j)$ with H_2 and N_2, J. Phys. Soc. Jpn. 55, 3831-3840 (1986).

58. G. Parlant and E. Gislason, Charge transfer in ion-molecule collisions: Theoretical studies by means of the vibrational semiclassical method, Electronic and Atomic Collisions, Invited Papers, XV-ICPEAC, pp.359-368, North-Holland, Amsterdam (1988).
59. K. Kadota and Y. Kaneko, Secondary electron ejection from contaminated metal surface by He and Ar atoms, Jpn. J. Appl. Phys. 13, 1554-1561 (1974).
60. T. Matsuo, N. Kobayashi, and Y. Kaneko, Time-of-flight study of low energy charge transfer reactions of Ar ions and metastable Ar ions with H_2 molecules, J. Phys. Soc. Jpn. 51, 1558-1565 (1982).
61. T. Matsuo, N. Kobayashi, and Y. Kaneko, Study of low energy charge transfer reactions of $Ar^+ + N_2$ and $Ar^+ + O_2$ by time-of-flight technique, J. Phys. Soc. Jpn. 55, 3045-3053 (1986).
62. Y. Kaneko, N. Kobayashi, and I. Kanomata, Low-energy ion-neutral reactions. I. $^{22}Ne^+ + {}^{20}Ne$, and $Ar^+ + N_2$, J. Phys. Soc. Jpn. 27, 992-998 (1969).
63. D. Smith and N.G. Adams, Charge-transfer reaction $Ar^+ + N_2 \rightarrow N_2^+ + Ar$ at thermal energies, Phys. Rev. A23, 2327-2330 (1981).
64. T. Tobita, N. Kobayashi, and Y. Kaneko, Charge transfer reaction $Ar^+(^2P_{3/2}, {}^2P_{1/2})$ and N_2 below 0.1 eV, Abstracts of Papers, XIII-ICPEAC (Berlin), pp.626 (1983).
65. T. Kato, K. Tanaka, and I. Koyano, State-selected ion-molecule reactions by TESICO technique. IV. importance of the two spin-orbit states of Ar^+ in the charge transfer reactions with N_2 and CO, J. Chem. Phys. 77, 337-341 (1982).
66. T. Kato, K. Tanaka, and I. Koyano, State-selected ion-molecule reactions by TESICO technique. V. $N_2^+(v) + Ar \rightarrow N_2 + Ar^+$, J. Chem. Phys. 80, 834-838 (1982).
67. B. Friedrich, W. Trafton, A. Rockwood, S. Howard, and J.H. Futrell, A crossed-beam study of the charge-transfer reaction of Ar^+ with N_2 at low and intermediate energies, J. Chem. Phys. 80, 2537-2542 (1984).
68. C.-L. Liao, J.-D. Shao, R. Xu, G.D. Flesch, Y.-G. Li, and C.Y. Ng, A state-to-state study of the electron transfer reactions $Ar^+(^2P_{3/2,1/2}) + N_2(X,v=0) \rightarrow Ar(^1S_0) + N_2^+(X,v')$, J. Chem. Phys. 85, 3874-3890 (1986).
69. M. Noll and J.P. Toennies, Vibrational state resolved measurements of differential cross sections for $H^+ + O_2$ charge transfer collisions, J. Chem. Phys. 85, 3313-3325 (1986).
70. K. Okuno, A. Fukuroda, N. Kobayashi, and Y. Kaneko, High resolution translational energy spectroscopy of one-electron capture in $Ne^{2+} - N_2$ collision, Abstracts of Papers, XV-ICPEAC (Brighton), pp.561 (1987).

RECOMBINATION PROCESSES

M. R. Flannery

School of Physics
Georgia Institute of Technology
Atlanta, Georgia 30332-0430
U.S.A.

1. SCOPE

The aim here is to survey the mechanisms basic to various types of recombination processes and to provide some recent results. Most assume significance in astrophysics (interstellar medium, stellar and planetary atomospheres) and some in laboratory (Tokamak) fusion plasmas and in various types of lasers. They span a wide range of physical conditions, e.g., the ranges $10 \leqq T \lesssim 10^6$ in temperature T (K) and $1 \lesssim N \leqq 10^{20}$ in particle density N (cm^{-3}). Recombination includes here not only electron-ion and ion-ion processes but also ion-atom (molecule) association. Most of the processes below may be characterized by the mechanism responsible for stabilization of an intermediate resonant collision complex. Typical two-body rates k (cm^3s^{-1}) for simple atomic and diatomic systems are indicated in parenthesis beside each process.

For termolecular association (TA)

$$A^+ + B + M \rightleftarrows (AB^+)^* + M \rightarrow AB^+ + M, \quad ((10^{-28}\text{-}10^{-32})N\ cm^3s^{-1}) \tag{1}$$

in a gas M of density N, stabilization of $(AB^+)^*$ occurs via $(AB^+)^*$-M collisions at a quenching frequency $\nu_q \lesssim 10^{-9}N\ s^{-1}$, while radiative association(RA)

$$A^+ + B \rightleftarrows (AB^+)^* \rightarrow AB^+ + h\nu\ , \quad (10^{-9}\text{-}10^{-17}\ cm^3s^{-1}) \tag{2}$$

occurs via photon emission (vibrational and electronic) at a radiative rate $\nu_r \sim 10^3\text{-}10^6\ s^{-1}$ depending on the type (vibrational or electronic) of stabilizing transition.

For dissociative recombination (DR),

$$e + AB^+ \rightleftarrows AB^* \rightarrow A^* + B\ , \quad (10^{-6}\text{-}10^{-7}\ cm^3s^{-1}) \tag{3}$$

stabilization of AB^* occurs by quantal predissociation along repulsive covalent excited molecular states at a dissociative frequency $\nu_d \sim 10^{15}\ s^{-1}$.

Emission of radiation provides the required stabilization in dielectronic recombination (DIR)

$$e + A^{z+}(i) \rightleftarrows [A^{z+}(k) - e]_{n'\ell'} \rightarrow A^{(z-1)}(j;n\ell) + h\nu\ , \quad (10^{-11}\ cm^3s^{-1}) \tag{4}$$

which occurs at (resonant) electron energies much higher than the lower threshold energies for which the direct (nonresonant) radiative recombination (RR)

$$e + A^{z+}(i) \rightarrow A^{(z-1)+}(i;n\ell) + h\nu\ , \quad (10^{-12}\ cm^3s^{-1}) \tag{5}$$

is more important. In contrast to the above formation of intermediate long-lived scattering resonances in processes (1)-(4), termolecular ion-ion recombination (TR)

$$A^+ + B^- + M \rightarrow AB + M\ , \quad ((10^{-24}\text{-}10^{-25})N\ cm^3s^{-1}) \tag{6}$$

of simple systems proceeds by nonresonant scattering since the Coulomb attraction cannot accomodate quasi-bound levels. The rates are fast since the third body M effectively utilizes the many (A^+-B^-) Coulombic superthermal encounters which occur at large ion-ion separations $R \leqq 370$ Å at room temperature. Elastic A^+-M and B^--M collisions are very efficient in removing most of the energy gained by A^+ and B^- from the Coulomb field so that the highly excited bound levels of AB so formed are then destroyed by multistep collisional cascades to stable levels. In parallel to the resonant scattering in process (1), TA can also proceed via nonresonant A^+-M collisions which change the energy and angular momentum of A^+-B relative motion.

Termolecular electron-ion recombination (TER)

$$e + A^+ + M \rightarrow A(n) + M\ , \quad ((10^{-26}\text{-}10^{-29})N\ cm^3s^{-1}) \tag{7}$$

also proceeds via collisions with the gas M, but at a much smaller rate, since elastic electron-atom M collisions cause only a small fraction

(~2m/M) of energy to be transferred to M. Rates become larger for molecular M which absorb a much larger fraction of energy via rotational and vibrational excitation, and for molecular ions when dissociative recombinations involving bound electrons can provide substantial enhancement.

As is well known, mutual neutralization (MN)

$$A^+ + B^- \rightarrow A^* + B \,, \quad (10^{-7}\text{-}10^{-8}\ cm^3s^{-1}) \tag{8}$$

proceeds by direct coupling of the diabatic ionic potential energy curve with the covalent curves, which however involve much smaller ion-ion separations R ~ 10-50 Å to yield rates smaller by an order of magnitude than for reaction (6). The fact that the Coulombic interaction between the ions is strong at large separations where the (Landau-Zener) probability for curve crossing is weak ensures the dominance of termolecular process (6) over bimolecular process (8), even at modest pressures. Since collisions with M can form bound $(A^+\text{-}B^-)$ states which in turn promote more efficient curve crossing, MN can be considerably enhanced by an ambient gas. It does not occur parallel to TR (6) so that the effective rate for neutralization is then not simply the sum $(k_{TR} + k_{MN})$ of the individual rates.

In an electron-ion plasma of intermediate density $n_e \sim 10^{11}\ cm^{-3}$, recombination

$$e + A^+ + e \rightarrow A(n) + e + h\nu \tag{9}$$

proceeds by collisions into high n-levels, which become de-excited by e-A(n) collisions and radiative emission. State-to-state rates for DIR (4), BR (5), DR (3), and TER (7) would all be relevant. Collisional-radiative recombination (CRR) then yields the familiar set of quasi-equilibrium (input = output) master equations to be solved for the individual excited state populations N_n in terms of the concentration of free electrons, ions and recombined atoms in the lowest stabilized states.

1.1. Current Status of Recombination

The present state of recombination is that theory (with reliable results) for most of the above processes involving simple atomic or diatomic systems is reaching maturity and is approaching a well-defined hi-tech state. In particular the recent theoretical developments[1-3] of DIR indicate that DIR cross sections may be calculated to within the same degree of accuracy (~10%) as electron-ion inelastic collisions. Termolecular ion-ion recombination[4-6] of simple ion systems in a gas has been solved as a universal function of mass species, and gas density and temperature. Results for simple systems of general mass are available at low density. Dissociative recombination[8] of simple diatomic systems is in principle well known but lack of relevant molecular potential energy curves and branching ratios to final products prohibit rigorous quantal calculation. Ion-neutral reactions and termolecular electron-ion recombination for complex systems remain by comparison in a more exploratory condition, although substantial progress[9,10] has recently occurred.

Reliable experiments[8,11-13] exist for DR, TA and MN which proceed with measurable rates $(10^{-7}\text{-}10^{-9}\ cm^3s^{-1})$. Technical breakthroughs have recently permitted measurements on DIR[14,15] and RA[16] which proceed at much slower rates (10^{-10}-10^{-15} cm^3s^{-1}), respectively. The influence of electric fields in the experiments is important, particularly for DIR and to a lesser extent for DR. Theories of recombination in external fields are currently under development.

Although TR (6) is now well understood theoretically and proceeds at the largest rate of any recombination processes involving simple systems, reliable experimental measurement, apart from some historical data[17], is as yet not forthcoming although some activity has recently emerged on the neutralization of $H_3O^+(H_2O)_n$ + NO_3^- ions in He[18]. There are at present no measurements from a given laboratory which span the full range of gas pressures studied theoretically and which monitor the identity of ions as the pressure changes. The task is difficult in that the ions may well be clustered to high orders.

1.2. Generic Kinetic and Resonant-Scattering Treatments

Identify the interacting species in processes (1)-(9) as A, B and M with concentrations n_A, n_B and N, respectively. The two-stage sequences common to TA(1), RA(2), DIR(3) and DR(5) are the formation of a long-lived unstable collision complex AB*, or scattering resonance, followed by an irreversible stabilization mechanism, whether radiative as in RA and DIR, collisional as in TA or dissociative as in DR. The complex with energy degenerate to and lying within the continuum of dissociated A(i) + B(j) states is formed when the excess energy and angular momenta of internal and relative motion of A and B become redistributed among the internal degrees of freedom of AB*. Following large perturbations in A-B close encounters, a quasi-equilibrium of these excited states of AB* is established. Thus processes TA(1), RA(2), DR(3) and DIR(4) above may be conveniently analyzed in terms of the macroscopic two-stage sequence

$$A + B \underset{\nu_d}{\overset{k^*}{\rightleftarrows}} AB^* \overset{\nu_s}{\rightarrow} \text{products} \tag{10}$$

which involves the stabilization at frequency ν_s of quasi-bound resonant scattering states of AB* formed at rate k* (cm^3s^{-1}) before AB* can redissociate (or autoionize) back to the initial or any other dissociated channel at frequency ν_d. For a quasi-steady-state density n^*_{AB} of the complex AB*, the overall process then proceeds at a rate (cm^3s^{-1})

$$k = n^*_{AB}\nu_s = \left[\frac{\nu_s}{\nu_d+\Sigma\nu_s}\right] k^* \equiv P_s\, k^* \tag{11}$$

where P_s is the probability of routing to a particular pair of stabilized products s. A negative temperature T dependence is anticipated for k since ν_d increases with T. As the density N of the gas M is raised, eq. (11) for collisional association TA predicts an initial linear variation of k with N (when $\nu_d >> \nu_s \sim k_sN$) increasing towards a saturation value k* (when $\nu_s >> \nu_d$) times the branching ratio $[\nu_s/\sum_s \nu_s]$ for that particular pair of products.

The reaction volume (cm^3)

$$K = \tilde{n}^*_{AB}/\tilde{n}_A\tilde{n}_B = k^*/\nu_d \tag{12}$$

where tildes (~) denote thermodynamic equilibrium values, is pivotal in determining the T-dependence of the overall rate

$$k = \left[\frac{\nu_s K(T)}{k^* + \sum_s \nu_s K(T)}\right] k^* = K(T)\left[\frac{\nu_s \nu_d}{\sum_s \nu_s + \nu_d}\right] \tag{13}$$

Note that K is not an equilibrium constant in the usual sense since AB* is distributed only among those states satisfying energy and angular momentum conservation above the dissociation limit. It is given in usual notation by

$$K(T) = \frac{h^3}{(2\pi M_{AB} kT)^{3/2}} \frac{q(AB^*)}{q(A)q(B)} \frac{\omega_{AB}}{\omega_A \omega_B} \tag{14}$$

where q is the internal partition function, or the number of quantum states available at temperature $T \sim \sum_i \exp(-E_i/kT)$, and where ω is the electronic statistical weight, associated with each reactant A and B and with the activated complex AB* of reduced mass M_{AB}. While q(A) and q(B) are generally known, q(AB*) must include only those rotational-vibrational-electronic states of AB* accessible at energies above the dissociation threshold of AB. It also includes states which satisfy conservation of total angular momentum produced from the orbital angular momentum for A-B relative motion and the combined internal angular momentum of the individual reactants.

The key quantities which characterize the T-dependence and rate limiting step of each of RA, TA, DIR and DR ate therefore K(T) and the stabilization frequency ν_s. For polyatomic species, not only is calculation of K difficult but ν_s is uncertain to the extent that the type of transition (vibration or electronic) may not be established. This uncertainty can involve at least two orders of magnitude difference in the rates[10].

For cases RA, DIR, DR and TA, a microscopic state-to-state generalization (phase-space or multichannel) of the basic premise underlying eq. (11) can be written down in terms of all the relevant electronic, vibrational and rotational quantum numbers for the internal degrees of freedom i and j of A(i) and B(j), for the translational energy and angular momentum of A-B relative motion and for the total conserved angular momentum and energy. The simplified expression (11) however not only serves as a guide to experimentalists in elucidating the role, and extracting the rate peculiar to various stabilization mechanisms but is also capable of providing order-of-magnitude rates and the associated dependence on temperature T fairly reliably.

The intimate connection of eq. (11), standard in chemical kinetics, with scattering theory is instructive. When the redissociation or autoionization channels in processes (1)-(5) are considered as a series of non-overlapping resonances and when the nonresonant background scattering is neglected, then Breit-Wigner resonance scattering theory with explicit inclusion of all multichannels, consistent with energy and angular momentum conservation, can be applied. To preserve a simple notation in order to isolate the key connection, and to illustrate the essential technique, let AB* exhibit only relative motion scattering resonances (quasi-bound states) at A-B relative energies $E = E_r^*$. The cross section for the resonant reaction of A and B with internal energies $E_{A,B}$ is

$$\sigma(E;E_A,E_B) = \frac{\pi}{E}\frac{\hbar^2}{2M_{AB}}\frac{\omega(AB^*)}{\omega(A)\omega(B)}\sum_r \frac{\Gamma_a\Gamma_s}{[(E_T-E_r^*)^2+\frac{1}{4}\Gamma^2]} \tag{15}$$

where the total energy of the system is $E_T = E_A + E_B + E$, where the energy widths for stabilization and re-dissociation (autoionization) are related to the corresponding frequencies by

$$\Gamma_s = \hbar\,\nu_s\,, \quad \Gamma_d = h\,\nu_d \tag{16}$$

where Γ is the total width $(\sum_d\Gamma_d+\sum_s\Gamma_s)$ for all dissociative (d) and stabilization channels (s). The electronic statistical weight of species X is $\omega(X)$. The rate of recombination for a Maxwellian distribution of relative energies E at temperature T is

$$k(T) = \left[\frac{8kT}{\pi M_{AB}}\right]^{1/2}\int_0^\infty \varepsilon\,\sigma(\varepsilon)\,\exp(-\varepsilon)d\varepsilon\ ;\quad \varepsilon = E/kT \tag{17}$$

where M_{AB} is the reduced mass of A-B. Since the Dirac delta function $\delta(x)$ is the limit of $\pi^{-1}h(x^2+h^2)^{-1}$ as $h\to0$, the rate (17) for sharp resonances $\Gamma \ll E_T-E_r^*$ then reduces to

$$k(T,E_A,E_B) = \frac{h^3}{(2\pi M_{AB}kT)^{3/2}}\frac{\omega_{AB}}{\omega_A\omega_B}\left[\sum_r\frac{\nu_d\nu_s}{(\sum_s\nu_s+\nu_d)}\exp\{-(E_r^*/kT)\}\right]\exp\{(E_A+E_B)/kT\} \tag{18}$$

On assuming that the frequencies are independent of the resonance positions E_r^*, then $\sum_r \exp(-E_r^*/kT)$ is simply the partition function q(AB*) arising from all the resonance states of AB*. On averaging over all internal states i and j of A and B and with the use of detailed balance, eqs. (12), (11) and (14) are then recovered since $\sum_{i,j}\exp(-E_A/kT)\times\exp(-E_B/kT)$ is the product q(A)q(B) of the reactant partition functions. This connection provides a basis for eq. (11) or (14) more quantitative than the earlier steady-state kinetic rate argument. The extension to include all multichannels directly is straightforward, but the case of overlapping resonances existing in various polyatomic systems requires attention, and may well under approximation provide the rate (13) in current use.

Because of the long-range Coulombic attraction in the entrance channels the remaining related processes (TR, TER), as indicated earlier, do not proceed via the resonating tight complex but rather by energy-changing collisions between M and A^+-B^- pairs. The collisions are effective for those pairs with separation $R \lesssim R_T = e^2/kT \sim 370$ Å at room temperature, which in a sense can be regarded to form an extremely large loose non-resonating complex with reaction volume $K = \frac{4}{3}\pi R_T^3$. At low gas density N, eq. (13) predicts

$$k_{TR} \sim \frac{4}{3}\pi\, R_T^3\, \nu_s = \frac{4\pi}{3} R_T^3 \langle v_{AB}\rangle\, N\, \sigma \tag{19}$$

where σ is the cross section for free-bound energy-changing $(A^+$-$B^-)$-M collisions, and emphasizes the characteristic linear N and the $T^{-5/2}$-T^{-3} dependencies. At high N, however, the rate does not converge to the saturation value k* predicted by eq. (13). The rate of approach of A^+ and B^- to R_T is limited by the transport rate, which decreases as N^{-1} and which becomes comparable to the reaction rate (19) within R_T at about 1 atm.

For TA(2) however the transport rate always remains much higher than the rate limiting step of reaction so that saturation to the thermal rate k* is eventually obtained.

2. RADIATIVE AND TERMOLECULAR ASSOCIATION

2.1. Simple Systems

The underlying physics of radiative recombination of a simple system such as

$$C^+(^2P_{1/2}) + H(^2S) \rightarrow CH^+(A\ ^1\Pi) \rightarrow CH^+(X\ ^1\Sigma) + h\nu \tag{20}$$

becomes transparent in a semiclassical treatment[19], where the cross section is

$$\sigma(E) = 2\pi \int_0^\infty P_r(E,\rho)\rho\ d\rho \tag{21}$$

at relative energy E. The probability of radiative emission during a collision at impact parameter ρ is

$$P_r(E,\rho) = \int_{-\infty}^{\infty} G(t)A(t)dt = \oint_R^{\infty} G(R)A(R)dR/v_R \tag{22}$$

where the radial speed at relative separation R is v_R with turning point R_T, and where G(R) is the probability that CH^+ during the collision is in state i(A $^1\Pi$), which radiates at a local rate

$$A(R) = \left[\frac{4}{3\hbar^4 c^3}\right]|M(R)|^2\ \Delta E^3(R) \tag{23}$$

to the stabilized state f(X $^1\Sigma$). Since the intermediate A $^1\Pi$ state i correlates adiabatically with the $C^+(^2P_{3/2})$ state (which is essentially unoccupied in the interstellar medium), state i is collisionally produced mainly by nonadiabatic radial coupling with the a $^3\Pi_1$ state which correlates adiabatically with the initial $C(^2P_{1/2})$ state. The molecular states, with wave functions $\psi_{i,f}$ and energy separation $E(R) = V_f(R) - V_i(R)$, are connected via the dipole matrix element $M(R) = \langle\psi_f(\underline{r},R)|e\underline{r}|\psi_i(\underline{r},R)\rangle$. Rates $k_{RA} = 1.3\times10^{-17}\ cm^3s^{-1}$ obtained[19] for reaction (20) over the temperature range $20 \leq T\ (K) \leq 1000$ do not, however, satisfactorily explain the discrepancy between the observed and theoretically deduced abundances of the radical CH^+ in diffuse interstellar clouds. A quantal treatment can in addition acknowledge the discrete vibrational levels of the intermediate electronic state i(A $^1\Pi$) and can include quasi-bound shape resonances formed within the centrifugal barrier. These effects enhance[20] the semiclassical rates for process (20) by ~25% to $\sim1.66\times10^{-17}\ cm^3s^{-1}$ mainly at lower T. Also state i may support predissociating levels between the fine structure states $C^+(^2P_{1/2})$ and $C^+(^2P_{3/2})$ of the reactants. Inclusion[20] of all these effects in a resonant-nonresonant quantum calculation with improved potentials and emission rates shows a strong temperature dependence in the rates which rise sharply to a maximum value $\sim8\times10^{-17}\ cm^3s^{-1}$ at 30 K and which decrease monotonically at larger T. These improved rates still do not satisfactorily explain the above discrepancy.

Termolecular association

$$A^+ + B + M \rightarrow AB^+ + M \tag{24}$$

for formation of simple diatomics as He_2^+, Ne_2^+, etc., can be considered[21] as proceeding via a multistep series of collisions between A^+-B pairs and M which change both the energy E and angular momentum L of relative He^+-He motion to such an extent that bound stabilized levels are formed. At lower energies E there is an additional contribution from quasi-bound resonances[22] formed at positive E within the centrifugal barrier.

A multichannel generalization of eq. (13) to simple (structureless) atomic systems yields the termolecular association rate[23]

$$k_{TA} = \int_0^\infty dE \int_0^{L^2_{max}} dL^2 \left[\frac{k_i^s k_i N}{k_i^* + k_i^s\, k_i N}\right] k_i^* \tag{25}$$

where subscripted-i rates refer to specific energy E and angular momentum L of A-B relative motion, where L_{max} is the maximum L of the complex at fixed E, and where

$$k_i^s\, N\, n_i^* = \int_{R_i^-}^{R_0} n_i(R)\ \nu_i(R)d\underline{R} \tag{26}$$

is the overall frequency for stabilization of all the $(A\text{-}B)_i$ pairs with internal separation R between the innermost turning point R_i^- of radial motion and the radial boundary $R_0(E,L)$ of the complex. The pair-distribution per unit interval $d\underline{R}\ dE\ dL_i^2$ is $n_i(R)$, and $\nu_i(R) = k_q N$ is the frequency of $(A\text{-}B)_R^*$ - M quenching collisions with rate k_q at fixed (R,E_i,L_i^2). At low gas densities N this distribution can be taken as its equilibrium value $\tilde{n}_i$, since ν_i in eq. (26) is already linear in N. When the quenching coefficient k_q is constant, and equal to some fraction β of the constant Langevin limiting rate for spiraling AB*-M collisions

$$k_L = 2\pi e(\alpha_M/M_s)^{1/2} \tag{27}$$

where α_M is the polarizability of M and M_s is the reduced mass of the (AB*-M) system, it then follows that

$$n^* = \int_0^\infty dE \int_0^{R_0(E)} d\underline{R} \int_0^{R^2p^2} \tilde{n}_i(R)dL^2$$

$$= \frac{2}{\sqrt{\pi}}(kT)^{-3/2} \int_0^\infty \exp(-E/kT)dE \int_0^{R_0(E)} [E-V(R)]^{1/2}\ d\underline{R} \tag{28}$$

For polarization attraction $V(R) \sim (-\alpha_B e^2/2R^4)$ between A^+ and B of polarizability α_B and orbiting radius $R_0(E) = (\alpha_B e^2/2E)^{1/4}$, eq. (28) yields

$$n^* = \frac{4}{3}\pi\ R_L^3\ \frac{8}{\sqrt{\pi}} \tag{29}$$

where R_L is $(\alpha_B e^2/2kT)^{1/4}$ and $(8/\sqrt{\pi})$ arises from both the focusing effect and enhancement of R_0 at small E. The association rate at low N is then

$$k_{TA} = n^*[\beta(T,N)k_L]N \sim (10^{-28}-10^{-31})N\ cm^3s^{-1} \qquad (30)$$

which exhibits the temperature dependence $\beta(T)T^{-3/4}$. The efficiency $\beta \sim 1$, but for He^+-He charge transfer collisions the quenching rate $k_q \sim \langle\nu_{AM}\sigma_{AM}\rangle$ involves an additional $(kT)^{1/2}$ factor from ν_{AM} and a factor $(kT)^{-1}$ from focusing effects so that $k_{TA} \sim T^{-5/4}$ at low temperature.

2.2. Complex Systems

Here, rates are much higher due to increase in the physical size and in the number of internal modes of the intermediate complex. For triatomic ionic systems as

$$C^+ + H_2 \underset{\nu_d}{\overset{k^*}{\rightleftarrows}} (CH_2{}^+)^* \overset{\nu_r}{\rightarrow} CH_2{}^+ + h\nu \qquad (31a)$$

which initiates carbon phase chemistry in diffuse and dense interstellar clouds[24], and for polyatomic complexes as in either

$$CH_3{}^+ + H_2 \rightleftarrows (CH_5{}^+)^* \rightarrow CH_5{}^+ + h\nu \qquad (31b)$$

which is a precursor[9] to the formation of methane (CH_4), or in

$$CH_3{}^+ + H_2O \underset{\nu_d}{\overset{k^*}{\rightleftarrows}} (CH_3{}^+\cdot H_2O)^* \overset{\nu_r}{\rightarrow} CH_3{}^+\cdot H_2O + h\nu \qquad (31c)$$

which can be photodissociated[24] to produce methanol (CH_3OH) in the interstellar medium, the "kinetic chemical" approach is as yet the only viable method. The collision duration is much longer than that for simple systems as process (20) and there are simply too many degrees of freedom in the intermediate complex to consider in a full quantal state-to-state fashion. Moreover the complex offers a near continuum of closely spaced vibrational (and electronic) energies, overlapping resonances and many intramolecular processes so that a state-to-state method could not be considered as providing the most feasible or realistic description.

In order to isolate radiative association (RA) from termolecular association (TA) extremely low neutral densities $\lesssim 10^{10}$ cm^{-3} and temperatures T (10-30 K) are required. The mechanisms often proceed in parallel so that, in the coupled sequence,

$$A + B + (M) \underset{\nu_d}{\overset{k^*}{\rightleftarrows}} AB^* \overset{\nu_r}{\rightarrow} AB + h\nu \qquad (32)$$

$$\underset{\nu_d}{\overset{k^*}{\rightleftarrows}} AB^* \overset{\nu_q}{\rightarrow} AB + M \qquad (33)$$

radiative association occurs at the rate

$$k_{RA} = n^*_{AB}\ \nu_r = \left[\frac{\nu_r}{\nu_d+\nu_s}\right] k^* \qquad (34)$$

and termolecular association at the rate

$$k_{TA} = n^*_{AB}\ \nu_q = \left[\frac{k_q M}{\nu_d+\nu_s}\right] k^* \tag{35}$$

The frequency of stabilization of the complex, against both natural and collisional disruption at frequency ν_d, is

$$\nu_s = \nu_r + k_q\ N \tag{36}$$

the sum of the radiative decay frequency ν_r, and the frequency $k_q N$ for collisional quenching. At low densities $N(H_2,He) \sim 10^3-10^{10}\ cm^{-3}$ in interstellar clouds, $\nu_s = \nu_r << \nu_a$, so that the overall association is radiative controlled and proceeds at rate

$$k_A = k_{RA} = K\ \nu_r \tag{37}$$

where the reaction volume is given by eq. (14). At intermediate densities $10^{10}-10^{16}\ cm^{-3}$, ν_s still remains much smaller than ν_a, and association proceeds at rate

$$k_A = k_{RA} + k_{TA} = K\ (\nu_r + k_q\ N) \tag{38}$$

which increases with gas density N, until eq. (35) saturates to the limiting rate $k_{TA} = k^*$ of collisional formation of the original complex. The rate (38) is determined by the character of the interaction between the transition channels within the complex and differences in temperature dependence are mainly controlled by the T-variation of the reaction volume K(T). Radiative stabilization rates ν_r for complex systems are also uncertain, but are expected to be $\nu_r \sim 10^3\ s^{-1}$ for vibrational transitions and $\nu_r \sim 10^5-10^6\ s^{-1}$ for electronic transitions. The large electronic rates ν_r permit association in interstellar clouds to proceed faster than originally supposed[9].

Typical values of the relevant rates are the Langevin limit $k^* \sim 10^{-9}\ cm^3s^{-1}$, $\nu_a \sim 10^7\ s^{-1}$, $\nu_r \sim 10^3-10^5\ s^{-1}$ and the Langevin limit $\nu_q \sim 10^{-9}\ N\ s^{-1}$. Radiative rates $k_{RA} \sim 10^{-13}-10^{-11}\ cm^3s^{-1}$ and termolecular rates $k_{TA} \sim 10^{-25}\ N\ cm^3s^{-1}$ are then expected for complex systems at low gas density N. Termolecular association therefore begins to compete with RA for $N \sim 10^{12}\ cm^{-3}$ while at higher $N \gtrsim 10^{15}\ cm^{-3}$, TA becomes dominant.

Few experiments exist on RA, mainly due to the smallness of the rate $\sim 10^{13}\ cm^3s^{-1}$ and difficulty in achieving low temperatures ($T \sim 10$ K) and densities ($N \lesssim 10^9\ cm^{-3}$) needed for isolation of RA. The trap technique of Dunn and associates[16] represents a spirited effort while at higher $N \sim 10^{11}-10^{13}\ cm^{-3}$, the ICR (ion-cyclotron resonance) experiment[25] measures the RA and TA combined (36). By contrast, many TA experimental studies at yet higher $N \geq 10^{15}\ cm^{-3}$ exist for atmospheric species - the SIFT (selected ion flow tube) technique[12] being the major source. For TA, reasonable (order-of-magnitude) agreement exists with theory, particularly in the temperature variation. For RA, the few measurements of (24) and (25) do not agree with available theory and do not furnish information on the type (vibrational or electronic) of radiative stabilization. Interesting discrepancies between experiment and theory based on (32) and (33) for polyatomic species are discussed by Bates and Herbst[10].

3. DISSOCIATIVE RECOMBINATION

3.1. Direct Process

In the direct two-stage mechanism (Fig. 1(a))

$$e + AB^+(v_i) \underset{\nu_a}{\overset{k^*}{\rightleftarrows}} (AB^*)_r \overset{\nu_s}{\rightarrow} A^*+B, \ k_{DR} \sim 10^{-7}(300/T)^{1/2} \ cm^3 s^{-1} \qquad (39)$$

the electron of energy ε excites an electron of the ion-core AB^+ and is then resonantly captured via a Franck-Condon (FC) vertical transition onto the repulsive state r of the double excited molecule (AB*). Competition between reverse autoionization at nonlocal frequency ν_a and predissociation at nonlocal frequency ν_s continue until the electronically excited neutral fragments accelerate past the stabilization point R_s. Beyond R_s the increasing energy of relative separation has reduced the total electronic energy to such an extent that autoionization is essentially precluded and the neutralization is then rendered permanent. The kinetic energy of the electron (in the field of AB^+) is effectively transferred here to motion of the nuclei not by direct collision but via a rearrangement in eq. (39) of the whole electronic cloud. DR is a "reactive" process in the sense that the reactants and products involve different collision partners.

The autoionization character of AB* for $R < R_s$ makes resonant capture originally possible, and the covalent repulsive character for $R > R_s$ makes neutralization finally permanent. For reasonable capture over a range of ε, the autoionization width $\Gamma_a \sim \hbar\nu_a$ must not be too small, while large stabilization probabilities P_s demand small widths. The requirement of resonant capture without any energy transfer between electronic and nuclear motion is that the vertical difference in the potential-energy curves (PE^+ and PE*) for XY^+ and XY* equals ε (Fig. 1(a)). For thermal-energy electrons this requirement is best fulfilled when PE* crosses PE^+ on the right side of its minimum (cf. Fig. 1(a)), as for most cases of doubly excited electronic states with more than four electrons. This energy-matching can consequently occur over the full range of ε.

Large capture rates depend therefore on good electron-electron communication (correlation) and on good vibrational overlap between the AB^+ bound and AB*-continuum nuclear wave functions, an overlap which is sensitive to the initial vibrational level v_i of the ion and to the crossing of PE* and PE^+. When the only crossings in Fig. 1(a) are provided by the upper repulsive PE* curves, then the capture probability remains small for v_i=0 ions and thermal electrons, and becomes large only when these curves are accessed by more energetic electrons $\varepsilon \gtrsim 0.5$ eV, which imply however smaller Coulomb focused scattering cross sections $\sigma \sim \varepsilon^{-1}$. This is the situation with H_2^+ (v_i=0) and He_2^+. Conversely, the overlap of AB^+ (v_i=2) in Fig. 1(a) with the lower PE* is poor, relative to the much larger overlap with the upper curves. Note that ε is measured from R_c on PE^+.

In keeping with eq. (11), the recombination cross section for simple systems may be factored as

$$\sigma_{DR}(\varepsilon) = \sigma_c(\varepsilon) \ P_s(\varepsilon) \qquad (40)$$

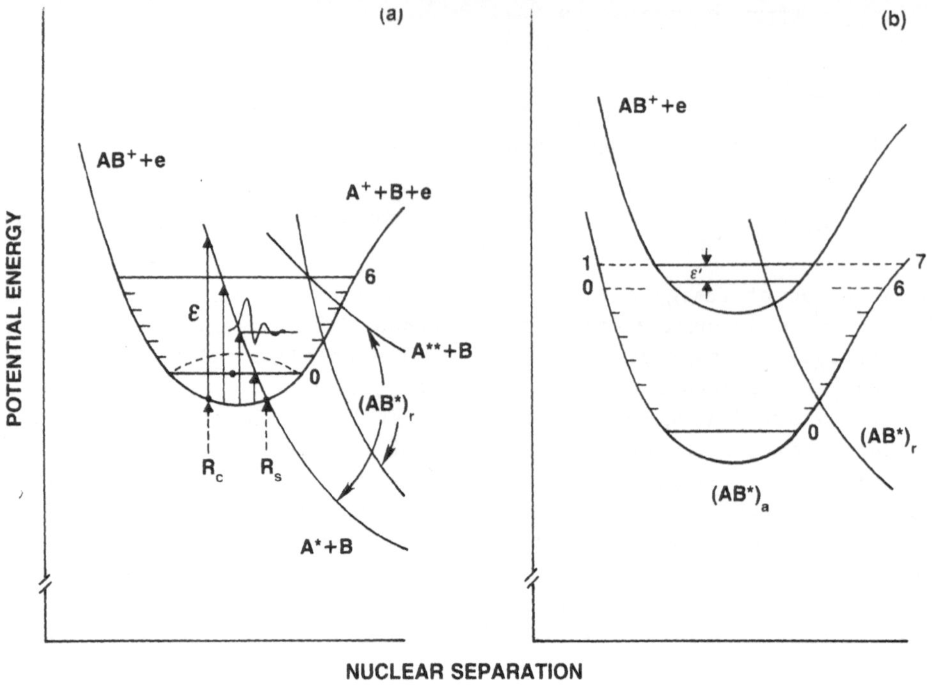

Fig. 1. Schematic representation of potential energy curves for dissociative recombination, $e + AB^+ \rightarrow A^* + B$, (a) via the direct (vertical transition) mechanism and (b) via the indirect (Rydberg) mechanism.

where the cross section for capture at R_c is[26]

$$\sigma_c(\varepsilon) = \frac{C}{\varepsilon}\left|V(R_c)\right|^2\left|\psi_v^+(R_c)\right|^2[dR/d(PE^*)]_{R_c} \tag{41}$$

Here V(R) is the electronically averaged interaction coupling the initial and intermediate molecular systems, ψ_v^+ is the vibrational wave function for $AB^+(v)$ and C is $(2\pi^3/mh)[\omega(AB)/\omega(AB^+)]$. The stabilization probability is given as in eq. (11) by $\nu_s/(\nu_a+\sum_s\nu_s)$. By analogy with dissociative attachment, it may also be approximated by[26],

$$P_s(\varepsilon) = \exp[-\int_{t_c}^{t_s}[\Gamma_a(R)/\hbar]dt] \tag{42}$$

where $\Gamma_a(R)$ is a local autoionization width (so that $\Gamma_a = \hbar\nu_a$ is the probability of electron ejection per unit time) and where the integration is over the interval from the time t_c at formation of AB* at R_c to the time t_s when stabilization at R_s is rendered permanent. This interval depends on the total energy and slope of PE*. Although the local $\Gamma_a(R)$ in eq. (42) is not strictly appropriate to recombination at thermal ε, eq. (42) remains useful as an estimate of the influence of autoionization. Thus P_s is reduced by an increase in number of open bound vibrational channels over which autoionization proceeds when electrons are emitted not only at energy $\varepsilon_a = \varepsilon$ but also at $\varepsilon_a \neq \varepsilon$ when the energy imbalance is absorbed by vibrational motion. It is enhanced not only by a reduction in the time interval, but also by an increase in the density of intermediate complexes and product channels, as with ion clusters. The ε^{-1}-dependence of σ_c results in recombination rates (11) which decreases at $T_e^{-0.5}$. For typical diatomic molecular ions as Ne_2^+ or NO^+, dissociation occurs at frequencies $\nu_a \sim 10^{15}$ s^{-1}, large compared with $\nu_a \sim 10^{14}$ s^{-1} for autoionization, so that P_s is close to unity. At thermal energies Coulomb focusing dominates the capture so that $\sigma_c \sim \varepsilon^{-1} \gtrsim 10^{-14}$ cm^2. Rates k_{DR} $(Ne_2^+) \sim 2\times10^{-7}$ $(300/T)^{0.5}$ cm^{3}s^{-1} are then quite typical. As one proceeds through an ion sequence ($Ne_2^+ \rightarrow Xe_2^+$), the natural increase in σ_c is due both to the stronger interactions and larger vibrational amplitudes and P_s remains substantial. Owing to the increasing steepness of PE*, it generally increases. Continued increase in σ_c however implies a corresponding increase in autoionization width so that P_s will eventually decrease, until it becomes limited to (ν_s/ν_a) as for the case of polyatomic systems.

3.2. Indirect Process

In the following indirect additional mechanism for DR[26],

$$e + AB^+(v) \rightleftarrows [AB^*(n,v)]_a \rightleftarrows (AB^*)_r \rightarrow A^* + B \qquad (43)$$

the electron is captured into attractive (a) vibrationally excited (v') Rydberg states (n) of AB* which converge to the initial electronic state of AB^+ (Fig. 1(b)) and which are then coupled by configuration interaction to the dissociative channels. The first stage involves energy transfer from the electron directly to vibrational excitation of the nuclei. In contrast to the broad ε-range enjoyed by the direct process (39), only selected energies ε' close to the Rydberg level (Fig. 1(b)) contribute to the indirect process which is therefore characterized by a series of narrow resonances (enhancements or dips) in the overall recombination cross section at the low electron energies $\varepsilon \leqq 1$ eV favored by this process.

The formal multichannel quantum theory of DR via the direct and indirect mechanisms can be constructed[8]. For full quantal calculations the following information is required as input: (a) identification and calculation of the relevant PE^+ and PE* curves for the capture cross section including those for the vibrationally excited Rydberg states, (b) the quantum coupling between the autoionization and dissociation channels for the widths Γ_a and Γ_s and (c) branching ratios to all possible products of dissociation. Since the coupling (b) appears as a resonance in the asymptotic phase of the electronic wave function the widths may be obtained either from direct electron-ion scattering calculations or from extrapolation of the properties of the Rydberg and valence bound states across the ionization threshold. The main theoretical problems are associated with the uncertainty of the role of the vibrationally excited Rydberg states and with the branching ratios which in turn involves solution of a set of coupled equations incorporating the interactions between the various products of dissociation. The "reactive" DR process combines therefore both electron-ion, ion-ion and neutral-neutral scattering technologies. Because of the sensitivity (as indicated above) of k_{DR} on the slopes, shapes and relative positions of PE^+ and PE* and the lack of accurate PE curves for most systems, rigorous calculation has been confined mainly[27,28] to H_2^+ and to some diatomic ions (N_2^+, O_2^+ and NO^+) of atmospheric significance[8].

DR for even the simplest diatomic system $e+H_2^+$, although not quite typical, is instructive. The sole candidate in the direct process for $e-H_2^+$ ($X\ ^2\Sigma_g^+$, v) recombination at low energies $\varepsilon \leq 1$ eV is the lowest doubly excited $^1\Sigma_g$ ($2p\sigma_u$) state of H_2 which crosses the $^2\Sigma_g^+$ ion state in the vicinity of the v = 2 level, and which dissociates into ionic fragments $H^+ + H^-$. Because of the propensity rule $\Delta v' = 1$ for vibrational autoionization in process (43), the recombination can be actively hindered by the higher vibrational levels v' of Rydberg states ($1s\sigma_g n\ell m$) with intermediate $n \leq 8$, and the contribution from these levels is weak. However, the sequence, coupling the direct and indirect processes,

$$e + H_2^+\ (v=0) \rightarrow H_2^{**} \rightarrow H_2(n,v\geq 2) \rightarrow H_2^{**} \rightarrow H_2^+ + e$$

does interfere destructively[27,28] with the direct process. The resulting resonant dips in the cross section have recently been observed[29]. Rates of the processes $e + H_2^+(v)$ can be given as $10^{-9}k_0(300/T)^\gamma\ cm^3s^{-1}$ where (k_0,γ) have been calculated[28] as (0.8,0.3), (6,0.5), (0.45,0.66), (0.66,0.32) and (1.1,0.77) for v = 0, 1, 2, 3 and 4, respectively.

The DR-rate for $CH^+(v=0)$ at 120 K was also calculated[28] to be $\sim 1.12\times10^{-7}\ cm^3s^{-1}$ in good agreement with a merged beam experiment[8].

Even though measured DR rates for many ions of planetary and astrochemical interest can be used with reasonable confidence, severe disagreement exists for the simplest triatomic H_3^+ important to the Jovian atmosphere and to interstellar chemistry. The rate is expected to be small since the $^2A_1$ repulsive part of the PE* curve of H_3 intersects the $^1A_1$ state of H_3^+ at 1 eV above the v = 0 level. Recent measurements[30] which vary from $2\times10^{-8}\ cm^3s^{-1}$ at 1000 K to $1\times10^{-8}\ cm^3s^{-1}$ at 1000 K for v = 0 and 1 ions are orders of magnitude higher than the revised upper-limit rate[31] of $\sim 10^{-10}\ cm^3s^{-1}$ in the range 90-500 K. Further work is clearly required to resolve this discrepancy.

Polyatomic ions and clusters offer many more additional degrees of freedom for capture of the electron, in both mechanisms. With increasing ion complexity, the multiplicity of readily excited internal modes of small energy separation makes the near resonant energy condition of the indirect process easier to attain by presenting a near continuum of closely spaced vibrational energies. Trapping then becomes more efficient over a broad range of ε'. This is confirmed by the large rates $k_{DR} \sim 2\times10^{-6}$ $(300/T)^{0.4}\ cm^3s^{-1}$ for dimer complexes $N_2^+\cdot N_2$, $O_2^+\cdot O_2$ and $CO^+\cdot CO$, important in atmospheric chemistry. That polar clusters $H_3O^+\cdot(H_2O)_n$ and $NH_4^+\cdot NH_3$ with rates $k_{DR} \sim 3\times10^{-6}\ cm^3s^{-1}$ appear fairly insensitive to T, has not been satisfactorily explained as yet.

As systems become more complex ($Ne_2^+ \rightarrow Xe_2^+$), the resulting increase in the capture cross sections σ_c tends to be offset by a corresponding decrease in the stabilization probability P_s from near unity until stabilization becomes the rate-limiting step. The rate from eq. (11) is then

$$k_{RA} = K(T)\ \nu_s \tag{44}$$

where the reaction volume has now the interesting form[9]

$$K(T) = \frac{h^3}{(2\pi mkT)^{3/2}} \frac{\omega(AB^*)}{2\omega(AB^+)} \left[\int |\psi_v(R)|^2 \left[\frac{dR}{d(PE^*)}\right] \exp(-E/kT)dE\right] \tag{45}$$

which contains an effective Franck-Condon factor which essentially selects only that portion of the full internal partition function of AB* that contributes to the capture by the vertical transition at $R = R_c$. Polyatomic systems relevant to interstellar cloud chemistry have recently been discussed by Bates[9].

4. DIELECTRONIC RECOMBINATION AND RADIATIVE RECOMBINATION

Dielectronic recombination (DIR) ar high temperatures ($\sim 10^6$ K)

$$e(\varepsilon\ell) + A^{z+}(i) \underset{\nu_a}{\overset{k^*}{\rightleftarrows}} [A^{z+}(k)-e]_{n\ell} \overset{\nu_r}{\rightarrow} A^{(z-1)+}_{n\ell}(j) + h\nu \tag{46}$$

is a resonant capture process into doubly excited Rydberg levels subsequently stabilized by radiative emission at frequency ν adjacent to, and usually on the lower frequency side of, the resonance transition

$$A^{z+}(k) \rightarrow A^{z+}(j) + h\nu_R \tag{47}$$

of the recombining ion of charge Ze. These satellite lines are observed in solar and in high temperature fusion plasmas and provide valuable diagnosis of electron temperature, electron density and the various stages of ionization. The frequency shift which originates from core perturbation by the $n\ell$-electron is small for high Rydberg $n\ell$-levels but would be quite large for low-lying n levels. Since the product ion may be subsequently re-ionized by interaction with its environment the stabilization mechanism is not quite as secure as that for dissociative recombination.

Although stabilization of the high Rydberg ion mainly occurs at high electron temperatures T_e by the inner-core transition (47) with the captured electron acting as a spectator, stabilization can also occur by a radiative transition $n\ell \rightarrow n'\ell'$ of the outer electron. This mechanism tends to be effective mainly at the much lower temperatures ($\lesssim 10^4$ K) characteristic of planetary nebulae. It is also effective for ions with low-lying metastable levels, as in

$$e + O^+(^4S) \rightarrow O^*[2p^3(^2D)n'\ell'] \rightarrow O^*(^2D,n\ell) + h\nu \tag{48}$$

The rate for dielectronic recombination (DIR) for an initial state i of the ion is, in the isolated resonance approximation (IRA), given by eq. (18) as

$$k_{DIR}(T;i) = \left[\frac{h^3}{(2\pi mkT)^{3/2}}\right] \frac{1}{2g_i} \sum_d \sum_f g_d \left[\frac{\nu_a(d\rightarrow i)\nu_r(d\rightarrow f)}{\nu_r(d)+\nu_a(d)}\right] \exp(-E_d^*/kT) \tag{49}$$

where g_i and g_d ($=2(2\ell+1)$) are the electronic statistical weights for state i of the recombining ion and for intermediate resonant state d ($=n\ell$) at energy E_d^* above state i. Each resonant state d may autoionize back (via an Auger transition) to state i with frequency ν_a ($d\rightarrow i$) or radiate with frequency ν_r ($d\rightarrow f$) to bound levels f. The total radiative and Auger rates from d to all states are $\nu_r(d)$ and $\nu_a(d)$, respectively. The total DIR rate is obtained by summing over all possible initial states i, intermediate states d and final bound states f. Note that the factor $h^3/(2\pi mkT)^{3/2}$ in eq. (49) is $(4\pi I_A/kT)^{3/2} a_0^3 = 4.1212\times10^{-16}\ T^{-3/2}\ cm^3$.

DIR within the past three years has been subjected to much theoretical[3] and experimental[15] study. The existing calculations are based on either the Coulombic model, the distorted-wave method and the relativistic configuration interaction method. For example, Chen[32], in a series of excellent papers, has used the multiconfiguration Dirac-Fock model to evaluate the detailed transition energies and Auger and radiative rates. The calculations not only include the Coulomb interaction r_{12}^{-1} but also the Breit interaction and other quantum-electrodynamic corrections. A considerable amount of theoretical data has now been accumulated[3] for many different isoelectronic sequences - for cases where the number N of electrons in the initial ion is 1-5, 8-12, 18 and 19.

The autoionization frequency decreases with (n,ℓ) as $\nu_a \sim n^{-3} \times \exp(-a\ell^2)$ owing to a decrease in communication between the core and Rydberg electrons, and is independent of Z. The radiative frequency is $\nu_r \sim \alpha^3 Z^p$ for core decay (p = 4 or 1 with or without a change in the core principal quantum number) and $\nu_r \sim \alpha^3 Z/n^3$ for outer electron decay. For small $n \ll 50$ and low ℓ, $\nu_r \ll \nu_a$ so that eq. (49) is radiatively limited. At nebular temperatures $T \sim 10^4$ K the exponential in eq. (49) restricts the summation to levels within ~ 0.15 eV of the ionization limit and ν_r is determined by outer electron decay. Since $\nu_a \ll \nu_r$ for large n, convergence can be obtained. Rates $k_{DIR} \sim (12\text{-}7)\times 10^{12}\ cm^3 s^{-1}$ for C^{2+}, N^{3+}, O^{4+} recombination at $T \sim 10^4$ K which exceed the direct radiative contribution are typical[3].

At high T (~ 10^7 K) ~ 1 keV characteristic of the solar corona, the full Rydberg series of autoionization levels must be included and core relaxation is the main radiative decay. For $n \gg 50$, $\nu_a \gg \nu_r$ so that eq. (49) is limited by autoionization. While the number of resonances increases as $2n^2$, only the low ℓ fraction are effective. Electric fields can however mix high ℓ-states with low ℓ-states so that DIR could be significantly enhanced. Typical rates[32] are ~ $3\times 10^{-11}\ cm^3 s^{-1}$ at 1 keV for F-like Se^{25+} - an X-ray laser candidate.

The separation ΔE (a.u.) between resonances of Rydberg series is $\sim Z^2/n^3$ which can become less than the radiative width $\Gamma_r = \hbar\nu_r$. The detailed resonance structure is then smeared out by interaction with the radiation field and IRA breaks down. Bell and Seaton[2] have solved this problem by quantum defect theory which because of its close connection with Rydberg series is ideally suited to DIR. Thus DIR cross sections can in principle be calculated to the same accuracy as electron-ion scattering cross sections (to within 10%).

For ions with low Z, Coster-Kronig (CK) channels, such as $1s2pn\ell \to 1s2s + e$ for He-like ions, become energetically accessible for large n. This effect of autoionization to excited states of the recombining ion has generally been neglected in the fluorescence yield $\nu_r(d\to f)\ [\nu_r(d) + \nu_a(d)]^{-1}$ in all calculations of eq. (49) until only recently. For example, the onset of the above CK transition for Be^{3+} ion is at $n^* = 3$, and n^* increases with Z (e.g., $n^* = 9$ for F^{7+}). Inclusion of CK transitions reduce[32] the peak values of the total DIR rates for B^{3+}, N^{5+} and F^{7+} by 60%, 13% and 4%, respectively. This trend is correct since the relative contributions to DIR from high n-state (important at low Z) decrease with n, while onset of CK transitions occurs at higher n as Z increases.

The CK effects are not, of course, included in the largely historical semiempirical formulas of Burgess[33] (for core decay $\Delta n = 0$) and of Merts et al.[34] (for $\Delta n = 1$). These formulas, although used quite generally by astrophysicists, have recently been shown to overestimate[32] small-Z rates by a factor of three and to underestimate large-Z rates by as much as a factor of two.

In addition to CK transitions for low Z, some remaining problems appear to be (a) effects of external fields on DIR, (b) three-body density effects on k_{DIR} and (c) fine-structure effects. For (c), fine-structure states of the excited ion core provide two Rydberg series of autoionization channels which can mutually interfere (as in the decay $3p_{3/2}(n\ell) \rightarrow 3p_{1/2}(\varepsilon_1\ell) + 3s_{1/2}(\varepsilon_2\ell)$ in Mg^+). A problem which appears to be solved is the coupling between resonant DIR and the following nonresonant radiative recombination (RR) which, while negligible for ions with low Z, becomes appreciable at high Z.

The subsequent chain of atomic processes in astrophysics was initiated by the basic $(e\text{-}H^+)$ radiative recombination (RR) process

$$e + A^{z+}(i) \rightarrow A^{(z-1)+}(n\ell) + h\nu \tag{50}$$

into level $n\ell$. Since RR is a direct inverse of photoionization with cross section $\sigma_I^{n\ell}$ $(h\nu)$, the RR rate bv detailed balance is

$$k_R^{n\ell}(T) = \left[\frac{8kT}{\pi m}\right]^{1/2}\left[\frac{kT}{mc^2}\right]\left[\frac{g_{n\ell}}{2g_i}\right]\exp(I_{n\ell}/kT)\int_{I_{n\ell}/kT}^{\infty}\left[\frac{h\nu}{kT}\right]^2\sigma_I^{n\ell}(h\nu)\exp(-h\nu/kT)\,d(h\nu/kT) \tag{51}$$

where g_i and $g_{n\ell}$ are the electronic statistical weights of the initial ion and the recombined ion in level n with ionization potential $I_{n\ell}$. Various analytical forms for σ_I can be adopted, e.g., when $(h\nu)^3\sigma_I(h\nu)$ equals its value $I_n^3\sigma_0^{n\ell}(I_n)$ at threshold then the rate is

$$k_R^{n\ell}(T) = 1.5\times10^{-13}\left[\frac{300}{T}\right]^{1/2}\left[\frac{I_n}{I_H}\right]^2\left[\frac{g_{n\ell}}{2g_i}\right]\bar{\sigma}_0^{n\ell}(T)\ \text{cm}^3\text{s}^{-1} \tag{52}$$

where, in terms of the exponential integral E_1, the averaged cross section is

$$\bar{\sigma}_I^{n\ell}(T) = \sigma_0^{n\ell}[x_n \exp x_n]E_1(x_n), \quad x_n = I_n/kT \tag{53}$$

which reduces at low temperatures $kT << I_n$ to

$$\bar{\sigma}_I^{n\ell}(T) = \sigma_0^{n\ell}\,[1 - (kT/I_n) + 2(kT/I_n)^2 - 6(kT/I_n)^3 + \dots] \tag{54}$$

The quantal cross section for photoionization of hydrogenic ions of charge Z by radiation of scaled energy ω $(= h\nu/I_n)$ is

$$\sigma_I^{n\ell}(h\nu) = \sigma_K^n(\omega)\,G_{n\ell}(\omega) \rightarrow \begin{cases} \omega^{-3}, & h\nu \rightarrow I_n \\ \omega^{-\ell-7/2}, & h\nu >> I_n \end{cases} \tag{55}$$

The departure from the (Kramer) semiclassical (high n and ℓ averaged) photoionization cross section[35]

$$\sigma_K^n(\omega) = \frac{2^6\alpha}{3\sqrt{3}}\left[\frac{n}{Z^2}\right]\left[\frac{I_n}{h\nu}\right]^3 \pi a_0{}^2 = 7.9\ n^2 Z^4 \omega^{-3}\ (\text{Mb}) \tag{56}$$

where α is the fine-structure constant ($e^2/\hbar c$). The $G_{n\ell}$ in eq. (55) is the bound-free Gaunt factor. The rate (14) is then

$$k_R^{n\ell}(Z,T) = \left[\frac{8kT}{\pi m}\right]^{1/2} \left[\frac{I_n}{kT}\right] \frac{1}{n} \left[\frac{2^5}{3\sqrt{3}}\right] (\alpha^3 \pi a_0{}^2) \left[\frac{g_{n\ell}}{2g_i}\right] F_{n\ell}(T^*) \tag{57}$$

Departures of eq. (57) from the above standard $(Z^2 n^{-3} T^{-1/2})$ low temperature rule is provided by the function

$$F_{n\ell}(T^*) = \frac{1}{T^*} \exp(1/T^*) \int_1^\infty \frac{G_{n\ell}(\omega)}{\omega} \exp(-\omega/T^*)\, d\omega \tag{58}$$

which decreases monotonically from $G_{n\ell}$ (1) as the scaled temperature T^* ($= kT/I_n$) increases. For interstellar clouds $kT \ll I_n$ and $F_{n\ell}(T^* \ll 1)$ tends to $G_{n\ell}$ (1), the threshold factor. Note that eq. (57) also provides the universal scaling law

$$k_R^{n\ell}(Z,T) = Z k_R^{n\ell}(1, T/Z^2) \tag{59}$$

Recombination rates are greatest into low n levels and the $\omega^{-\ell-1/2}$ variation of $G_{n\ell}$ in eq. (58) preferentially populates states with low $\ell \sim 2$-5. Highly accurate analytical fits for $G_{n\ell}$ (ω) have been obtained[36] for $n \leq 20$ so that eq. (57) is expressed in terms of known functions of fit parameters. This procedure (which does not violate the S_2 sum rule) has been extended[36] to non-hydrogen systems of neon-like Fe XVII, where $\sigma_I^{n\ell}(\omega)$ is a monotonically decreasing function of ω.

Variation of the ℓ-averaged values, $n^{-2} \sum_{\ell=0}^{n-1} (2\ell+1) F_{n\ell}(T^*)$, is close[36] in both shape and magnitude to the corresponding semiclassical function $S(T^*)$ given by eq. (58) with $G_{n\ell}(\omega) = 1$. Hence the ℓ-averaged recombination rate is

$$k_R^n(Z,T) = 1.1932\times10^{-12} \left[\frac{300}{T}\right]^{1/2} \left[\frac{Z^2}{n}\right] F_n(T^*)\ \text{cm}^3\text{s}^{-1} \tag{60}$$

where F_n can be calculated directly from eq. (58) or be approximated as $G_n(1)\times S(T^*)$. A computer program based on a three-term expansion of G_n is also available[37].

Tables exist[38] for the effective rate

$$k_E^{n\ell}(T) = \sum_{n'=n}^{\infty} \sum_{\ell'=0}^{n-1} k_R^{n'\ell'}\ C_{n'\ell',n\ell} \tag{61}$$

of populating levels $n\ell$ of hydrogen via radiative recombination into all levels $n' \geq n$ with subsequent radiative cascade $(i \rightarrow f)$ with probability $C_{i,f}$ via all possible intermediate paths. Tables[38] also exist for the total rate

$$k_R^N = \sum_{n=N}^{\infty} n^{-2} \sum_{\ell=0}^{n-1} k_R^{n\ell} \tag{62}$$

of recombination into levels N and above of hydrogen. They are useful in deducing time scales of radiative recombination and rates from eq. (59) for complex ions.

When effective at higher temperatures, dielectronic recombination proceeds in general faster than RR. Since $k_R \sim Z^2$, RR can however become competitive for highly charged ions. A unified treatment of DIR and RR has recently been presented[39]. The mutual interference of the corresponding amplitudes and continuum-continuum coupling is expected to be most important for individual transitions involving low-lying autoionization levels and is probably negligible for DIR arising from highly excited levels. If the photoionization cross section $\sigma_I^{n\ell}(h\nu)$ already includes the effects of autoionizing resonances, no further correction for DIR to RR may be necessary.

5. MUTUAL NEUTRALIZATION

Until fairly recently (1984), lack of agreement of various curve-crossing and Landau-Zener type theories with experiment for such a simple system as

$$H^+ + H^- \rightarrow H^*(n) + H \tag{63}$$

remained embarassing, and agreement between the two main experiments remained very good. Then a 1983-theory[40] which included couplings (neglected in previous theories) to the n = 3 level still did not agree with measurement, until new experiments[41,42] were performed in 1984 and 1985. The process (63) is now apparently well understood, but careful quantum mechanical calculations and experiment are required.

In dense interstellar clouds, mutual neutralization (MN) of complex systems can be important and can produce qualitative changes[24] in the chemistry sequence. For example, when polycyclic aromatic hydrocarbons (PAH) exist in high abundance, the negative charge is carried not by electrons but by PAH^- so that MN, as in $C^+ + PAH^- \rightarrow C + PAH$, replaces dissociative recombination (DR) so that the C-abundance is enhanced[24].

6. TERMOLECULAR RECOMBINATION

6.1. Ion-Ion Recombination

The theory of termolecular ion-ion recombination

$$A^+ + B^- + M \underset{k}{\overset{\alpha}{\rightleftarrows}} AB + M \tag{64}$$

of positive and negative atomic ions of concentrations $N_{A,B}$ (t) at time t in a gas M is also well established[43], and is also suitable as a case study. The effective two-body association rate $\alpha(N,T)$ cm^{3}s^{-1} and the dissociation frequency $k(N,T)$ s^{-1} are functions of gas density. At low gas density they are given by[43]

$$\alpha \, \tilde{N}_A \tilde{N}_B = \int_{-D}^{\infty} P_i^S \, dE_i \int_{-D}^{\infty} (P_i^S - P_f^S) \, C_{if} \, dE_f = k \, \tilde{n}_s \tag{65}$$

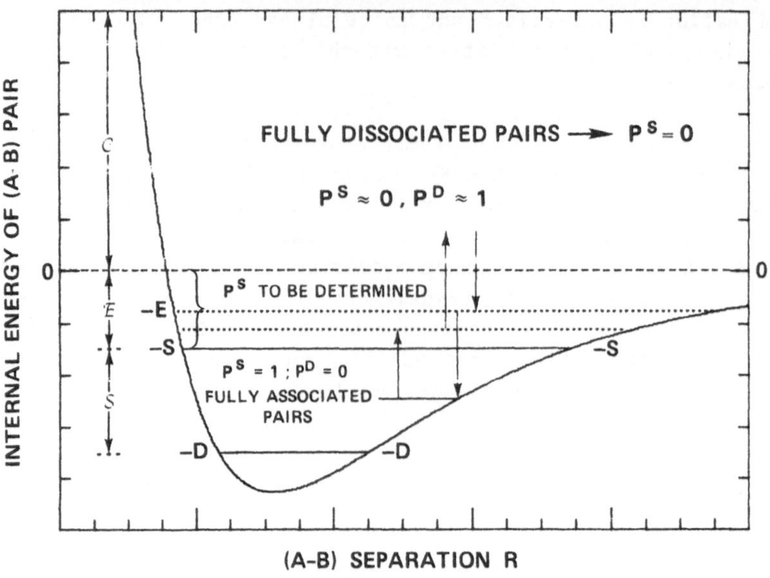

Fig. 2. Schematic diagram of energy blocks $\mathcal{C}$, $\mathcal{E}$ and $\mathcal{S}$ pertinent to recombination at low gas densities.

where P_i^S, which measures the departure from equilibrium, is the stochastic probability that a pair A^+-B^- with energy distribution n_i over internal relative energy E_i of the pair is connected via a multistep series of energy- (state-)changing collisions to a stabilized sink $\mathcal{S}$ of low-lying fully associated pairs of concentration n_s (cf. Fig. 2). The sink $\mathcal{S}$ extends over the energy range $-S \geq E_i \geq -D$ where $-D$ is the lowest energy level and where $-S$ is that bound level below which P_i^S is unity. The one-way equilibrium rate C_{if} for $E_i \to E_f$ collisional transitions per unit interval $dE_i dE_f$ is $n_i \nu_{if}$, and the distribution n_i at low gas densities satisfies the input-output collisional master equation[43]

$$\frac{dn_i}{dt} = [\gamma_c(t) - \gamma_s(t)] \int_{-D}^{\infty} (P_i^S - P_f^S)\, C_{if}\, dE_f \qquad (66)$$

at low gas densities where energy relaxation is the rate limiting step. The departures from their steady equilibrium (tilde) values of the total time-dependent concentrations of fully dissociated pairs (in block $\mathcal{C}$, $0 \leq E_i \leq \infty$ where $P_i^S \approx 0$) and of fully associated pairs (in block $\mathcal{S}$ where $P_i^S \approx 1$) are

$$\gamma_c(t) = N_A(t) N_B(t) / \tilde{N}_A \tilde{N}_B \ ; \quad \gamma_s(t) = n_s(t)/\tilde{n}_s \qquad (67)$$

respectively. For quasi-steady state (QSS) of the intermediate block $\mathcal{E}$ $(0 \geq E_i \geq -S)$ of highly excited levels at time t, eq. (66) vanishes so that eq. (65) reduces to

$$\alpha\, \tilde{N}_A \tilde{N}_A = \int_{-E}^{\infty} dE_i \int_{-D}^{-E} (P_f^S - P_i^S)\, C_{if}\, dE_f \qquad (68)$$

for arbitrary energy $-E$ in block $\mathcal{E}$.

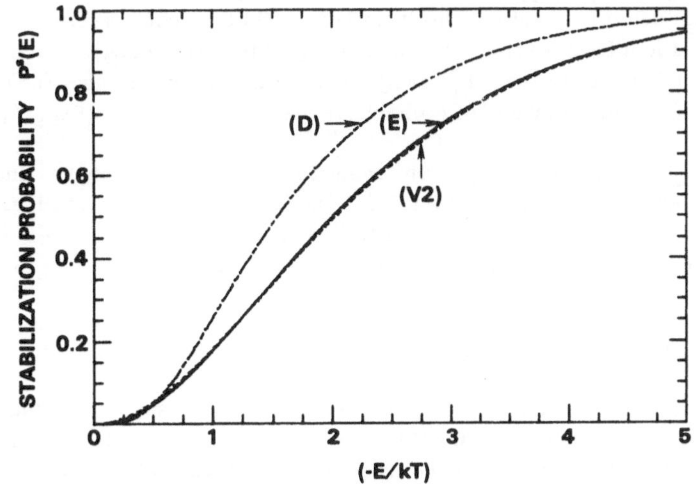

Fig. 3. Stabilization probabilities. (E): quasi-steady-state[43]. (V2): two-parameter variational[44]. (D): Diffusion[45].

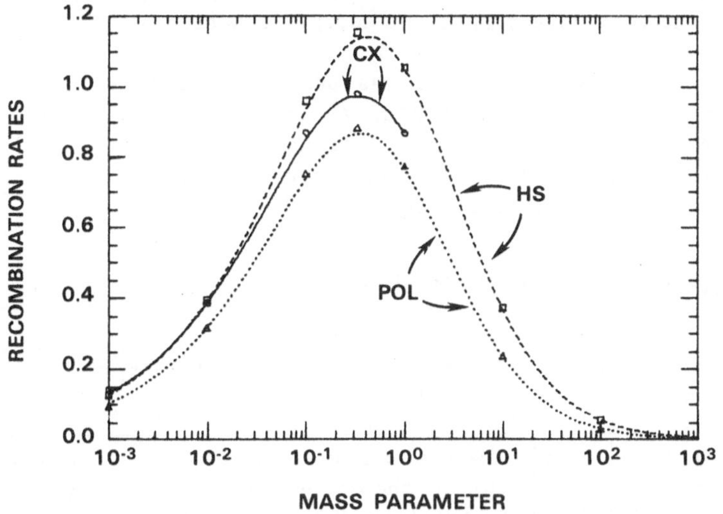

Fig. 4. $A^+ + B^- + M$ partial recombination rates $\left[\frac{M_A}{M_{AB}}\right]\alpha(a)$ normalized to Thomson's rate $\alpha_T(a)$ as a function of mass parameter $a = M_B M_g / M_A(M_A+M_B+M_g)$ for various A^+-M or B^--M interactions. (CX: symmetrical resonance charge transfer. HS: hard-sphere. POL: polarization attraction.) The full rates are $\alpha(a)\alpha_T(a) + \alpha(b)\alpha_T(b)$, where $b = (M_A/M_B)^2 a$ and where Thomson's rate is $\alpha_T(a) = \frac{4}{3}\pi R_e^3(3kT/M_{AB})^{1/2}\sigma_{AM}N$[43].

6.1.1. Variational principle. It has been recently proposed[44] that P_i^S are so distributed that the rate (65) is a minimum. This distribution leads exactly to the QSS distribution given by eq. (66) set to zero. Thus eq. (65) provides a variational expression for the QSS condition, so that P_i^S may be determined (Fig. 3) variationally or from the direct solution of the integral equation (66). The variational[44] and QSS[43] rates obtained are of course identical.

6.1.2. Diffusion method. By performing a Fokker-Planck conversion of the integral equation (66), the resulting (but approximate) differential equation is identical with a diffusion equation in energy space which can be solved analytically for P_i^S (Fig. 3). Insertion in eq. (65) yields a proposed diffusional method[45] which is highly accurate (Fig. 4).

6.1.3. Bottleneck limit. On assuming that pairs above and below a bound level -E are in equilibrium with that fully dissociated and associated (blocks $\mathcal{C}$ and $\mathcal{S}$, respectively, i.e., $P_i^S = 0$ for $E_i > -E$ and $P_i^S = 1$ for $E_i < -E$) then either eq. (65) or eq. (68) yields,

$$\alpha(-E)\ \tilde{N}_A\tilde{N}_B = \int_{-E}^{\infty} dE_i \int_{-D}^{-E} C_{if}\ dE_f \tag{69}$$

the one-way equilibrium collisional rate across -E, which is then an upper limit to the exact rate. Variation of α with -E yields the least upper limit at the bottleneck energy E* (see Refs. 23 and 43).

Other approximations such as coupled nearest-neighbor (CNN) limit and uncoupled intermediate $\mathcal{E}$-block levels (UIL), based on analogy of eqs. (65) and (66) with electrical networks recently proposed[46], have also elucidated the most likely paths of energy reduction.

6.1.4. Gas density. As the gas density N is raised nonequilibrium effects in internal separation R of A^+ and B^- must be considered. The appropriate input-output collisional-transport master equation satisfied by the distribution $n_i(\underline{R})$ of A^+-B^- pairs per unit interval $d\underline{R}\ dE_i$ has been shown to satisfy the continuity equation[47]

$$\frac{d}{dt} n_i(R,t) = \frac{\partial n_i}{\partial t} + \frac{1}{R^2}\frac{\partial}{\partial R}[R^2 j_i^d(R)]_{E_i}$$

$$= - \int_{V(R)}^{\infty} [n_i(R)\nu_{if}(R) - n_f(R)\nu_{fi}(R)]dE_f \tag{70}$$

where $j_i^d(R)$ $(= j_i^+ - j_i^-)$ is the net outward transport current of pairs expanding at R, where $\nu_{if}(R)$ is the frequency per unit interval $d\underline{R}dE_i dE_f$ for $E_i \to E_f$ collisional transitions for ion-pairs at fixed separation R and where V(R) is the energy of interaction between A and B. Integration of eq. (70) over all accessible R yields the standard master equation (66).

The question of reproducing the cumulative effects of multistep energy-changing collisions by an accumulative strong collision within a loose collision complex of radial extent R_T can now be examined[43]. The rate of recombination within a sphere of radius R_T and the overall probability $P_i^A(R_T)$ of association within R_T are related for low gas density by

$$\alpha(R_T)\ \tilde{N}_A\tilde{N}_B = \int_0^{\infty} [4\pi R_T^2 \tilde{j}_i^{\,-}(R_T)] P_i^A(R_T) dE_i \tag{71}$$

which is expressed via eq. (70) in terms of the stabilization probabilities P_f^S by[43]

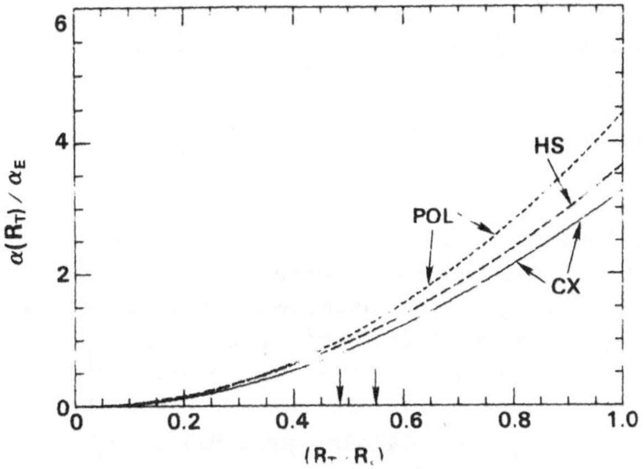

Fig. 5. Variation of $\alpha(R_T)$, eq. (72) with $P_f^S = 1$, to exact rate, eq. (72) with $R_T \to \infty$, for ion-ion recombination of equal-mass species under various A^+-M interactions (cf. Fig. 4).

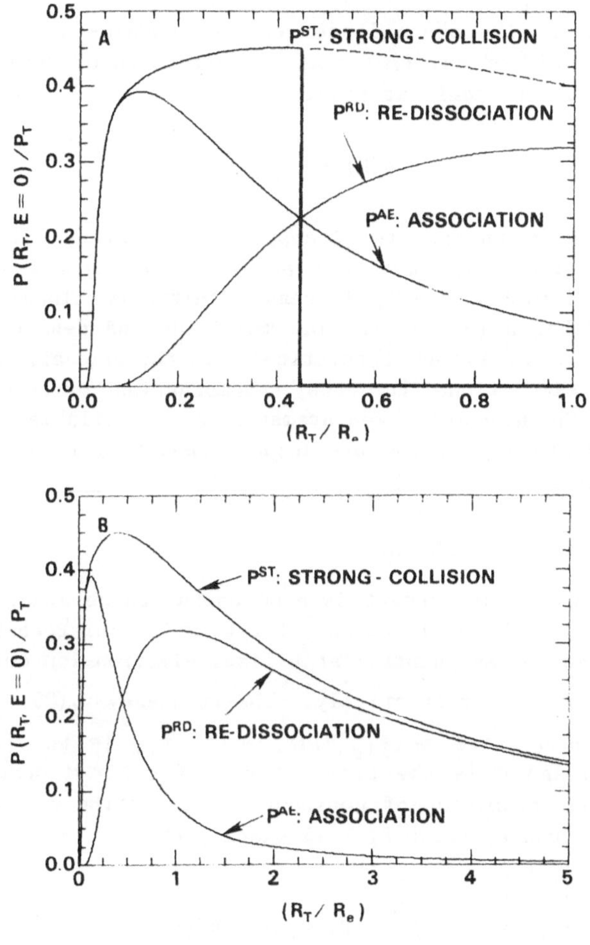

Fig. 6. Probability for eventual association and re-dissociation of A^+-B^- pairs with zero internal energy. P^{AE} and P^{RD}: exact association and redissociation. P^{ST}: strong collision. The probabilities are normalized to Thomson's low density probability $P_T = R_T(\sigma_{AM}N)$[43].

$$\alpha(R_T)\,\tilde{N}_A\tilde{N}_B = \int_0^\infty dE_i \int_0^{R_T} dR \int_{-V(R)}^{0} C_{if}\, P_f^S\, dE_f \tag{72}$$

A strong-collision (or classical) treatment refers to the assignment P_f^S = 1 in eq. (72).

Figure 5 illustrates the ratio of the effective strong-collision rate to α_E, the exact rate α ($R_T \to \infty$). Agreement can be obtained by assigning (de-facto) $R_T \approx 0.5\ R_e$. The underlying reason becomes apparent from Fig. 6. The exact probability P_i^{AE} that ($E_i = 0$) dissociated pairs ultimately associate dominates the probability P_i^{RD} for ultimate re-dissociation (after bound levels are formed) for smaller $R_T << R_e = e^2/kT$, so that P_i^{AE} and the strong-collision probability P_i^{ST} (from eq. (72) with $P_f^S = 1$) are essentially equal. Pairs with larger $R_T >> R_e$ are however mainly re-dissociated (Fig. 6). The strong collision rate at $R_T \sim 0.45$ comes out to be twice the rate $\alpha(R_T)$ of eq. (72). The remaining contribution from $R \geq R_T$ to the exact rate then provides agreement with the strong collision rates[43].

The concept of the above loose reaction complex is useful in showing[23] with the aid of eq. (70) that the variation of recombination with gas density yields the familiar result

$$\alpha(N) = \frac{\alpha_{RN}\,\alpha_{TR}}{\alpha_{RN}+\alpha_{TR}} \tag{73}$$

where $\alpha_{TR}(N)$, the known rate for transport of pairs by diffusional drift to within separation R_T, decreases as N^{-1}. The reaction rate α_{RN} by collision with M within $R \leq R_T$ increases initially linearly with N and saturates at higher N (~ 1 atm). The magnitude and density variation of eq. (73), with accompanying theoretical procedures, agree with Monte-Carlo computer simulations for the recombination of rare gas halide systems[4,6]. No benchmark measurements are available, but the two historical measurements at low and high N respectively in general agree with eq. (73)[17].

6.2. Electron-Ion Recombination

The trapping radius concept is also useful to obtain classical rates (i.e., eq. (72) with $P_f^S = 1$) not only for termolecular recombination (64) but also for electron and neutral stabilized electron-ion collisional recombination (9) and (7) respectively. The frequency $\nu_i(R) = \int_{V(R)}^{0} \nu_{if}(R)\,dE_f$ formation of bound pairs is $v_{12}(R)\sigma N$, where v_{12} is the speed of $A^+ + B^-$ relative motion and σ is the cross section for AB*-M deactivating collisions. On the assumption of constant cross section σ_0 for such collisions, eq. (72) reduces (with $P_f^S = 1$) exactly to

$$\alpha\,\tilde{N}_A\tilde{N}_B = \left[\int_0^\infty dE_i \int_0^{R_0} \tilde{n}_i(R)\; v_{12}(R)\,d\underline{R}\right](\sigma_0 N) \tag{74}$$

$$= \left[\frac{8kT}{\pi M_{AB}}\right]^{1/2} \left[\frac{4}{3}\pi R_0^3\right] \left[1 + \frac{3}{2}\frac{R_e}{R_0}\right] \sigma_0\ N \tag{75}$$

where R_e is e^2/kT. A classical version of the semiquantal bottleneck treatment (Section 6.1.3 above) yields, a priori, the trapping radius to be $R_0 = 0.41\ R_e$. The rates α of termolecular recombination (64), and of e-ion collisional recombination $(e + A^+ + e)$ at electron temperature T_e and electron density n_ε are therefore,

$$\alpha_{TR}(T) = 0.32\left[\frac{8kT}{\pi M_{AB}}\right]^{1/2}\left[\frac{4}{3}\pi R_e^3\right](\sigma_0 N) \sim 2.3\times10^{-25}(300/T)^{2.5}\ N\ (cm^3s^{-1}) \quad (76)$$

and

$$\alpha_{ee}(T) = 0.32\left[\frac{8kT}{\pi m}\right]^{1/2}\left[\frac{4}{3}\pi R_e^3\right](\frac{1}{9}\pi R_e^2)n_e \sim 2.7\times10^{-20}(300/T)^{4.5}\ n_e\ (cm^3s^{-1}) \quad (77)$$

respectively. In eq. (77), σ_0 for electron-electron collisions is taken as the Coulomb cross section $(\frac{1}{9}\pi R_e^2)$ for energy changes $\geq \frac{3}{2}kT$. These expressions provide the correct order of magnitude and temperature dependence, and, in general, agree with experiment. In particular eq. (76) agrees with the expression of Mansbach and Keck[48] derived from more elaborate analysis. At higher T_e and lower n_e, the highly excited levels collisionally formed within kT of the ionization limit become increasingly stabilized by radiative transitions. The resulting rate for collisional-radiative recombination can then be approximated as[49]

$$\alpha_{CR} = [3.8\times10^{-9}\ T_e^{-4.5}n_e + 1.55\times10^{-10}\ T_e^{-0.63} + 6\times10^{-9}\ T_e^{-2.98}n_e^{0.37}]\ cm^3s^{-1} \quad (78)$$

where the first term is eq. (77), the second term is the radiative correction and the third term arises from collisional-radiative coupling. This expression agrees with the experimental data[49] to within 10% for a Lyman optically thick plasma with n_e and T_e in the range $10^9 \leq n_e\ (cm^{-3}) \leq 10^{13}$ and $2.50 \leq T_e\ (K) \leq 4000$.

For termolecular $(e + A^+ + B)$ collisional recombination only a small fraction $\delta = 2m/M_B$ can be transferred in e-B elastic collisions so that the E_i-integration in eq. (74) must be so restricted to give

$$\alpha_{eB}(T_e) = \left[\frac{8kT}{\pi m}\right]^{1/2} 4\pi R_0^3\ (\sigma_0 N) \int_0^{r_0} r^2 dr \int_0^{\delta/r} (1+\frac{1}{r\varepsilon})\ \varepsilon e^{-\varepsilon} d\varepsilon \quad (79)$$

with $R = rR_e$ and $\varepsilon = E_i/kT$. Hence

$$\alpha_{eB}(T_e) = 4\pi\delta\left[\frac{8kT_e}{\pi m}\right]^{1/2} R_e^2 R_0\ (\sigma_0 N) \sim [10^{-26}/M_B(AMU)](300/T)^{2.5}\ N \quad (80)$$

which agrees exactly with the diffusion result of Pitaevskii[50] when R_0 is taken as the Thomson radius $\frac{2}{3}R_e$, and which is linear[51] in the trapping radius R_0. This result $(\sim10^{-28}\ cm^3s^{-1})$ is in general agreement with experimental data for $(e + Cs^+ + He)$ but is much smaller than that $(\sim10^{-26})$ for $(e + He^+ + He)$, which proceeds far more effectively[52] via formation of an intermediate complex He_2^* which then dissociates into neutral fragments.

The rate for $(e + A^+ + B)$ is greatly increased[51] for a molecular gas B where energy reductions are effected mainly by rotational and vibrational transitions. Allowance for the discreteness of e-A^+ Rydberg levels reduces α_{eM} and produces a sharper decrease with temperature[53]. When A^+ is a molecular ion XY^+ a dissociative recombination channel opens. Here the e-XY^+ pairs formed in highly excited Rydberg molecular levels XY* by collision with M, in addition to being collisionally and radiatively quenched to stable bound states of AB, may predissociate along repulsive curves X*+Y, i.e., by dissociative recombination involving bound electrons - the second half of the indirect mechanism(43). The contribution from this collisional dissociative recombination[53] can dominate the contribution from direct collisional relaxation. That quantal curvecrossing is involved makes it similar to the enhancement[6] of mutual neutralization (A^++B^-) by third bodies. In the limit of high gas density N, the recombination rate α_{eM} becomes transport limited, as in eq. (73) for ion-ion recombination and decreases as N^{-1}. Because of the higher electron mobilities, its onset however occurs at much higher N. Between the linear low-density region and the transport limited N^{-1} region only Monte Carlo simulations have been performed[54]. For e + A^+ + M recombination in a molecular gas the rotational and vibrational cross sections of Takayanagi[55] and of Takayanagi and Itikawa[56] and the recommended molecular constants[56] are invaluable.

Acknowledgment This research is supported by the U.S. Air Force Office of Scientific Research under Grant No. AFOSR-84-0233.

REFERENCES

1. M.J. Seaton and P.J. Storey, in: Atomic Processes and Applications (P.G. Burke and B.L. Moiseiwitsch, eds.), pp. 133-197, North-Holland, Amsterdam (1976).
2. R.H. Bell and M.J. Seaton, J. Phys. B18, 1589-1629 (1985).
3. Y. Hahn and K.J. LaGattuta, Physics Reports 166, 195-268 (1988).
4. M.R. Flannery, in: Applied Atomic Collision Physics (E.W. McDaniel and W.L. Nighan, eds.), Vol. 3, pp. 141-172, Academic, New York (1982).
5. M.R. Flannery, Phil. Trans. Roy. Soc. London. Ser. A 304, 447-497 (1982).
6. D.R. Bates, in: Adv. Atom. Mol. Phys. 20, 1-40 (1985).
7. M.R. Flannery, J. Chem. Phys. 88, 4228-4241 (1988).
8. See contributions in: Proceedings of International Workshop on Dissociative Recombination: Theory, Experiment and Applications (J.B.A. Mitchell and S. Guberman, eds.), World Scientific, New Jersey (1988).
9. D.R. Bates, in: Recent Studies in Atomic and Molecular Processes (A.E. Kingston, ed.), pp. 1-27, Plenum, New York (1987).
10. D.R. Bates and E. Herbst, in: Reaction Rate Coefficients in Astro-Physics (T.J. Miller and D.A. Williams, eds.), in press, Kluwer, Dordrecht (1989).
11. J.B.A. Mitchell, in: Atomic Processes in Electron-Ion and Ion-Ion Collisions (F. Brouillard, ed.), pp. 185-222, Plenum, New York (1986).
12. D. Smith, in: Adv. Atom. Mol. Phys. 24, 1-49 (1988).
13. K. Dolder and B. Peart, in: Adv. Atom. Mol. Phys. 22, (1986).
14. G.H. Dunn, in: Atomic Processes in Electron-Ion and Ion-Ion Collisions (F. Brouillard, ed.), pp. 93-116, Plenum, New York (1986).
15. See contributions in: Atomic Excitation and Recombination in External Fields (M.H. Nayfeh and C.W. Clark, eds.) pp. 439-452, Gordon and Breach, New York (1985).
16. S.E. Barlow, G.H. Dunn, and K. Schauer, Phys. Rev. Lett. 52, 902-905; 53, 1610 (1984).

17. M.R. Flannery, in: Atomic Processes and Applications (P.G. Burke and B.L. Moiseiwitsch, eds.), pp. 407-466, North-Holland, Amsterdam (1976).
18. H.S. Lee and R. Johnsen, J. Chem. Phys., in press (1989)..
19. A. Giusti-Suzor, E. Roueff, and H. van Regemorter, J. Phys. B9, 1021-1034 (1976).
20. H. Abgrall, A. Giusti-Suzor, and E. Roueff, Astrophys. J. 207, L69-L72 (1976); M.M. Graff, J.T. Moseley, and E. Roueff, Astrophys. J. 269, 796-802 (1983).
21. D.R. Bates and C.S. McKibbin, Proc. Roy. Soc. A 33, 13-28 (1974).
22. A.D. Dickinson, R.E. Roberts, and R.B. Bernstein, J. Phys. B5, 355-365 (1976).
23. M.R. Flannery, in: Recent Studies in Atomic and Molecular Processes (A.E. Kingston, ed.), pp. 167-197, Plenum, New York (1987).
24. A. Dalgarno, in: Recent Studies in Atomic and Molecular Processes (A.E. Kingston, ed.), pp. 51-61, Plenum, New York (1987).
25. B.G. Anicich and W.T. Huntress, Jr., Astrophys. J. Suppl. Ser. 62, 553-672 (1986).
26. J.N. Bardsley and M.A. Biondi, in: Adv. Atom. Mol. Phys. 26, 1-57 (1970).
27. A. Giusti-Suzor, J.N. Bardsley, and C. Derkits, Phys. Rev. A28, 682-691 (1983).
28. H. Nakamura, H. Takagi, and K. Nakashima, in: reference 8.
29. H. Hus, F. Yousif, C. Noren, A. Sen, and J.B.A. Mitchell, Phys. Rev. Lett. 60, 1006-1009 (1988).
30. H. Hus, F. Yousif, A. Sen, and J.B.A. Mitchell, Phys. Rev. A38, 658-663 (1988).
31. N.G. Adams and D. Smith, in: Rate Coefficients in Astrochemistry (T.J. Millar and D.A. Williams, eds.), pp. 173-192, Kluwer, Dordrecht (1988).
32. M.H. Chen, Phys. Rev. A34, 1079-1083 (1986).
33. A. Burgess, Astrophys. J. 141, 1588-1590 (1965).
34. A.L. Merts, R.D. Cowan, and N.H. Magee, Los Alamos Scientific Laboratory Report No. LA-6220-MS, 1976.
35. H.A. Bethe and E.E. Salpeter, Quantum Mechanics of One- and Two-Electron Atoms, pp. 308-322, Plenum, New York (1977).
36. B.F. Rozsnyai and V.L. Jacobs, Astrophys. J. 327, 485-501 (1988).
37. D.R. Flower and M.J. Seaton, Comp. Phys. Commun. 1, 31-34 (1969).
38. P.G. Martin, Astrophys. J. Suppl. 66, 125-138 (1988).
39. V.L. Jacobs, J. Cooper, and S. L. Haan, Phys. Rev. A36, 1093-1113 (1987).
40. V. Sidis, C. Kubach, and D. Fussen, Phys. Rev. A27, 2431-2446 (1983).
41. S. Szucs, M. Karema, M. Terao, and F. Brouillard, J. Phys. B17, 1613-1622 (1984).
42. B. Peart, M.A. Bennett, and K. Dolder, J. Phys. B18, L439-L444 (1985).
43. M.R. Flannery and E.J. Mansky, J. Chem. Phys. 88, 4228-4241 (1988).
44. M.R. Flannery, J. Chem. Phys. 89, 214-222 (1988).
45. M.R. Flannery, J. Chem. Phys. 87, 6947-6956 (1987).
46. M.R. Flannery, J. Chem. Phys. 89, 4086-4091 (1988).
47. M.R. Flannery, J. Phys. B20, 3929-4938 (1987).
48. P. Mansbach and J. Keck, Phys. Rev. 181, 275-289 (1965).
49. J. Stevefelt, J. Boulmer, and J.F. Delpech, Phys. Rev. A12, 1246-1251 (1975).
50. L.P. Pitaevskii, Sov. Phys. - JETP 15, 919-921 (1962).
51. D.R. Bates, J. Phys. B13, 2587-2599 (1980).
52. D.R. Bates, J. Phys. B12, L35-L38 (1979).
53. D.R. Bates, J. Phys. B14, 3525-3534 (1981).
54. W.L. Morgan, in: Recent Studies in Atomic and Molecular Processes (A.E. Kingston, ed.), pp. 149-166, Plenum, New York (1987).
55. K. Takayanagi, in: Adv. Atom. Mol. Phys. 1, 149-194 (1965).
56. K. Takayanagi and Y. Itikawa, in: Adv. Atom. Mol. Phys. 6, 105-153 (1970).

MOLECULAR PROCESSES IN THE UPPER ATMOSPHERES OF THE EARTH AND OTHER PLANETS

Tatsuo Shimazaki

Ames Research Center, NASA
Moffett Field, California 94035
U.S.A.

1. INTRODUCTION

Five of the nine known planets of the solar system, i.e., Mercury, Venus, Earth, Mars, and Pluto, resemble each other in size, mean higher density, and abundance of heavier elements. They are called the terrestrial planets, and with the exception of Pluto, these planets lie nearest to the sun. They have a solid surface and an atmosphere except for Mercury which has essentially no atmosphere. The atmosphere of Pluto, the outermost planet, is not well known. The four "giant" planets - Jupiter, Saturn, Uranus, and Neptune - are much larger than the terrestrial planets and have a lower mean density. They have no solid surface and these planets are composed mainly of hydrogen (H_2) and helium (He) gases similar to the solar atmosphere, although Uranus and Neptune contain large quantities of water, methane, and ammonia.

The main objectives of this chapter are to discuss the atomic and molecular processes in the upper atmospheres of three terrestrial planets, i.e., Earth, Venus, and Mars. The most detailed discussion is devoted to the Earth, however, since our knowledge about this planet is naturally the most advanced.

We will first discuss the absorption of solar radiation by atmospheric molecules and the neutral and ionic composition of the upper atmosphere. Subjects of discussion include molecular absorption of the solar ultraviolet (UV) and X-ray radiation, photodissociation and photoionization processes, and some physical and chemical processes important for the distribution of molecules such as eddy and molecular diffusion, plasma (or ambipolar) diffusion, and ion-neutral and charge transfer reactions as well as neutral-neutral reactions.

We will then discuss impact of charged particles on molecules in the upper atmosphere. Topics include secondary ionization by photoelectrons, their transport and thermalization (electron and ion temperatures), and impact of the corpuscular radiation (protons and electrons) impinging upon the upper atmosphere from space.

Atoms, molecules, and their ions produced in the upper atmosphere are sometimes in excited states, and those excited (metastable) species play very important roles in the atomic and molecular processes and in the heating of the upper atmosphere. Production and loss mechanisms of each important excited species are discussed and the significance of these metastable species for upper atmosphere chemistry and physics is explained.

When the excited species make transitions to lower energy levels, various line and band emissions occur in the spectrum range from UV, through visible to infrared (IR). Even forbidden transitions can occur strongly sometimes under the physical conditions of the upper atmosphere. These emissions can be observed as the airglow or the aurora at high latitudes. Excitation mechanisms for various airglow and auroral emissions and their characteristics (the intensity, the height range where the emissions occur, etc.) are summarized.

2. SOLAR RADIATION AND CHEMICAL COMPOSITION OF THE UPPER ATMOSPHERE

Chemical composition of the atmosphere is essential to any discussion on molecular processes. Absorption of solar photon radiation is the main energy source for producing various chemically active species including ions in the upper atmosphere. The radiation of wavelengths below about 2000 Å is the source of energy for most of the upper atmospheric phenomena. The solar radiation below 2000 Å originates near the surface of the photosphere and the spectrum is close to that of black body radiation at the equivalent temperature ~4500 K. The spectrum also has many strong atomic emission lines including H Lyman α (1215.7 Å), H Lyman β (1025.7 Å), Si III (1206.5 Å), C III (977 Å), O V (629.7 Å), He I (584.3 Å), and He II (303.8 Å).

Upon absorbing solar UV radiation, atmospheric molecules can be decomposed into two or more fragments. This process is called photodissociation if the products are all neutrals, and photoionization if the products include ions and one or more electrons. Photodissociation occurs mainly at wavelengths longer than ~1300 Å (or photons of energy less than ~9.53 eV), whereas photoionization occurs mainly at wavelengths shorter than ~1000 Å (or photons of energy larger than ~12.4 eV).

The solar radiation attenuates mainly by molecular absorption as it penetrates through the upper atmosphere. The degree of penetration of the solar radiation is determined by the optical depth

$$\tau = \sum_i \sigma_i \int_z^\infty n_i(z)dz \tag{1}$$

where σ_i is the absorption cross section and n_i is the number density of the ith constituent. The main molecule absorbing the solar UV radiation is O_2 for the Earth's upper atmosphere and CO_2 for the atmosphere of Venus and Mars. The strongest absorption (or energy deposition) occurs at the height where the optical depth becomes equal to $\cos\chi$, where χ is the solar zenith angle. We call this height the penetration height. Below this height, the intensity of solar radiation decreases exponentially, at a very rapid rate. Figure 1 shows the penetration height as a function of wavelength for the case of vertical incidence of solar radiation (χ = 0 or $\cos\chi$ = 1); it should be somewhat higher when the solar radiation impinges obliquely ($\chi \neq 0$ or $\cos\chi < 1$).

Different molecules absorb the radiation at different wavelengths. Absorption and the subsequent photodissociation and photoionization of molecules occur at wavelengths shorter than the threshold of that molecule

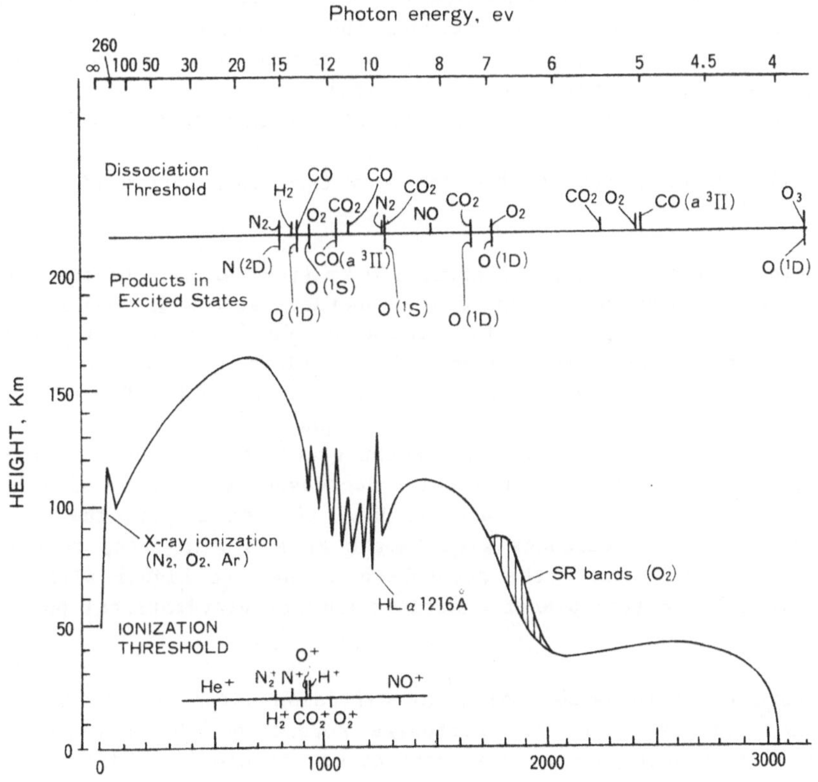

Fig. 1. Penetration height to e^{-1} of the solar UV and X-ray intensity for the case of vertical incidence on Earth. Dissociation and ionization thresholds of various atmospheric molecules are shown. Excited molecules produced by photodissociation are also indicated.

or ion. Figure 1 shows the dissociation and ionization thresholds of various atmospheric atoms and molecules. We can learn from Fig. 1 which molecules can be dissociated or ionized in different wavelength ranges and the height ranges where they occur.

2.1. Neutral Composition

The lower atmospheres of the terrestrial planets are composed mainly of N_2 and O_2 on Earth and CO_2 on Mars and Venus (we call them primary or parent constituents). However, the compositions of the upper atmospheres are significantly different from them; various kinds of atoms and molecules are produced by the actions of solar UV and X-ray radiations upon the primary constituents (we call them secondary constituents). Additional secondary constituents are produced through the dissociation of some secondary constituents and also through chemical reactions between a secondary and a parent constituent or between secondary constituents. The chemical reactions are particularly enhanced if the atoms or molecules participating in the reaction are in the excited states. The products of photodissociation are sometimes in excited states; they include $O(^1D)$, $O(^1S)$, $N(^2D)$, and $CO(a\ ^3\pi)$ as are shown below the dissociation threshold of the corresponding dissociating molecules in Fig. 1.

If a molecule is photochemically very active, its concentration is determined by the equilibrium between the production and loss rates of that molecule. Excited species are generally in photochemical equilibrium. However, if molecules are not chemically active (or the photochemical life time is longer than the characteristic transport time), they can be transported a long distance without major chemical changes. In such cases, the dynamics essentially controls the distribution of that molecule.

Below a certain height of about 110 km (for Earth), the atmosphere is well mixed by turbulence produced by planetary waves and tidal waves propagated from lower heights. This critical height is called the <u>turbopause</u>. Below the turbopause, the height distributions of chemically inactive molecules are determined by the atmospheric scale height, $H = kT/mg$, where k is the Boltzmann constant, T the atmospheric temperature, m the mean mass of atmospheric molecules, and g the gravitational acceleration. The concentration of atmospheric molecules decreases by a factor of e^{-1} as one goes up the distance of one scale height. If H is larger, the concentration of a molecule decreases with height at a slower rate; this happens either when T is larger (the atmosphere expands to higher altitudes) or when g is smaller (the planet exercises smaller gravitational pull to the molecule).

Above the turbopause, molecular diffusion under gravity becomes the dominant transport process. Diffusive separation should occur between heavier and lighter gases. The coefficient of molecular diffusion of the ith constituent of mass m_i through an atmosphere of molecular mass m can be calculated rigorously by the well-established mathematical formula[1]

$$D_{ij} = \frac{3}{8ns_i^2}\left\{\frac{kT(m+m_i)}{2\pi m m_i}\right\}^{1/2} \qquad (2)$$

where s_i is the collisional cross section. It is inversely proportional to the total concentration of molecules (n) and increases rapidly with height as the concentration decreases. Thus, the effect of molecular diffusion and diffusive separation becomes increasingly marked at higher altitudes. In fact, most atoms and molecules tend to distribute according to their own scale height, $H_i = kT/m_i g$. Since lighter gases such as H and O atoms have a larger H_i because of a smaller m_i, they tend to decrease in concentration much more slowly with height and become more abundant than heavier molecules (N_2, O_2, etc.) as we proceed upwards.

The detailed physics of the turbulent mixing process is not well known, but since its effect becomes dominant over molecular diffusion below the turbopause, we assume that the coefficient of mixing (or eddy diffusion) is about same as the molecular diffusion coefficient at the turbopause. Thus, the eddy diffusion coefficient (K) in the Earth's upper atmosphere is about $(1\text{-}5)\times10^6$ cm^2s^{-1}; the values for Mars and Venus are about an order of magnitude larger (see Table 1).

The height distributions of various neutral species predicted with model calculations taking into account all physical (dynamical) and photochemical processes discussed above are shown in Figs. 2(a), (b) and (c) for Earth, Venus and Mars, respectively. The biggest difference between the three planets is that atomic oxygen is much more abundant on Earth than on Venus and Mars. The main source for atomic O in the upper

Table 1. Comparison of three terrestrial planets

	Venus	Earth	Mars
Average distance from sun ($\times10^8$ km)	1.082	1.496	2.279
Radiation flux received from sun (Earth=1)	1.91	1.00	0.43
Mass ($\times10^{26}$ g)	48.7	59.8	6.43
Diameter ($\times10^3$ km)	12.112	12.756	6.78
Rotation period (day)	242.9	1.00	1.026
Surface pressure (mb)	10^5	1.013×10^3	6.0
Gravitational acceleration (cm s^{-2})	870	980	360-400
Surface temperature (K)	720±20	280±20	180±30
Thermospheric temperature[a] (K)	160-350	800-1500	120-400
Eddy diffusion coefficient (cm^2 s^{-1})	10^7-10^8	$(1\text{-}5)\times10^6$	$(1\text{-}5)\times10^7$
Atmospheric composition (%)	CO_2(96.4) N_2(3.4) $H_2O(10^{-3})$ $SO_2(10^{-3})$ O_2(70ppm) Ar(20ppm)	N_2(78.1) O_2(20.9) Ar(0.93) CO_2(0.03) H_2O(0.1-10)[b] O_3(0.5ppm)	CO_2(95.3) N_2(2.7) Ar(1.6) O_2(0.13) CO(0.07) H_2O(0.03)[b] O_3(0.03ppm)[b]
Mean molecular weight	42.94	28.96	43.78

[a] Changes largely with solar cycle.
[b] Variable in considerable range.

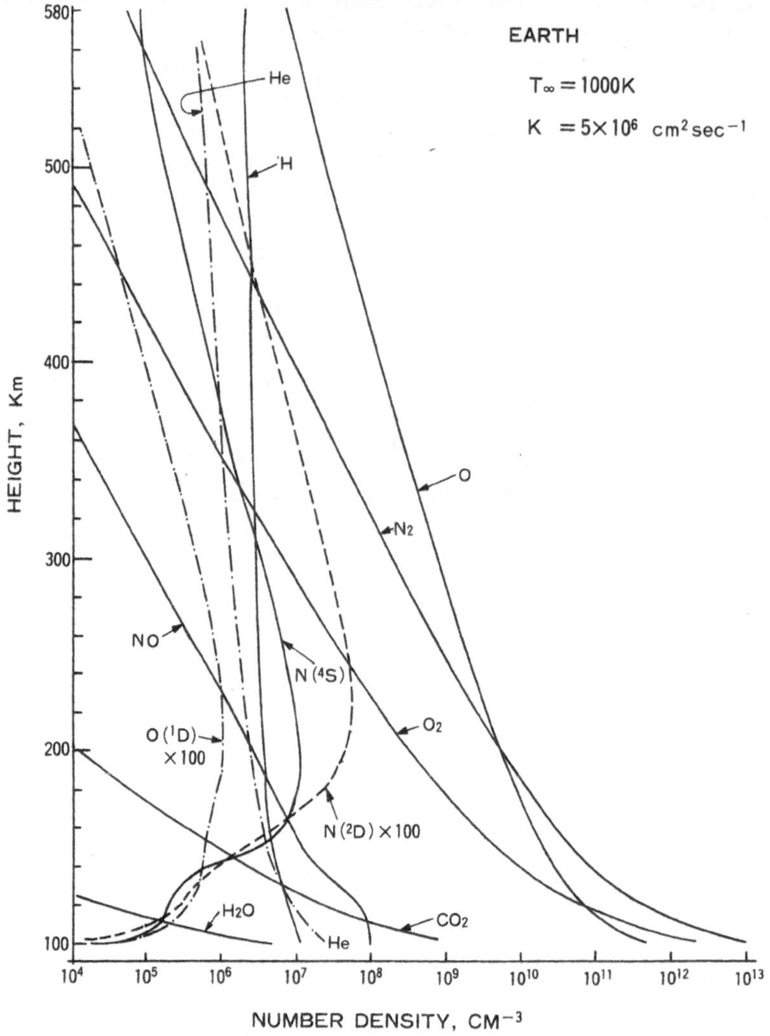

Fig. 2(a). Vertical distribution of neutral constituents predicted with a numerical model for the upper atmosphere of Earth (exospheric temperature 1000 K, eddy diffusion coefficient 5×10^6 cm^2s^{-1}, and solar zenith angle 65°).

atmosphere of Earth is of course the photodissociation of a larger amount of O_2. There is very little O_2 on Mars and Venus, but photodissociation of CO_2 should have produced a large amount of O in the upper atmosphere of these planets.

There are some physical and chemical mechanisms which can reduce effectively the O density in the upper atmosphere of Mars and Venus. Strong vertical eddy diffusion would be able to remove O atoms from the upper atmosphere by transportation down into the lower atmosphere. Earlier studies[2,3] indicated that a large eddy diffusion coefficient (> 5×10^8 cm^2s^{-1}) was necessary for explaining the observed O densities in the upper atmosphere of Mars. A recent re-evaluation of the problem, however, suggests a smaller coefficient ~5×10^7 cm^2s^{-1}[4].

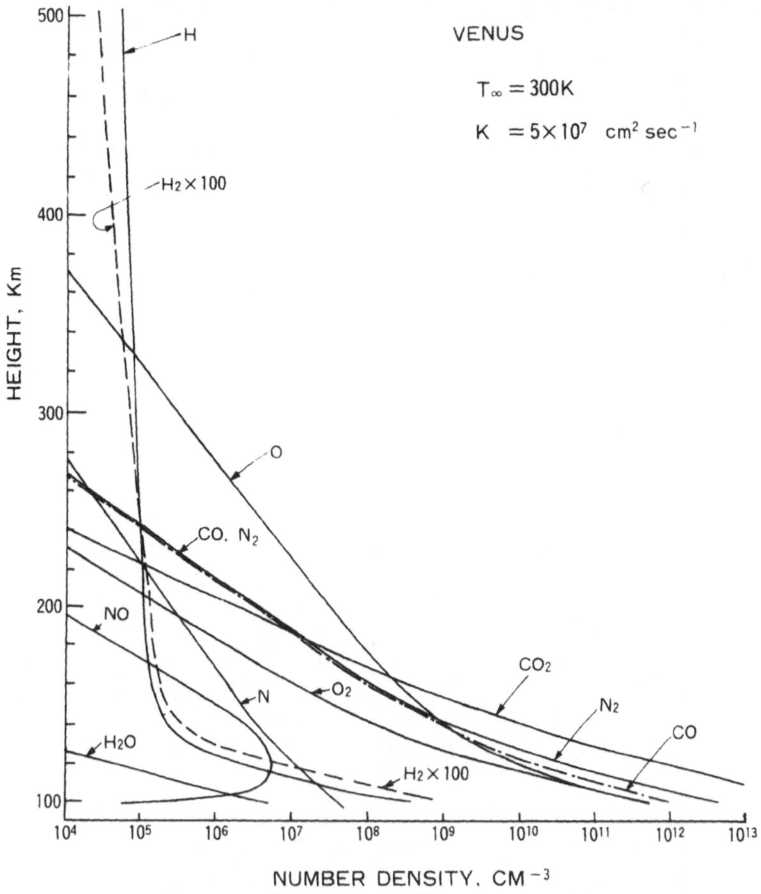

Fig. 2(b). The same as Fig. 2(a) but for Venus (exospheric temperature 300 K, eddy diffusion coefficient 5×10^7 cm^2s^{-1} are assumed).

In the lower atmosphere of Mars, oxygens would recombine with CO through catalytic reactions of HO_x either through the following sequence of reactions[3]

$$H + O_2 + CO_2 \rightarrow HO_2 + CO_2$$

$$HO_2 + O \rightarrow OH + O_2$$

$$CO + OH \rightarrow CO_2 + H$$

net: $CO + O \rightarrow CO_2$

or through the reaction series, via production of H_2O_2[5]

2×) $H + O_2 + CO_2 \rightarrow HO_2 + CO_2$

$$HO_2 + HO_2 \rightarrow H_2O_2 + O_2$$

$$H_2O_2 + h\nu \rightarrow 2\ OH$$

2×) $CO + OH \rightarrow CO_2 + H$

net: $2\ CO + O_2 \rightarrow 2\ CO_2$

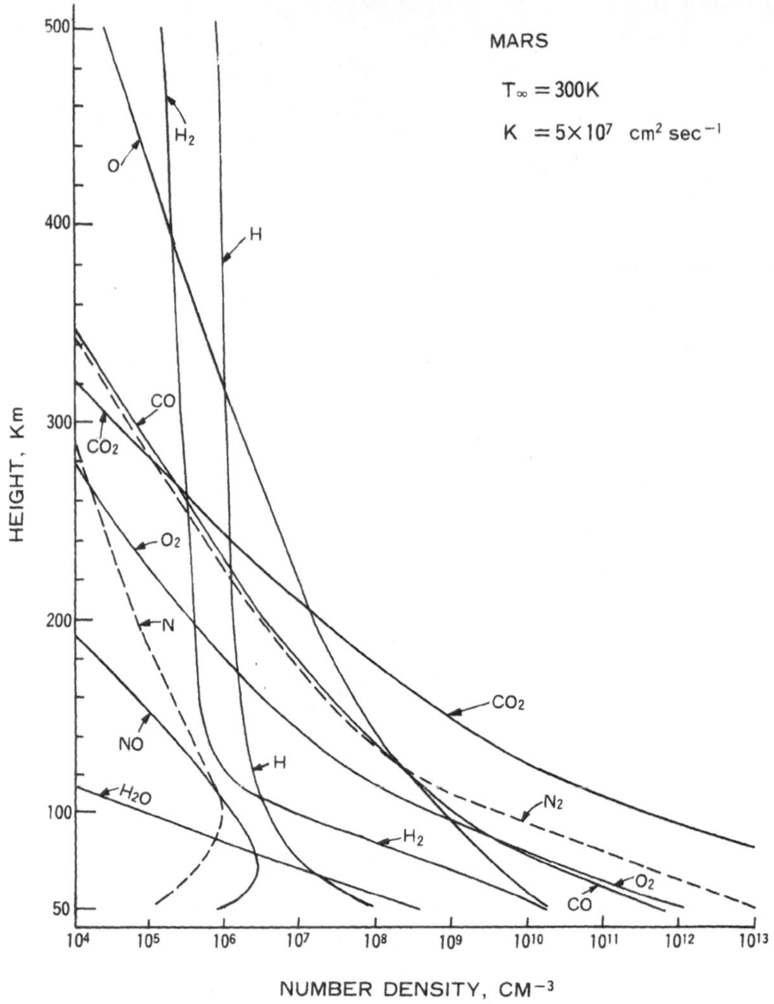

Fig. 2(c). The same as Fig. 2(b) but for Mars.

In both cases the net effect is the recombination of oxygen with CO to reproduce CO_2. Odd hydrogen needed for catalyses can be provided from H_2O and H_2 through their photodissociation and reactions with $O(^1D)$.

The result of a one-dimensional model study indicates that HO_x generated from a small amount of water vapor on Mars (less than the observed global average ~10 μm) is sufficient to reproduce the existing CO_2, so that photochemical stability of CO_2 against their decay by photodissociation may be maintained in the Martian atmosphere. However, water vapor on Mars is highly variable and the mechanism by which the observed level of CO_2 is maintained and its relationship to the water vapor distribution must be considered on the global scale[4].

The higher thermospheric temperature on Earth would make the scale height larger, which explains the large extension of the O density to higher altitudes. Thermospheric temperatures on Mars and Venus are much smaller than that on Earth; this is partly because of the radiative characteristics of the CO_2 atmosphere (a smaller heating efficiency by EUV absorption and a strong cooling by emission at the 15 μm band) and partly because of the effect of the downward transport of heat by large eddy diffusion. Assuming the representative thermospheric temperature to be 1000 K for Earth and 300 K for Venus and Mars (see Table 1), the scale height of O atoms is calculated to be ~52.6 km for Earth, ~19.3 km for Venus, and ~40.7 km for Mars. The O density in the upper atmosphere of Venus decreases rapidly with height due to a small scale height (T is smaller, whereas g is about the same as on Earth). Since Mars has a smaller gravitational acceleration, its scale height is larger than that of Venus, although the two planets have about the same thermospheric temperature. Because of its far distance from the sun, however, Mars receives less solar radiation, about a quarter that of Venus and about half that of Earth (see Table 1); thus, the production of O atoms through CO_2 photodissociation could be much smaller and the O density becomes extremely small on Mars.

2.2. Ionic Composition

The vertical distributions of various ions are predicted with model calculations for three terrestrial planets as shown in Figs. 3(a)-(c). The model includes the effect of ion-diffusion due to ion-neutral particle and ion-ion collisions. The ambipolar (plasma) diffusion plays a particularly important role for the distribution of O^+ ions above ~200 km, as will be discussed later. In general, molecular ions are dominant in the lower region and they are controlled mainly by photochemistry (photochemical equilibrium), whereas in the upper region atomic ions dominate and the dynamics plays the important role for their distribution. The boundary between the two regions are ~200 km for the Earth and Venus and ~300 km for Mars.

In the lower region, the initial molecular ions produced by photoionization are mainly N_2^+ on Earth and CO_2^+ on Mars and Venus. However, the most dominant ions observed are NO^+ and O_2^+ on Earth and O_2^+ on Mars and Venus. These phenomena are caused mainly by rapid conversion of N_2^+ and CO_2^+ into NO^+ or O_2^+ through the chemical reactions

$$N_2^+ + O_2 \rightarrow O_2^+ + N_2, \quad k = 5.1\times10^{-11}(T_i/300)^{-0.8}\ cm^3s^{-1} \tag{3}$$

$$N_2^+ + O \rightarrow NO^+ + N, \quad k = 1.4\times10^{-10}(T_i/300)^{-0.44}\ cm^3s^{-1} \tag{4}$$

and

$$CO_2^+ + O \rightarrow O_2^+ + CO, \quad k = 1.6\times10^{-10}\ cm^{-3}s^{-1} \tag{5}$$

where T_i is the ion temperature in degrees Kelvin. Loss of molecular ions occurs mainly by dissociative recombination

$$O_2^+ + e \rightarrow O' + O'', \quad k = 1.6\times10^{-7}(T_e/300)^{-0.55}\ cm^3s^{-1} \tag{6}$$

$$NO^+ + e \rightarrow N' + O', \quad k = 4.5\times10^{-7}(T_e/300)^{-0.83}\ cm^3s^{-1} \tag{7}$$

$$N_2^+ + e \rightarrow N' + N'', \quad k = 1.8\times10^{-7}(T_e/300)^{-0.39}\ cm^3s^{-1} \tag{8}$$

and

$$CO_2^+ + e \rightarrow CO' + O, \quad k = 3.8\times10^{-7}(T_e/300)^{-1.0}\ cm^3s^{-1} \tag{9}$$

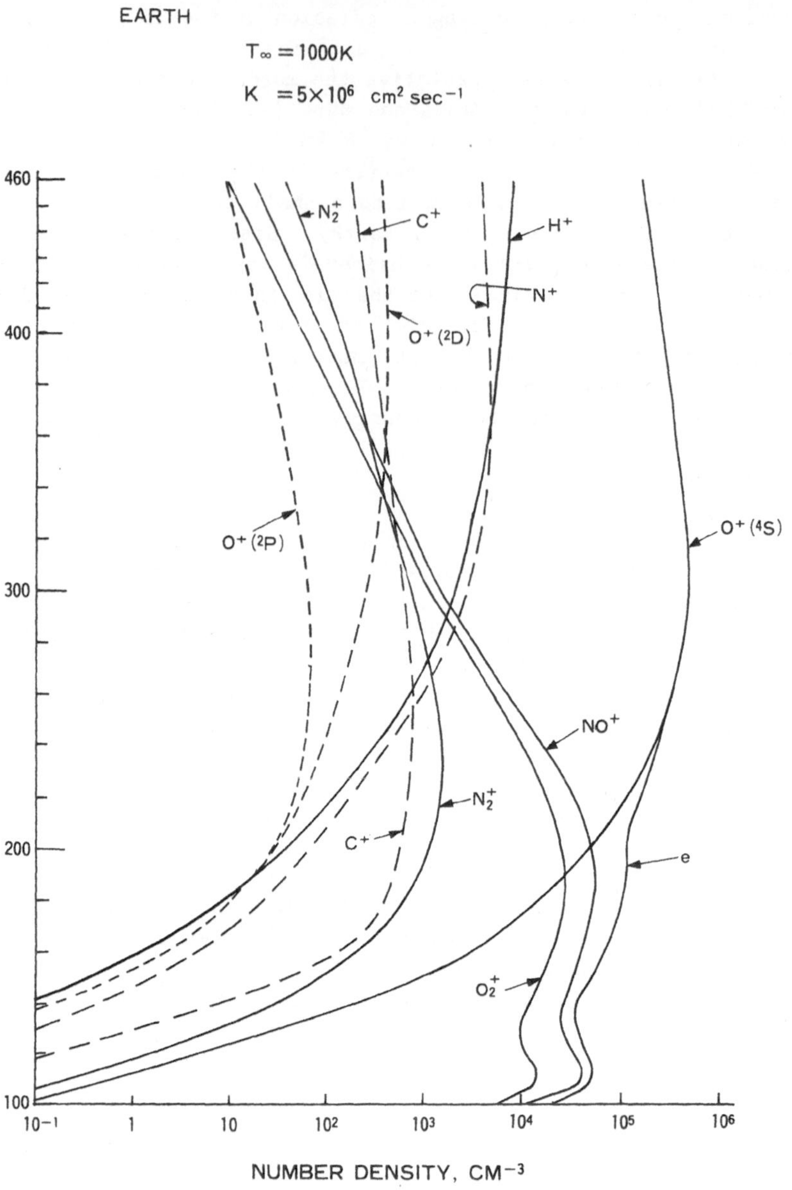

Fig. 3(a). Vertical distribution of ionic constituents predicted with a numerical model for Earth for the same condition as Fig. 2(a).

where T_e is the electron temperature in degrees Kelvin. Photoionization of molecules N_2, O_2, and CO_2 and reactions (3)-(9) control concentrations of various molecular ions in the lower region. Products from the reactions (6)-(9) could be in excited states, which could then be sources for airglow emissions as will be discussed later.

The maximum electron density appears at ~110 km on Earth (see Fig. 3 (a)), which constitutes the ionospheric E layer. The major ions involved are NO^+ and O_2^+, and their production originates from the photoionization of N_2 or O_2 by the solar radiations around 1000 Å, including H Lyman-β (1025.7 Å) and C III(977 Å) emissions, and the X rays below 100 Å (note that these radiations can penetrate to the height ~110 km, see Fig. 1).

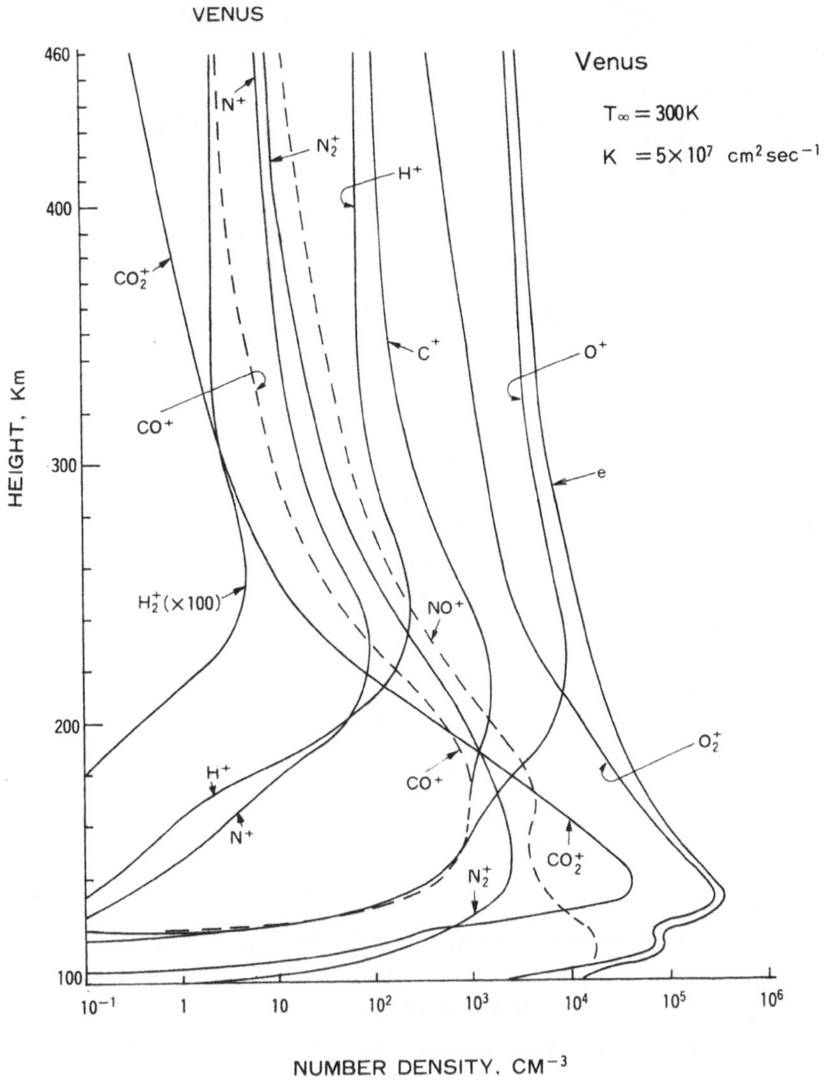

Fig. 3(b). The same as Fig. 3(a) but for Venus.

The nitric oxide (NO) molecule has a large ionization threshold (1337.8 Å) and can be ionized by the strong solar H Lyman-α radiation (1215.7 Å). This radiation can penetrate deep into the atmosphere because of the existence of a window of O_2 absorption near this wavelength (Lyman-α). The ionization of NO is responsible for the ionospheric D region below 100 km. Below ~60 km, the cosmic ray particles become the major ionizing agents.

Additional maxima of NO^+ and O_2^+ densities occur at ~190 km on Earth; these correspond to the electron density peak of the ionospheric F_1 layer. Actually, the F_1 layer often appears as a region of reduced rate of the increase with height in the distribution of electron density rather than a clear maximum. The solar radiation at ~500-800 Å is primarily responsible for ionization near the F_1 layer peak (see Fig. 1), and the most important sources for NO^+ and O_2^+ ions in this region are the photoionization of O followed by reactions

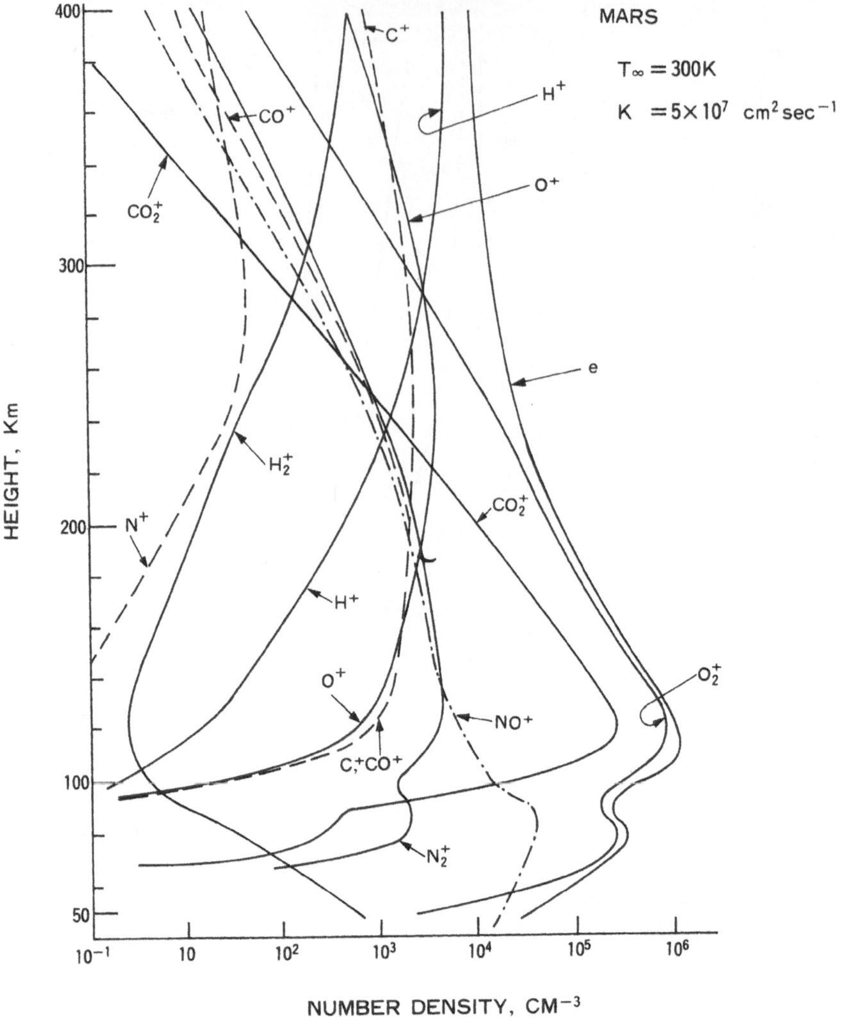

Fig. 3(c). The same as Fig. 3(a) but for Mars.

$$O^+ + N_2 \rightarrow NO^+ + N, \qquad k = 6\times10^{-13}\ cm^3s^{-1} \tag{10}$$

and

$$O^+ + O_2 \rightarrow O_2{}^+ + O, \qquad k = 2\times10^{-11}(T_i/300)^{-0.4}\ cm^3s^{-1} \tag{11}$$

The production rate of O^+ by the photoionization of O is proportional to the O density, whereas the loss rate is proportional to the N_2 (or O_2) density since the loss of O^+ ions occurs by chemical reactions (10) and (11). The chemical equilibrium density of O^+ is then proportional to the ratio of O to N_2 (or O_2) densities. Because the N_2 (or O_2) density decreases with height much more rapidly than the O density (see Fig. 2(a)), the ratio of O to N_2 (or O_2) densities, and therefore, the chemical equilibrium density of O^+ must increase indefinitely with height.

In the region above ~300 km, however, the dynamical effect due to ambipolar diffusion should affect the O^+ density significantly. The chemical time constant of O^+ loss due to reaction (10) is $\tau_c=1/[k_{10}\ n(N_2)]$

and increases from ~5 min at 200 km to ~2.5 hrs at 300 km, whereas the time constant of transport due to ion-diffusion ($\tau_d = H^2/K$) decreases from ~4 hrs at 200 km to ~18 min at 300 km. Thus $\tau_c \ll \tau_d$ at 200 km and $\tau_c \gg \tau_d$ at 300 km, which means that the chemical effect dominates over the dynamical effect at 200 km but the reverse is the case at 300 km. Electrons diffuse faster than ions because of their high mobility, but a polarization field is set up preventing the charge separation. Thus, ions and electrons diffuse with the same velocity; this process is called ambipolar diffusion. The downward transport of O^+ ions by ambipolar (plasma) diffusion would suppress the increase of O^+ density above 300 km, and a maximum of O^+ density would be produced at around that height. This maximum ion density corresponds to the electron density peak of the ionospheric F_2 layer. The F_2 peak is the most pronounced peak of the electron density in the Earth's ionosphere.

The most pronounced electron density peak on Venus and Mars appears at around 120 - 130 km (see Fig. 3(b) and (c)); these layers are composed mainly of O_2^+ ions and correspond to the F_1 layer of the Earth's ionosphere. Mars and Venus lack the F_2 layer because of their scarcity of neutral atomic oxygen. A much smaller electron density peak is formed just below the major peak for both Venus and Mars. This corresponds to the E layer of the Earth's ionosphere.

3. IMPACT OF CHARGED PARTICLES

Charged particles are known to be an important additional energy source for many physical and chemical phenomena in the upper atmosphere. In this section, we will first discuss some details of the behavior of the photoelectrons produced by the interaction of atmospheric molecules with solar radiation. We will then discuss various charged particles impinging upon the upper atmosphere from space and their impact on atmospheric molecules.

3.1. Photoelectrons

Takayanagi and Itikawa[6] have given a detailed review of the production of photoelectrons and their slowing-down processes in the ionosphere.

3.1.1. Secondary ionization by photoelectrons

Photoelectrons ejected from molecules as a result of photoionization by solar UV and X-ray radiations have energies from a fraction of one up to a few hundreds of electron volts. The energy of the ejected photoelectron is expressed as

$$e^* = h\nu - I - E_x \qquad (12)$$

where the first term of the right-hand side is the energy of the photon absorbed (h is Planck's constant and ν is the wave frequency), I is the first (lowest) ionization energy of the molecule, and E_x is the excitation energy of the residual ion.

After the first ionization and excitation, if the energy of the photoelectron is still large enough, additional (secondary) ionization can

Table 2. Products of ionization by electron impact on molecules and their ionization potentials

Direct Ionization		Dissociative Ionization	
[N_2 + e* reactions]			
$N_2^+(X\ ^2\Sigma_g^+)$	15.59 eV	$N^+(^3P) + N(^3S)$	24.32 eV
$N_2^+(A\ ^2\Pi_u)$	16.87	$N^+(^3P) + N(^2D)$	26.65
$N_2^+(B\ ^2\Sigma_u^+)$	18.75	$N^+(^3P) + N(^2P)$	28.0
$N_2^+(C\ ^2\Sigma_u^+)$	23.86	$N^+(^3P) + N^+(^3P)$	39.0
[O_2 + e* reactions]			
$O_2^+(X\ ^2\Pi_g)$	12.075	$O^+(^4S) + O(^3P)$	18.8
$O_2^+(a\ ^4\Pi_u)$	16.11	$O^+(^2D) + O^-(^2P)$	20.7
$O_2^+(A\ ^2\Pi_u)$	16.82	$O^+(^4S) + O(^1S)$	23.4
$O_2^+(b\ ^4\Sigma_g^-)$	18.17		
$O_2^+(c\ ^4\Sigma_u^-)$	24.56		
[$O(^3P)$ + e* reactions]			
$O^+(^4S)$	13.61		
$O^+(^2D)$	16.94		
$O^+(^2P)$	18.63		
[$O(^1D)$ + e* reactions]			
$O^+(^4S)$	11.64		
$O^+(^2D)$	14.97		
$O^+(^2P)$	16.66		

be produced by the photoelectron. The ions that can be produced through electron impact on various atmospheric molecules and the ionization potential (energy) needed for each ionization process are listed in Table 2. The occurrence of dissociative ionization needs much higher electron energies than the ordinary (or direct) ionization of molecules.

Ion production rates by initial ionization by solar photons (hν) and by secondary ionization by photoelectrons (e*) calculated for N_2, O, and O_2 in the Earth's ionosphere are shown in Fig. 4. It is interesting to note that photoelectron impact on N_2 is a dominant ionization source below ~140 km, where photoelectrons are produced mainly by photons of wavelengths less than 300 Å (or energies greater than 41.3 eV). Thus, photoionization of N_2 by these photons will produce an electron of energy ~25.7, ~24.4, ~22.6, or ~17.4 eV depending upon which transition is involved (see Table 2). Electrons of comparable energies can also be

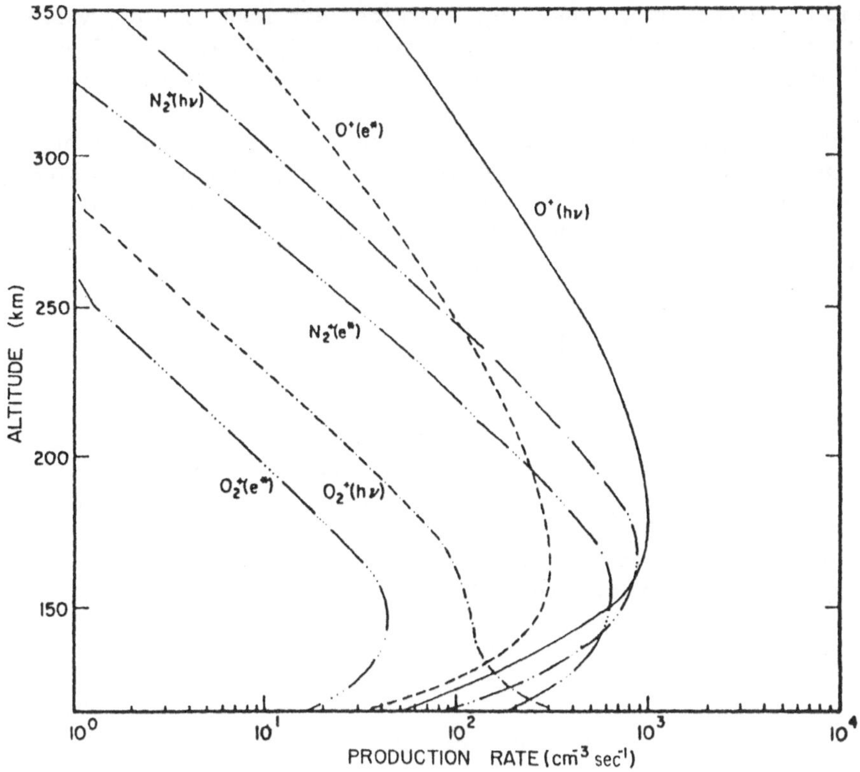

Fig. 4. Ion production rates by direct solar photons ($h\nu$) and by secondary electron (e^*) computed for solar zenith angle 26° [21].

produced by photoionization of O and O_2. The energies of these electrons are large enough for further (secondary) ionization of N_2, the major constituent below ~180 km.

3.1.2. Energy loss (slowing-down) of photoelectrons

As the photoelectron travels through the atmosphere, it loses the kinetic energy through inelastic collisions, mainly ionization and excitation of neutral gases. Inelastic collisions with ions have no important contribution to the photoelectron slowing-down. Electrons of energies greater than 20 eV or so can both ionize and excite molecules and will lose their energies quickly. The mean energy lost per collision is ~15 eV. Below ~20 eV, energy loss through excitation of the metastable levels becomes increasingly more important as the energy decreases. When the energy decreases to less than ~10 eV, the photoelectrons can no longer ionize molecules.

Excitation processes responsible for the energy loss in the Earth's upper atmosphere include electronic excitation of N_2, O_2, and O and vibrational and rotational excitation of N_2 and O_2. There are a large number of electronically excited levels of N_2 produced in electron impact. For electrons below ~7 eV, however, no significant electronic excitation of N_2 occurs, and vibrational excitation of low lying electronic states of

N_2 and O_2 becomes important. For energies less than ~1.5 eV, vibrational excitation is inefficient, and near thermal energies rotational excitation of N_2 is probably the most efficient mechanism for electron energy loss.

When the photoelectrons lose their energies and the number of excitation processes available to them becomes small, elastic collisions with ambient electrons, neutral molecules, and ions become the important processes for energy loss by photoelectrons. The rate of energy loss through elastic collisions is given by

$$- \frac{dE}{dx} = \left[\frac{2m_e\ m}{(m_e + m)^2}\right]\sigma_e\ nE \tag{13}$$

where n and m are the number density and the mass of the collision partner, and m_e and σ_e are the mass and the collision cross section of electrons. Because of the mass dependence, collisions with ambient electrons should play the most important role in the energy loss of photoelectrons; elastic collisions with neutral particles and ions are negligible in comparison with collisions with ambient electrons.

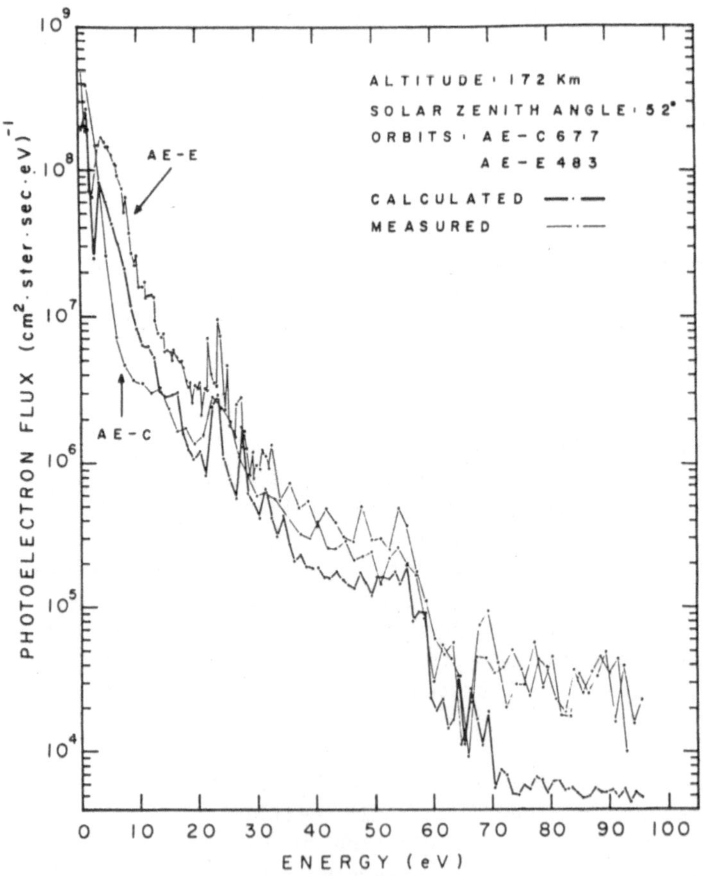

Fig. 5. Calculated and measured spectra of ambient photoelectron fluxes in the 0-100 eV range for AE-C and AE-E satellites [8].

Equilibrium photoelectron fluxes are calculated bv employing an accurate description of photoelectron production processes and by taking explicit account of the discrete nature of the electron energy losses[7,8]. The energy spectra of photoelectrons in the Earth's thermosphere have been measured in the photoelectron spectrometer experiments of the Atmospheric Explorer (AE) satellite. Figure 5 compares the results of measurements from AE-C and AE-E with theoretical calculations for the same geophysical conditions (solar zenith angle 52° and altitude 172 km). Agreement between the calculated and observed relative shape of the spectra is excellent, but the calculated fluxes are lower by approximately a factor of 1.5-2.0 in absolute magnitude than the measured fluxes. The particularly low AE-C fluxes below ~20 eV are caused by the fact that the sensor is "shadowed" by the spacecraft when the body of the spacecraft interferes with the spiral electron trajectories in the geomagnetic field.

3.1.3. Electron and ion temperatures

When energetic electrons are thermalized through various collision processes in the upper atmosphere where collision frequencies are small,

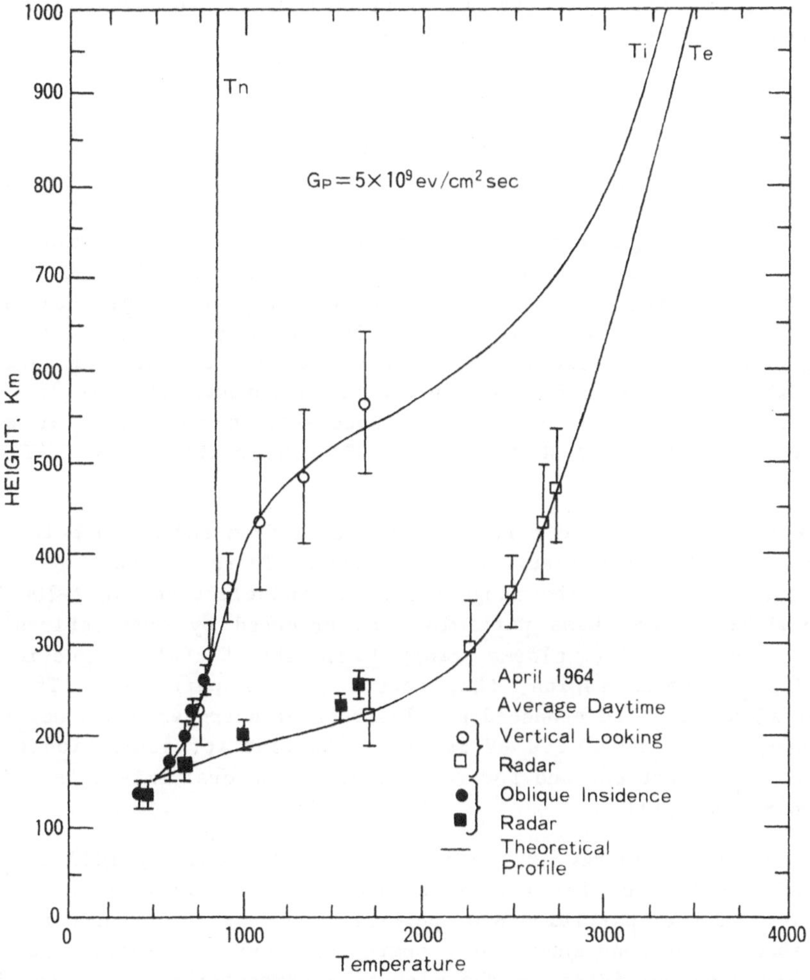

Fig. 6. Computed electron and ion temperature profiles compared with the observed daytime averaged profiles [12].

the energies of electrons and ions should be distributed with Maxwellian distributions of temperatures significantly different from that of the neutral temperature (T_n). The temperature disequilibrium among neutral particles, electrons, and ions in the upper atmosphere has been confirmed by measurements of electron and ion temperatures (T_e and T_i) by rocket-borne Langmuir probes and by ground-based Thomson-scatter radar experiments. These measurements have shown that T_e is different from T_n by 1000 K or more above ~150 km and that, in regions above ~300 km, T_i begins to rise above T_n slowly approaching the T_e profile. It has also been found that the profiles of T_e and T_i change significantly depending on parameters such as electron density, solar cycle, and geomagnetic activity.

Dalgarno et al.[9,10] have calculated T_e profiles by considering the partition of the photoelectron energy among various processes involving photoelectron-neutral and photoelectron-thermal-electron interactions. Geisler and Bowhill[11] have shown that heat conduction is important for determining the shape of the T_e profile at high altitudes. The importance of the heating rate in controlling the shape of T_e was demonstrated by Mantas and Evans[12]. Figure 6 illustrates the comparison of the calculated temperature profiles and the observed daytime-averaged profiles. The calculation uses the measured electron density and assumes a heat flux from the protonosphere of $\sim 5\times10^9$ eV $cm^{-2}s^{-1}$.

3.2. Corpuscular Radiations

Charged particles (protons and electrons) incident from outside of the planetary atmosphere could be responsible for ionization in the lower ionosphere and also for the production of aurora at high latitudes. Cosmic rays are high energy charged particles originated in galaxies; they have energies greater than 100 MeV and can reach below 30 km of the Earth's atmosphere. Solar cosmic ray particles have energies in the range of 1-100 MeV and can penetrate to the altitude of 90-30 km; they produce a strong ionization effect at polar caps leading to polar cap absorption (PCA) which affects radio wave propagation through that region. These particles are produced by particle acceleration in the solar magnetic plasma at the time of solar flare explosion, particularly during solar maximum.

High energy charged particles intruding from radiation belts include protons (> 1 MeV) and electrons (> 40 keV). There are some lower energy particles (protons in the range 0.1-1 MeV and electrons in 1-100 keV) in the magnetosphere; these particles are produced by interactions of the solar wind with the plasma trapped in the Earth's magnetic field. Invading the polar region, they become auroral particles. The optical (visible) auroras are produced by electrons of energies about several keV. The energy range below 100 eV is very important for ionization processes by secondary electrons and for many important auroral emissions excited by electron impact[13].

The kinetic energy of the solar wind with velocity 1000 km s^{-1} corresponds to 2.84 eV for electrons and 5.25 keV for protons. If their energies are equi-partitioned by polarization near the magnetosphere, both electrons and protons should have energy ~2.5 keV. The direct penetration of solar wind particles to the planetary atmosphere is limited, but the solar wind could be a significant energy source for the upper atmosphere of a planet such as Venus which has no magnetic field.

The impact of charged particles is not limited to ionization and production of aurora. For instance, dissociative excitation or ionization of N_2 by electron impact in aurora would produce a reasonable amount of hot (energetic) N atoms, which may then contribute to the heating of the auroral atmosphere and enhanced NO production at high latitudes. The important ionospheric reaction $O^+ + N_2 \rightarrow NO^+ + N$ is sensitive to the vibrational temperature of N_2. If N_2 is heated to a high vibrational temperature by excitation due to electron impact, this reaction could be enhanced significantly. This situation may occur particularly under the conditions of stable auroral red (SAR) arcs when the electron temperature is enhanced.

4. ACTIVATION OF MOLECULES BY EXCITATION

4.1. Electronic Excitation and Metastable Species

Molecules and ions can be excited to metastable states in the upper atmosphere through interactions with solar photons or corpuscular radiations (mainly slow electrons) and chemical reactions. They are formed in excited states corresponding to forbidden electronic transitions and have long lifetimes in comparison with the species excited via permitted transitions. The existence of these metastable species plays an important role in photochemistry and energy budget in the upper atmosphere.

Because of their long lifetime, the metastable species may undergo collisions with other constituents before radiative decay occurs and deactivates such states. For electron impact, metastable species generally have much larger ionization cross sections than the corresponding species in the ground state. They also have lower ionization thresholds; for instance, the threshold potential of $O_2(^1\Delta_g)$ is ~11.02 eV (or ~1118 Å), whereas that of ground state O_2 is ~12.06 eV (or ~1028 Å). Thus, some ionization may occur in the mesosphere through the ionization of $O_2(^1\Delta_g)$ by solar radiation of wavelengths greater than 1000 Å. However, that ionization should be limited to some extent because solar radiation in the same wavelength range is also absorbed by CO_2.

Because of the long lifetime and high internal energy of metastable excited species, chemical reactions in which they are participants usually have much larger reaction rate constants than the corresponding reactions with the ground state species. For instance, the reaction rate constant of the reaction $N(^2D) + O_2 \rightarrow NO + O$ is about 35 times larger than that of the reaction $N(^4S) + O_2$ for the temperature 1000 K, and is an important source for themospheric NO. Reactions of $O(^1D)$ with N_2O and H_2O (and H_2) are the main sources for NO_x and HO_x, respectively, in the Earth's middle atmosphere, whereas the reactions of $O(^3P)$ with these molecules are very slow. It is well known that NO_x and HO_x play important roles in modifying the ozone density in the Earth's stratosphere.

The processes for production of excited species in the planetary upper atmosphere include photodissociation (PD), photoionization (PI), simultaneous photodissociation and ionization (PDI), excitation by electron impact (EE), dissociation by electron impact (ED), simultaneous dissociation and ionization by electron impact (EDI), simultaneous ionization and excitation by electron impact (EIE), dissociative recombination (DR), chemical reaction (CR), ion-atom interchange reaction (IA), radiative scattering (RS), and excitation transfer (ET). The main loss processes are quenching (QE), radiative decay (RD), chemical reactions (CR), and charge exchange reaction (CE).

The knowledge of metastable species in the thermosphere has been reviewed by Torr and Torr[14]. The electronically excited matastable species that are thought to be important in the Earth's upper atmosphere are listed in Table 3. Their excitation energy (or threshold dissociation energy), lifetime due to radiative decay or transitions, and major source and sink reactions are summarized. The radiative transition of excited species should cause atmospheric cooling as well as airglow emission. The 6300 Å emission from the $O(^1D)$ - $O(^3P)$ transition is the most significant contribution to the energy loss due to emissions from metastable species above ~250 km, while the Vegard-Kaplan band emission from $N_2(A\ ^3\Sigma)$ is the major contribution at lower heights[14].

4.2. Vibrational and Rotational Excitation

Vibrational excitation of atmospheric molecules may also play an important role in upper atmosphere physics and chemistry. The heating efficiency of the upper atmosphere depends, amongst other factors, upon the degree of vibrational excitation produced in photodissociation and in chemical reactions.

Vibrationally excited molecular oxygen could be produced by chemical reactions such as $O + O + M \rightarrow O_2(v) + M$ and $OH + O \rightarrow O_2(v) + H$, collisions of O_2 with translationally hot ions, in particular O_2^+, electron impact ($O_2 + e \rightarrow O_2(v) + e$), and fluorescent scattering of solar radiation $O_2 + h\nu \rightarrow O_2'$ followed by $O_2' \rightarrow O_2(v) + h\nu$ during the daytime[15,16]. Production of $O_2(v)$ in the reaction $O + O + M$ may increase production of $O(^1S)$ in the lower thermosphere (80-115 km) in the Barth's mechanism (see Table 3). However, loss of $O_2(v)$ occurs effectively through deactivation by direct quenching by O (or atom-atom interchange reaction) $O_2(v) + O \rightarrow O_2(v' < v) + O$. Thus, high vibrational temperature of $O_2(v)$ is possible only if the O quenching rate were small and vibrational excitation of O_2 has little effect on the chemistry of the nocturnal atmosphere.

The main source for vibrationally excited molecular nitrogen is quenching of $O(^1D)$ by N_2, i.e., $O(^1D) + N_2 \rightarrow O(^3P) + N_2(v)$. Exothermic energy of this reaction could provide a major source for vibrational excitation of N_2. Other sources include quenching of $N_2(A\ ^3\Sigma_u^+)$ by O, photoelectron impact ($N_2 + e \rightarrow N_2(v) + e$) and chemical reactions (e.g., $N + NO \rightarrow N_2(v) + O$). If the bulk of energy were transferred to $N_2(v)$ in these processes, the vibrational temperature of $N_2(v)$ could be enhanced up to ~3000 K at high altitudes unless significant quenching takes place. As was discussed before, the high vibrational temperature of $N_2(v)$ would increase the rate of reaction (10), an important ionospheric reaction. An increase of 30-40 % would be possible at the F layer peak at solar maximum, although the effect is small at solar minimum particularly at lower heights. The vibrational energy of $N_2(v)$ will be lost by collisional quenching with O and CO_2 at lower heights (< ~120 km) and by electron quenching at higher altitudes. Collisions of $N_2(v)$ would also transfer the vibrational energy to other molecules resulting in their excitation (excitation transfer).

Table 3. Electronically excited metastable species in the Earth's upper atmosphere

Species	Excitation Energy	Life-time	Main Sources	Main Sinks	Remarks
$O(^1D)$	1.97 eV	110 s	PD: $O_2 + h\nu(1350\text{-}1717\ Å) \rightarrow O(^1D) + O$		
			PD: $O_3 + h\nu(<3100\ Å) \rightarrow O(^1D) + O_2$		
			EE: $O + e^* \rightarrow O(^1D) + e$	QE: $O(^1D) + M \rightarrow O(^3P) + M$	$M = N_2$ or O_2
			CE: $O^+ + O_2 \rightarrow O_2^+ + O(^1D)$	RD: $O(^1D) \rightarrow O(^3P) + 6300\ Å$	
			DR: $O_2^+ + e \rightarrow O(^1D) + O$		Nighttime source
			DR: $NO^+ + e \rightarrow N + O(\ D)$		
$O(^1S)$	4.19 eV	0.74 s	CR: $\left[\begin{array}{l} O + O + M \rightarrow O_2^* + M \\ O_2^* + O \rightarrow O(^1S) + O_2 \end{array}\right.$		Lower region (80-115 km): Barth's mechanism
			CR: $O + O + O \rightarrow O(^1S) + O_2$		Chapman mechanism
				QE: $O(^1S) + O_2(a^1\Delta_g) \rightarrow O + O_2$	
			PD: $O_2 + h\nu(<1334\ Å) \rightarrow O(^1S) + O$		
			EE: $O + e^* \rightarrow O(^1S) + e$	RD: $O(^1S) \rightarrow O(^1D) + 5577\ Å$	Higher region (> 200 km)
			DR: $O_2^+ + e \rightarrow O(^1S) + O$	RD: $O(^1S) \rightarrow O(^3P) + 2972\ Å$	and aurora
			IA: $O_2^+ + N \rightarrow O(^1S) + NO^+$		
			ET: $O + N_2(A^3\Sigma_u^+) \rightarrow O(^1S) + N_2$	QE: $O(^1S) + O_2(a^1\Delta_g) \rightarrow O + O_2$	Aurora
$N(^2D)$	2.37 eV	26 h	DR: $NO^+ + e \rightarrow N(^2D) + O$	QE: $N(^2D) + O \rightarrow N + O$	
			IA: $N_2^+ + O \rightarrow N(^2D) + NO^+$	CR: $N(^2D) + O_2 \rightarrow NO + O$	
			DR: $N_2^+ + e \rightarrow N(^2D) + N$	QE: $N(^2D) + e \rightarrow N + e$	
			EDI: $N_2 + e^* \rightarrow N(^2D) + N^+(^3P) + 2e$	RD: $N(^2D) \rightarrow N(^4S) + 5200\ Å$	
			PD: $N_2 + h\nu(800\text{-}1000\ Å) \rightarrow N(^2D) + N$		
			PDI: $N_2 + h\nu(<510\ Å) \rightarrow N(^2D) + N^+ + e$		

Table 3. (Cont.)

Species	Excitation Energy	Life-time	Main Sources	Main Sinks	Remarks
$N(^2P)$	3.56 eV	12 s	EDI: $N_2 + e \rightarrow N(^2P) + N^+ + 2e$ ED: $N_2 + e \rightarrow N(^2P) + N + e$ DR: $N_2^+ + e \rightarrow N(^2P) + N$	RD: $N(^2P) \rightarrow N(^2D) + 10{,}400$ Å RD: $N(^2P) \rightarrow N(^4S) + 3466$ Å QE: $N(^2P) + O \rightarrow N(^2D) + O$ CR: $N(^2P) + O_2 \rightarrow NO + O$	Auroral studies
$O_2(^1\Delta_g)$	0.98 eV	~50 m	PD: $O_3 + h\nu(<6110$ Å$) \rightarrow O_2(^1\Delta_g) + O$ ET: $O(^1D) + O_2 \rightarrow O + O_2(^1\Delta_g)$ CR: $O + O + M \rightarrow O_2(^1\Delta_g) + M$ CR: $OH + O \rightarrow O_2(^1\Delta_g) + H$	QE: $O_2(^1\Delta_g) + M \rightarrow O_2 + M$ CR: $O_2(^1\Delta_g) + N \rightarrow NO + O$ RD: $O_2(^1\Delta_g) \rightarrow O_2 + 1.27\mu m$	$M=O_2$, N_2, CO_2, H_2O and air
$O_2(^1\Sigma_g^+)$	1.63 eV	~7 s	PD: $O_3 + h\nu(<4630$ Å$) \rightarrow O_2(^1\Sigma_g^+) + O$	RD: $O_2(^1\Sigma_g^+) \rightarrow O_2$ + Atmospheric band	
$N_2(A^3\Sigma_u^+)$	6.5 eV	~2 s	EE: $N_2 + e^* \rightarrow N_2(A^3\Sigma_u^+) + e$ Cascading from higher excited states	QE: $N_2(A^3\Sigma_u^+) + M \rightarrow N_2 + M$ RD: $N_2(A^3\Sigma_u^+) \rightarrow N_2$ + Vegard Kaplan	$M=N_2$, O_2, N, O, NO
$O^+(^2D)$	3.31 eV	3.6 h	PI: $O + h\nu(<732$ Å$) \rightarrow O^+(^2D) + e$ RD: $O^+(^2P) \rightarrow O^+(^2D) + h\nu$	CE: $O^+(^2D) + N_2 \rightarrow N_2^+ + O$ QE: $O^+(^2D) + O \rightarrow O^+(^4S) + O$	Important source for N_2^+ ions
$O^+(^2P)$	5.60 eV	5 s	PI: $O + h\nu(<665$ Å$) \rightarrow O^+(^2P) + e$	QE: $O^+(^2P) + N_2 \rightarrow O^+(^4S) + N_2$ QE: $O^+(^2P) + O \rightarrow O^+(^4S) + O$ RD: $O^+(^2P) \rightarrow O^+(^2D) + 7320$ Å RD: $O^+(^2P) \rightarrow O^+(^4S) + 2470$ Å	
$N^+(^1D)$	1.89 eV	4.13 m	PDI: $N_2 + h\nu \rightarrow N^+(^1D) + N + e$ RD: $N^+(^1S) \rightarrow N^+(^1D) + h\nu$	QE: $N^+(^1D) + e \rightarrow N^+(^3P) + e$ RD: $N^+(^1D) \rightarrow N^+(^3P) + 6584$ Å	

$N^+(^1S)$	4.04 eV	0.9 s	PDI: $N_2 + h\nu \rightarrow N^+(^1S) + N + e$	RD: $N^+(^1S) \rightarrow N^+(^1D) + 5755$ Å	
				RD: $N^+(^1S) \rightarrow N^+(^3P) + 3063$ Å	
$N_2^+(A^2\Pi_u)$			CE: $N_2 + O^+(^2D) \rightarrow N_2^+(A^2\Pi_u) + O$	CE: $N_2^+(A^2\Pi_u) + O \rightarrow O^+ + N_2$	
$N_2^+(B^2\Sigma_u^+)$			RS: $N_2^+ + h\nu \rightarrow N_2^+(B^2\Sigma_u^+)$	DR: $N_2^+(B^2\Sigma_u^+) + e \rightarrow N + N$	
			PI: $N_2 + h\nu \rightarrow N_2^+(B^2\Sigma_u^+) + e$		
			EIE: $N_2 + e \rightarrow N_2^+(B^2\Sigma_u^+) + 2e$		
$O_2^+(a^4\Pi_u)$	16.11 eV		PI: $O_2 + h\nu \rightarrow O_2^+(a^4\Pi_u) + e$	QE: $O_2(a^4\Pi_u) + M \rightarrow O_2 + M$	$M = N_2, O_2$, NO
			Cascading		

PD=photodissociation, PI=photoionization, PDI=simultaneous photodissociation and photoionization, EE=excitation by electron impact, ED=dissociation by electron impact, EDI=simultaneous dissociation and ionization by electron impact, EIE=simultaneous ionization and excitation by electron impact, DR=dissociative recombination, CR=chemical reaction, IA=ion-atom interchange reaction, RS=radiative scattering, ET=excitation transfer, QE=quenching, RD=radiative decay, CE=charge exchange reaction.

A significant percentage of N_2^+ may be in vibrationally excited levels. Some N_2^+ ions may be produced by the charge exchange reaction

$$O^+(^2D) + N_2 \rightarrow N_2^+ + O \quad (14)$$

A large rate coefficient ($\sim 8\times 10^{-10}$ $cm^3 s^{-1}$) has been obtained for this reaction in laboratory experiments by Rowe et al.[17] and Johnsen and Biondi[18]. Such a high rate coefficient should have produced too much N_2^+ and NO^+ ions and too few $O^+(^4S)$ and O_2^+ ions in the thermosphere as compared with observations. Torr and Torr[14] and Abdou et al.[19] have shown that the problem can be solved if N_2^+ ions produced in reaction (14) are in a vibrationally excited state of $N_2^+(A\ ^2\Pi_u)$. Then the reaction

$$N_2^+(v) + O \rightarrow O^+(^4S) + N_2 \quad (15)$$

may be able to effectively convert the excess N_2^+ to $O^+(^4S)$, if the vibrational temperature of $N_2^+(v)$ is as high as ~3000 K.

It has been observed in the spectra of sunlit aurora at high altitudes (400-650 km) that the strong first negative band system due to the $N_2^+(B\ ^2\Sigma_u^+) - N_2^+(X\ ^2\Sigma_g^+)$ transition has an unusually great development of vibrational structure. Vibrational structure of the N_2^+ first negative system has also been detected in the non-auroral sunlit ionosphere by rocket observations. The presence of excited $N_2^+(B\ ^2\Sigma_u^+)$ ions in the upper atmosphere is due mainly to resonance absorption and scattering of sunlight by ground state $N_2^+(X\ ^2\Sigma_g^+)$. Additional sources include simultaneous ionization and excitation of N_2 by EUV photons ($\lambda < 661$ Å) and photoelectrons ($E > 18.7$ eV). The vibrational temperature of $N_2^+(B\ ^2\Sigma_u^+)$ would be of the order of 4500 K. They also have rotational structure which is of the Boltzmann form at $T_{rot} = 1600$ K.

The O_2^+ ions in the upper atmosphere may possess a high degree of vibrational excitation. They could be produced by photoionization of O_2, charge exchange reactions of O_2 with N^+, N_2^+, or O^+ ions, and second negative band system due to the $O_2^+(A\ ^2\Pi_u) - O_2^+(X\ ^2\Pi_g)$ transition. The excited $O_2^+(A\ ^2\Pi_u)$ ions may be expected via resonance absorption and scattering of sunlight by the ground state O_2^+. Deactivation by collisions with O_2 and O would effectively remove vibrational quanta particularly at lower heights.

Nitric oxide (NO) is an important infrared radiator in the lower thermosphere because of its fundamental vibrational band at 5.3 μm and first overtone band at 2.7 μm. Up to $v = 6$ excitations have been detected in an aurora[20]. The main source for vibrational excitation of NO is the chemiluminescent reaction of $N(^2D)$ with O_2 and collisions between thermal atomic oxygen and enhanced NO in the aurora. The dominant loss processes of NO(v) are radiative relaxation and collisional quenching most likely with O.

Vibrational-rotational excitation of OH is a very marked phenomenon in the mesosphere and lower thermosphere. Strong OH Meinel band emission occurs in the visible - IR range through $v' \rightarrow v''$ transitions. The observed spectra show that all the emissions occur through transitions from levels $v' \leq 9$. This implies that the main emission mechanisms are

$$H + O_3 \rightarrow OH(v' \leq 9) + O_2 \quad (16)$$

and

$$HO_2 + O \rightarrow OH(v' \leq 6) + O_2 \quad (17)$$

The vibrational temperature of OH(v) varies in the range 5000 - 10000 K with large seasonal and diurnal variations. It is generally higher when the solar zenith angle is larger, i. e., in winter than in summer and at midnight than at twilight. The large variation of $T_v(OH)$ may imply that the partition of OH(v) among different vibrational levels would change significantly, as well as the reaction rates of (16) and (17), because of large seasonal and diurnal variations in the concentrations of O and O_3.

It has also been observed that each OH(v) line is accompanied by many rotational lines. The result of an analysis of observed spectra indicates that the OH rotational temperature is in the range 180-240 K which is close to the atmospheric temperature at the height of OH emission (85 ± 5 km). This may imply that the collisional relaxation of rotational levels occurs only from lower vibrational levels after the higher levels are transfered to lower levels by cascading.

5. AIRGLOW AND AURORA

The airglow is the emissive, fluorescent or scattered radiation originated from the molecules in the upper atmosphere. Atoms and molecules activated or excited by solar irradiance, chemical reactions or electron impact would emit the energy of activation or excitation as airglow. The aurora occurs with the same mechanism as airglow, but the primary excitation mechanism of aurora is incoming charged particles from the sun. These particles are trapped in the geomagnetic field and intrude with spiral motions towards the poles, where they excite the atoms and molecules in the air. The aurora is up to several hundred times brighter and also has a higher degree of excitation than the airglow.

Forbidden transitions hardly occur in laboratory experiments because of the small transition probability. However, the probability is not exactly zero, and usually forbidden transitions can occur under the conditions of the upper atmosphere where collisional deactivation is small because the air is very thin and there is no wall effect which could quench excited molecules very quickly. Actually, there are some forbidden transitions occurring very strongly in the upper atmosphere; remarkable examples of such transitions can be seen in the atomic oxygen green line at 5577 Å that occurs by the $^1S \rightarrow {}^1D$ transition, the atomic oxygen red lines at 6300 and 6346 Å that occur by $^1D \rightarrow {}^3P_J$ transitions, and the carbon monoxide Cameron band at 2000-2400 Å that occurs by the a $^3\Pi \rightarrow X\ {}^1\Sigma^+$ transition and has been observed in the Martian upper atmosphere.

The airglow usually is measured from the ground at night. The excitation mechanism of molecules leading to nightglow emission originates mainly from chemiluminescence. The main sources are chemical reactions in the mesosphere and lower thermosphere and ion-electron reactions in the thermosphere. The former includes two-body chemical reaction (CR, an example is the reaction $O_3 + H \rightarrow OH^*(v) + O_2$ which produces the OH Meinel band emissions at 0.55 - 4.4 μm) and three-body excitation (TR, an example is the reaction $O + O + O \rightarrow O^*(^1S) + O_2$ from which the O 5577 Å emission occurs), while the latter includes dissociative recombination (DR, an example is the reaction $O_2^+ + e \rightarrow O^*(^1D) + O$ from which the O 6300 Å emission is produced), particle excitation including electron impact, and radiative recombination (RR, an example is the reaction $O^+ + e \rightarrow O^*(^3S) + h\nu$ which is a source for the O 1300 Å emission in the sub-tropical glow).

The airglow during the daytime is rather difficult to observe from the ground because of competition with strong Rayleigh scattering of sunlight by atmospheric molecules. For instance, the average intensity of the strong NaD line of dayglow is 2×10^4 R, whereas the intensity of Rayleigh scattering reaches 3×10^6 R $Å^{-1}$. Hence, observations of dayglow are limited to the specially strong lines or during the solar eclipse. This situation was much improved when it became possible to raise the instruments to higher altitudes by balloons, rockets, and even by satellites. The intensity of Rayleigh scattering decreases with height in inverse proportion to the air pressure. Thus, we now can measure more accurately the dayglow intensity and determine the altitudes of the emitting layers.

Twilight glow is observed when the ground layers are already in the Earth's shadow while the upper atmosphere is still illuminated. Although no special emission occurs in twilight, we can measure the airglow emission which is difficult to observe during the daytime. In addition, since the terminator changes quickly during twilight, we can measure the precise height of the emitting layer produced by solar irradiance.

There are three main emission mechanisms for dayglow (and twilightglow). The first is resonance scattering (RS) and resonance fluorescence (RF) which occur through irradiance of atmospheric molecules by solar radiation. Examples of RS are the hydrogen L_α (1216 Å) emission in the geocorona and the NaD (5890-96 Å) emission in the twilightglow, and an example of RF includes the hydrogen H_α (6563 Å) emission in the geocorona due to solar L_β radiation. The second sources for the dayglow emission are simultaneous dissociation and excitation (SDE) and simultaneous ionization and excitation (SIE). Both processes can occur either by solar radiation or charged particles (electrons). For example, the dissociation of O_2 by solar radiation at Schumann-Runge would produce $O(^1D)$, from which the O 6300 Å emission should occur. The photodissociation of CO_2 in the Martian upper atmosphere would produce an excited $CO(a\ ^3\Pi)$, from which the CO Cameron band (2000 - 2500 Å) emission takes place. The excitation of $N_2^+(B\ ^2\Sigma_u^+)$ and the subsequent first negative band emission (3914 Å) can occur by auroral electrons as well as by solar radiation. The third sources for the dayglow emissions are inelastic collisions by photoelectrons and auroral electrons. During the process of thermalization of photoelectrons produced by photoionization, the collisions of electrons with atmospheric molecules should cause optical excitation. The excitation by photoelectrons in dayglow happens generally at higher potential than the excitation in nightglow. Examples of the high potential excitation in dayglow include the O 1300 Å, the O 1356 Å, and the N_2 second positive band (blue-UV). Particle excitation (PE) can occur, particularly in aurora, through direct excitation by primary electrons and through excitation by secondary electrons produced by ionization due to electron impact.

Table 4. Airglow and auroral emissions for the Earth

Species	Transitions	Wavelength (Å)	Height (km)	Nightglow Intensity (R)	Nightglow Excitation Mechanism	Twilightglow Intensity (R)	Twilightglow Excitation Mechanism	Dayglow Intensity (R)	Dayglow Excitation Mechanism	Aurora Intensity (R)
Ca^+	$^2P - {}^2S$	3934,3968	>100			~50	RS			
CO_2	ν_2	15 μm	Strato.	10^8	TE					
	ν_3	4.5 μm	Strato.	10^7	ET					
H	2-1 (L_α)	1215.67	>120	$(3-10)\times10^3$	RS,CE	4×10^3	RS	10^4	RS	
	3-1 (L_β)	1025.72	>120	10	RS			40	RS	
	3-2 (H_α)	6562.80	>120	2	RF	10	RF			
	4-2 (H_β)	4861.32	>120	<1	CE					
He	$^1P - {}^1S$	584	800	2	CE	100	RS	<300	RS	
	$^3P - {}^3S$	10830	600			10^3	RS	4×10^3	RS+PE	
He^+	2-1 (L_α)	304	>500	10	CE	10	RS+CE	30	RS+CE	
K	$^2P - {}^2S$	7665	90-95			100	RS			
Li	$^2P - {}^2S$	6708	80-90			30	RS			
Mg	$^1S - {}^1P$	2952	~100					10	RS	
Mg^+	$^2S - {}^2P$	2796,2803	>100					10-100	RS	
N	$^4P - {}^4S$	1200	150-300					10^3	PE	
	$^2P - {}^4S$	3466	Aurora							10^4
	$^2D - {}^4S$	5199	150-300	<1	CE			200	DR+PE	
	$^2P - {}^2D$	10400	Aurora							10^5

Na	$^2P - {}^2S$	5890,5896	90-95	50	CR	10^3	RS	2×10^4	RS	
N_2	$a^1\Pi_g - X^1\Sigma_g$ (LBH)									
		1200-2500	>120					10^4	PE+RS	
	$B^3\Pi_g - A^3\Sigma_u{}^+$ (1st pos.band)									
		red-infra	>120					40	PE	
			Aurora							2×10^6
	$C^3\Pi_u - B^3\Pi_g$ (2nd pos.band)									
		UV-violet	>120					10^3	PE	
			Aurora							10^5
	$A^3\Sigma - X^1\Sigma$	UV	Aurora							2×10^5
$N_2{}^+$	$B^2\Sigma_u{}^+ - X^2\Sigma_g{}^+$ (1st neg.band)									
		>3914	150-200			~30	RS	4×10^3	RS+SIE	
			Aurora							1.5×10^4
	$A^2\Pi_u - X^2\Sigma_g{}^+$ (Meinel band)		150-200				RS+SIE		RS+SIE	
		>9000								
		Infra	Aurora							3×10^6
NO	$A^2\Sigma^+ - X^2\Pi$ (γ band)									
		>2264	~110			10	CR	$(0.5-3)\times10^3$	RS	
	$C^2\Sigma^+ - X^2\Pi$ (δ band)									
		>1915	~160			10	CR			
NO_2	Continuum	4000-7000	90-110	IR $Å^{-1}$	CR					
O	$^3P - {}^3P$	63 μm	>100	1.5×10^6	TE					
	$^1D - {}^3P$	6300,6364	150-350	~50	DR	~500	DR+SDE	10^3-10^4	SDE	
			Aurora							5×10^4
	$^1S - {}^1D$	5577	95-100	300	TR			300	TR+DR	
			150-200	10	DR			10^3	DR+PE	
			Aurora							10^5
	$3^3S - 2^3P$	1302,5,6	100-350	<50	RR			$(2-5)\times10^3$	PE+RS	

	$^5S - {}^3P$	1356-59	100-350	<10	RR			400	PE	
	$^1S - {}^3P$	2972	150-300					<100	DR+TR	
			Aurora							3×10^3
	$4^3P - 3^3S$	4368	100-350	0.5	RR					
	$3^5P - 3^5S$	7772-4-5	100-350	10-100	RR			10^3	PE	
	$3^3P - 3^3S$	8446				10	RR			
O^+	$^4P - {}^4S$	833	150-400					10^3	PE+RS	
	$^2P - {}^4D$	7319,7330	200-400			100	SIE	10^3	SIE+PE	
O_2	$B^3\Sigma_u{}^- - X^3\Sigma_g{}^-$	(Schumann-Runge)								
		1800-3000	90-120					1×10^4	RS	
	$A^1\Sigma_u{}^- - X^3\Sigma_g{}^-$	(Herzberg)								
		3000-5000	70-100	10^3	TR			300	SDE+TR	
	$b^1\Sigma_g{}^+ - X^3\Sigma_g{}^-$	(Atmospheric)								
		8645,7619	90-95	$(1-6)\times10^3$	TR			10^5	RS+SDE	
	$a^1\Delta_g - X^3\Sigma_g{}^-$	(IR atmospheric)								
		1.27 μm	40-95	$\sim10^5$	TR	10^6	SDE	10^7	SDE	
		1.58 μm								
	$C^3\Delta_u - a^1\Delta_g$	UV-violet			CR					
OH	$X^2\Pi$ vib.-	5500 -	85-95	5×10^6	CR					
	rot.	4.4 μm								
	$A^2\Sigma - X^2\Pi$	~3090	40-80					10^3	RS	
O_3	ν_3	9.6 μm	Strato.	10^7-10^8	TR					

RS=resonance scattering, RF=resonance fluorescence, CR=chemical reaction, TR=three body reaction, CE=charge exchange reaction, DR=dissociative recombination, RR=radiative recombination, PE=particle excitation, SDE=simultaneous dissociation and excitation, SIE=simultaneous ionization and excitation, ET=excitation transfer, TE=thermal electron (inelastic collision).

Additional excitation of molecules would occur by inelastic collisions of thermal electrons (TE) and excitation transfer (ET). Examples of TE include the O 63 μm emission in the thermosphere, the O 6300 Å emission in the sub-auroral red (SAR) arc, and the infrared emission from the stratospheric CO_2(15 μm) and O_3(9.6 μm). An example of ET is the excitation of the CO_2(4.3 μm) band by the vibrationally excited $N_2(v = 1)$.

Table 4 summarizes known airglow emissions for Earth. This table is a revised version of those in Nagata and Tohmatsu (1973). The transition involved, the wavelength of the emission, and the altitude of the emitting layer are given for each atom and molecule. The intensity of the emission and the excitation mechanism for nightglow, twilightglow, and dayglow are also given. Table 4 also includes the intensity of auroral emission for typical aurora on Earth.

REFERENCES

1. S. Chapman and T. G. Cowling, The Mathematical Theory of Non-Uniform Gases, 2nd ed., 415 pp., Cambridge Univ. Press (1953).
2. M. B. McElroy, and J. C. McConnell, Dissociation of CO_2 in the Martian atmosphere, J. Atmosph. Sci. 28, 879-884 (1971).
3. M. B. McElroy and T. M. Donahue, Stability of the Martian atmosphere, Science 177, 986-988 (1972).
4. T. Shimazaki, Photochemical stability of CO_2 in the Martian atmosphere: Re-evaluation of eddy diffusion coefficient and the role of water vapor, J. Geomag. Geoelectr., in press (1989).
5. T. D. Parkinson and D. M. Hunten, Spectroscopy and aeronomy of CO_2 on Mars, J. Atmosph. Sci. 29, 1380-1390 (1972).
6. K. Takayanagi and Y. Itikawa, Elementary processes involving electrons in the ionosphere, Space Sci. Rev. 11, 380-450 (1970).
7. G. A. Victor, K. Kirby-Docken, and A. Dalgarno, Calculations of the equilibrium photoelectron flux in the thermosphere, Planet. Space Sci. 24, 679-681 (1976).
8. S. P. Hernandez, J. P. Doering, V. J. Abreu, and G. A. Victor, Comparison of absolute photoelectron fluxes measured on AE-C and AE-E with theoretical fluxes and predicted and measured N_2 2PG 3337Å volume emission rates, Planet. Space Sci. 31, 221-233 (1983).
9. A. Dalgarno, M. B. McElroy, and R. J. Moffett, Electron temperatures in the ionosphere, Planet. Space Sci. 11, 463-484 (1963).
10. A. Dalgarno, M. B. McElroy, and C. G. Walker, The diurnal variation of ionospheric temperatures, Planet. Space Sci. 15, 331-345 (1967).
11. J. E. Geisler and S. A. Bowhill, Ionospheric temperatures at sunspot minimum, J. Atmos. Terr. Phys. 27, 457-474 (1965).
12. G. P. Mantas and J. V. Evans, The thermal structure of the ionosphere, Aeronomy Rep., No.28, Univ. Illinois, Urbana-Champaign, Ill., U.S.A. (1968).
13. K. Takayanagi and T. Yonezawa, Collision processes in the aurorae and ionization in the auroral zone, Rep. Ionos. Space Japan 15, 51-69 (1961).
14. M. R. Torr and D. G. Torr, The role of metastable species in the thermosphere, Rev. Geophys. Space Phys. 20, 91-144 (1982).
15. A. Dalgarno, Vibrationally excited molecules in atmospheric reactions, Planet. Space Sci. 10, 19-28 (1963).
16. A. Dalgarno, Metastable species in the ionosphere, Ann. Geophys. 26, 601-607 (1970).
17. B. R. Rowe, D. W. Fahey, F. C. Fehsenfeld, and D. L. Albritton, Rate constants for the reactions of metastable O^{+*} ions with N_2 and O_2 at collision energies 0.04 to 0.2 eV and the mobilities of these ions at 300 K, J. Chem. Phys. 73, 194-205 (1980).

18. R. Johnsen and M. A. Biondi, Laboratory measurements of the $O^+(^2D)$ + N_2 and $O^+(^2D) + O_2$ reaction rate coefficients and their ionospheric implications, Geophys. Res. Lett. 7, 401-405 (1980).
19. W. A. Abdou, D. G. Torr, P. G. Richards, and M. R. Torr, The effect on thermospheric chemistry of a resonant charge exchange reaction involving vibrationally excited N_2^+ ions with atomic oxygen, J. Geophys. Res. 87, 6324-6330 (1982).
20. W. T. Rawlins and G. E. Caledinia, Infrared emission from NO(Δv=1) in an aurora: Spectral analysis and kinetic interpretations of HIRIS measurements, J. Geophys. Res. 86, 1313-1324 (1981).
21. D. G. Torr, The photochemistry of the upper atmosphere, Chap.5 in The Photochemistry of Atmosphere, (J. S. Levine ed.), pp.165-278 (1985).

INTERSTELLAR MOLECULES

Norio Kaifu

Nobeyama Radio Observatory
National Astronomical Observatory
Minamisaku-gun, Nagano-ken 384-13
Japan

1. INTERSTELLAR MOLECULES AND MATTER CIRCULATION IN THE GALAXY

1.1. Radio Observations of Cold Interstellar Matter

The radio observations of interstellar molecules, initiated by the detection of OH[1] (1963) and by the successive detection of NH_3 (1968), H_2O and H_2CO (1969), provided a very wide window to the "cold universe." Now a number of huge and elaborate mm-wave telescopes and interferometers are revealing new exotic molecules in dark clouds, rotating gas disks which produce stars in their central regions, energetic molecular flows ejected from protostars, and bursts of star formation in distant galaxies. These phenomena occur in the cold interstellar matter, most of which forms a molecular gas cloud invisible with optical telescopes.

Thousands of spectral lines of nearly seventy molecular species were detected in interstellar matter, star-forming regions and expanding

Fig. 1. A dark cloud (left half) with the surface shone by a nearby OB star (right) in Orion. famous 'Horse head' nebula is at center. (By courtesy of Kiso Observatory, University of Tokyo.)

envelopes of red giant stars during the past twenty years (see the recent reviews by Irvine et al.[2,3]). Most of the lines arise from pure rotational transitions which lie in the mm wavelength region. They are easily excited in an interstellar cloud by the collision with H_2 molecules even with very low temperature.

Early observations with relatively small mm-wave telescopes have revealed that the interstellar molecular lines, especially the CO J=1-0 transition at 115 GHz, are observed mostly in regions called dark cloud. The dark clouds are widely distributed in the Milky Way. Nearby clouds can be seen in a long-exposured photographs (Fig. 1) as dark silhouettes against the light of background stars. Solid particles of small size (with a diameter of 0.1 μm or less) in the clouds (interstellar dust or grain) absorb or scatter the star light, and hence the visual extinction of individual cloud sometimes reaches 100 mag.

By observing the molecular spectral lines emitted from the dark clouds, we can make an extensive study of the structure, physical conditions and dynamics of the clouds, which were totally unknown before. The important and basic results obtained so far are:

(1) The main component of dark clouds is molecular gas. Most of them is H_2. CO is the second most abundant molecule, whose fraction is about 10^{-4} of H_2. Therefore the dark cloud is often called a molecular cloud.
(2) Molecular clouds are the dominant phase of the interstellar medium (about 90%).
(3) Interstellar dusts play an essential role in molecular clouds. First the dust acts as a screening filter against those stellar UV photons which break chemical bonds of the molecule. Second the H_2 molecule, which is the starting point of all chemical reactions in the dark cloud, should be formed on dust surfaces.
(4) Due to the very low temperature (≒10 K) and the very low gas density (10^2 to 10^5 cm^{-3}) in the molecular clouds, gas-phase chemical reactions in the clouds are mostly two-body collisions between an ion and a molecule. For the ion-molecule reaction the initial energy of ionization comes from the cosmic ray or from the penetrating stellar UV photons.
(5) Stars are formed in the dense core of the molecular clouds. The star-formation process could not be observed before because of the high obscuration of the molecular clouds against visible light, but now it can be studied in detail with mm-wave and infrared telescopes.

1.2. Molecular Clouds and Formation of Stars

Strong molecular line emissions were found first in the clouds adjacent to the HII region, i.e., the site of recent massive star formation. An HII region is an extended high temperature plasma which is heated by strong UV photons from an O-type or a B-type massive star that is newly formed. The HII region is normally associated with a huge, cold molecular cloud; the mother molecular cloud of the UV-emitting OB star and the HII region. In the molecular cloud forming stars, HCN, HCO^+, NH_3, H_2CO, CS, CN, and many other molecules are found. As described in Section 4.1, such molecules as SiO, SO, SO_2, H_2O and CH_3OH are often an indication of a very active site of massive star formation.

Those molecular clouds with a core of massive star formation have a relatively high temperature and high density. The physical parameters of these clouds are: hydrogen number density $n(H_2) \fallingdotseq 10^5$ to 10^6 cm^{-3}, kinetic temperature $T_k \fallingdotseq 30$ to 100 K, linear size $L \fallingdotseq 1$ pc, mass $M \fallingdotseq 10^2$ to 10^3 Mo.

On the other hand, a "quiet" molecular cloud, which do not exhibit active star formation inside (often called "dark cloud" in its limited sense) is characterized by a different set of physical conditions: $n(H_2) = 10^3$ to 10^4 cm^{-3}, $T_k = 10$ K, L = 1 to 10 pc, $M = 10^2$ to 10^5 $M_\odot$.

These quiet (or dark) molecular clouds should preserve more or less the "pure" or basic nature of molecular clouds, that is the condition in which a cloud exists before being affected by shock or radiation heating due to the star forming process. The chemistry of the dark cloud is characterized by carbon chain molecules. This will be described in Sections 3 and 5.

Table 1. Interstellar molecules observed in radio frequencies

[Diatomic Molecules and Simple Inorganic Species]										
OH	CH	CN	CO	NO	CS	NS	SO	SiC	SiO	SiS
PN	NaCl	AlF	AlCl	KCl						
H_2O	H_2S	OCS	SO_2	NH_3						
[Carbon Chain Molecules]										
(CH)	CCH	CCCH	C_4H	C_5H	C_6H					
(CN)		CCCN								
(CO)		CCCO								
(CS)	CCS	CCCS								
HCN		HCCCN		HC_5N		HC_7N		HC_9N		$HC_{11}N$
	CH_3CN		CH_3C_3N							
		CH_2CCH		CH_3C_4H						
(SiC)			C_4Si							
[Aldehydes, Ketones, Alcohols, Ethers, and Related Molecules]										
HCO	H_2CO	H_2CS	CH_3CO	HCOOH	HC_3HO	NH_2CHO	CH_3OH	CH_3SH		CH_3CHO
$HCOOCH_3$	$(CH_3)_2O$	CH_3CH_2OH								
[Other Nitriles, Amines, and Related Molecules]										
HNC	HNCO	HNCS	CH_2NH	CH_3NH_2	CH_2CN	CH_3CN	CH_2CHCN	CH_3CH_2CN		
[Ring Molecules]										
SiC_2	C_3H	C_3H_2								
[Molecular Ions]										
HCO^+	HOC^+	HCS^+	HN_2^+	$HOCO^+$						

Circumstellar shells, which are not interstellar matter in an exact definition, are also known to be rich sources of molecular line emissions. These expanding envelopes of red giant/supergiant stars are important because they eject a lot of grains to the surrounding interstellar space. From the point of view of matter circulation in space, most of the molecules emitted are destroyed by UV radiation in interstellar space, but the grains have much longer lives and form dust clouds. Molecules are formed in turn in the dust cloud and the grains also grow by adsorption of atoms and molecules on their surface.

Table 1 lists the molecular species detected so far using the radio observation in the astronomical objects described above.

Molecular clouds are gravitationally unstable. The density, temperature and size of the cloud observed suggest that they would contract due to self gravity within 10^6 years and form stars. However, the life of the cloud, estimated from observations, is longer by at least one order of magnitude, i.e., 10^7-10^8 years. This is probably due to the internal pressure caused by a magnetic field or turbulent motion. Nevertheless the molecular cloud contracts to form stars inside. A part of the cloud core can form a star (sometimes many stars), possibly associated with a planetary system formed mainly from dust grains in the cloud. The remaining fraction, though considerable, of the cloud is ejected or pushed away into interstellar space by the energetic activity associated with the star formation (Sections 4.1 and 4.2).

Stars proceed along their evolutional paths with various time scales depending on their mass. When a star passes through a red giant phase, after the relatively stable main sequence phase, it ejects a considerable amount of their mass towards the space to form an expanding envelope. As is well known the massive stars progress very fast and finally reach the stage of gravitational collapse which is then followed by a supernova explosion. The supernova explosion, together with the red giant phase, is the main mechanism of the mass recurrence to interstellar matter. With these mechanisms a considerable amount of heavy and superheavy elements are added to the interstellar matter. Very roughly, a half of the total mass would recur from the stellar phase to the interstellar phase. Thus the matter circulates with a time scale of the order of 10^8 years, at least in the solar neighborhood in our Galaxy. The matter circulation is schematically illustrated in Fig. 2.

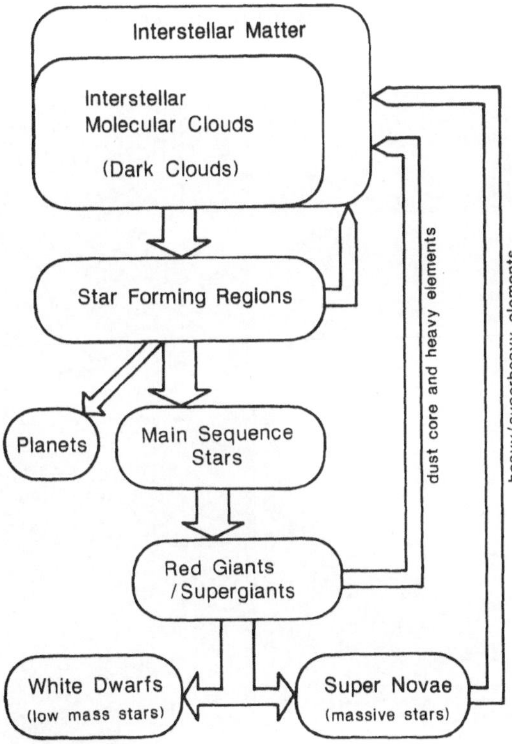

Fig. 2. A schematic diagram showing the circulation of matter in the Galactic space. Heavy elements and dust cores are gradually accumulated in the form of a dark cloud and condense to form the stars and planets. Note that about half of the matter go to the 'dead ends'.

2.1. Methods of Observation

The detection of various interstellar molecular lines has strongly stimulated the development of mm-wave telescopes and receiver systems, because the most important transitions of interstellar molecules lie in the short mm-wavelength region. A radio telescope for mm-wavelength requires high surface accuracy of better than $\lambda/4\pi$, where λ is the observed wavelength. Deformation due to gravity, thermal distortion and the difficulty in the pointing accuracy are the extremely serious problems for a large telescope. The mm-wave telescopes recently built, such as the Nobeyama 45-m telescope (NRO, Japan) and the Pico Velta 30-m telescope (IRAM, France - West Germany - Spain), are protected by thick panels of thermal insulator and employ the "homologous deformation" design method. The latter method allows the main dish to deform only "from a paraboloid to another paraboloid" when it changes the elevation angle[4] (see Fig. 3(a)).

Receiver frontends which detect and amplify the extremely weak signal from interstellar clouds have to be very sensitive and stable. The GaAs Schottky diode mixers cooled down to about 15 K and the cooled FET IF pre-amplifiers are widely used for mm-wave observations. The recent development of superconducting devices (SIS mixers) and cooled HEMT amplifiers has achieved a considerable improvement in receiver sensitivity.

(a) (b)

Fig. 3. (a) 45-m mm-wave telescope of the Nobeyama Radio Observatory (NRO). (b) An acousto-optical radiospectrometer (AOS) of NRO with 16,000 frequency channels for 2-GHz instantaneous bandwidth.

Radio spectrometers are also important tools for cosmic radio spectroscopy. To receive as much signal and information as possible, the spectrometer should have a wide instantaneous bandwidth and a high frequency resolution, i.e., a large number of parallel frequency channel outputs. The conventional filter-bank system is not suitable in this respect, and the AOS (acousto-optical radiospectrometer) with 1,000 or more frequency channels is more commonly used in the higher frequency observation. A digital correlator type spectrometer is also widely used for relatively low frequency observations.

The AOS developed in the Nobeyama Radio Observatory (NRO) is the largest radiospectrometer which has 32,000 channel frequency outputs (Kaifu and Chikada[5]). Half of the system is the wideband part which provides the total instantaneous bandwidth of 2 GHz and output of 16,000 channels. The other half of the system is the high-resolution part with 37 kHz frequency resolution and 16,000 channel outputs. This is a powerful tool for multi-transition observations and especially for a molecular line survey in a wide frequency range. (Fig. 3(b))

Observations are normally done by tracking the telescope toward the object for the time span required and by integrating the spectral data in a computer system to reduce the random noise and to achieve an enough S/N ratio. The integration time varies from an order of one second to days, depending on the intensity of the line and the value of S/N required. A sample of spectral observations is shown in Fig. 4.

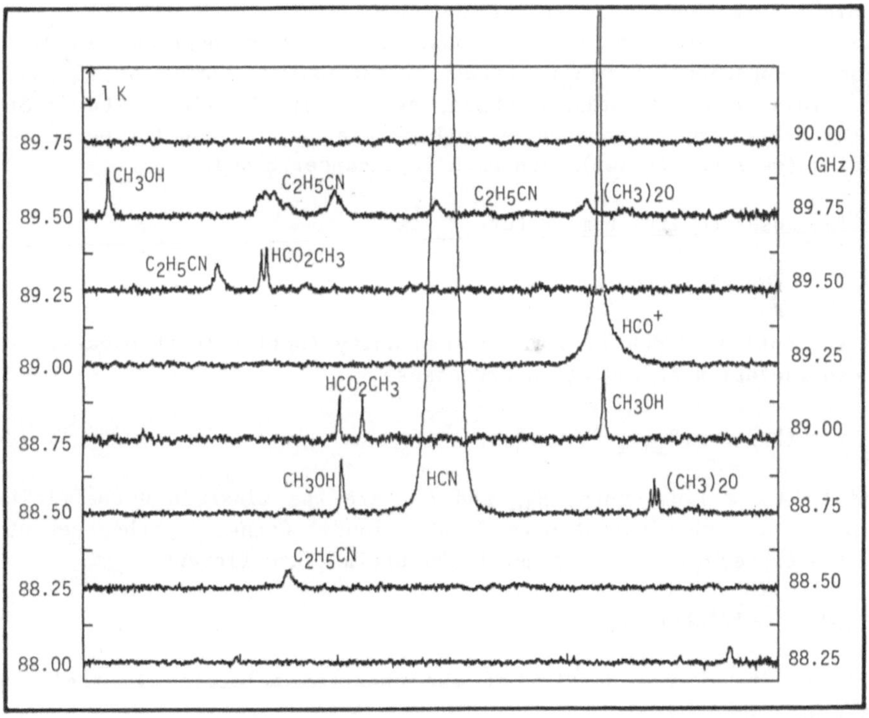

Fig. 4. A 88-90 GHz spectra obtained by a single observation toward Ori-KL, using the NRO 45-m telescope and wide-band 16,000 channel AOS (from Kaifu et al.[6]).

2.2. Line Intensity and Abundance of Molecules

The observed intensity of a molecular line depends on the excitation condition as well as the column density of the molecule. Rotational transitons of molecules in interstellar clouds are excited by collisions with H_2 molecules, which hold about 90 % of the number density of the cloud particles. An excitation temperature T_{ul} between an upper(u) and a lower(l) energy level of the molecule is defined by

$$n_u/n_l = (g_u/g_l)\exp\{\Delta E_{ul}/kT_{ul}\}$$

where n_u and n_l are the number densities of the two states, g_u and g_l are their statistical weights and $\Delta E_{ul} = h\nu$ is the energy difference between the two levels. T_{ul} normally lies between T_k and T_r, i.e., the cloud kinetic temperature and the radiation temperature at the transition frequency;

$$T_k > T_{ul} > T_r$$

When the collisional excitation exceeds the radiation process, T_{ul} approaches the kinetic temperature T_k. The critical density of H_2 which excites the molecule is roughly given by

$$n(H_2)_c = A_{ul}/\langle\sigma v\rangle$$

where A_{ul} is the Einstein coefficient of the spontaneous emission, σ is the collision cross section, and v is the average velocity of the H_2. In a typical dark cloud, $\langle\sigma v\rangle$ is of the order of 10^{-10} cm^3s^{-1} and A_{ul} has a value of the order of 10^{-5} for most molecular transitions in 100 GHz region. Consequently $n(H_2)_c$ takes an approximate value of 10^5 to 10^6 cm^{-3}. This gives an estimate of gas density in the line-emitting cloud, but the excitation often becomes much more effective due to the radiation trapping (self excitation) mechanism in a denser cloud.

The observed line temperature T_b is

$$T_b = (h\nu/k)[\exp\{h\nu/kT_{ul}\}-1]$$

when the optical depth is smaller than unity (optically thin case). From the consideration of radiation processes

$$\tau = (h\nu/c\Delta\nu)(B_{lu}n_l-B_{ul}n_u)L$$

where $\Delta\nu$is the line width, B_{lu} and B_{ul} are the Einstein B coefficients, and L is the line-of-sight size of the cloud. Connecting these equations with the expression of T_b using the Einstein A coefficient

$$T_b = (hc^2/8\pi k\nu\Delta\nu)A_{ul}n_uL$$

we obtain the column density of the observed molecules in their upper level, $N_u = n_uL$, by

$$N_u = (1.9\times10^3\nu^2[GHz]/A_{ul}[s^{-1}])\int T_b dv[K\ km\ s^{-1}]$$

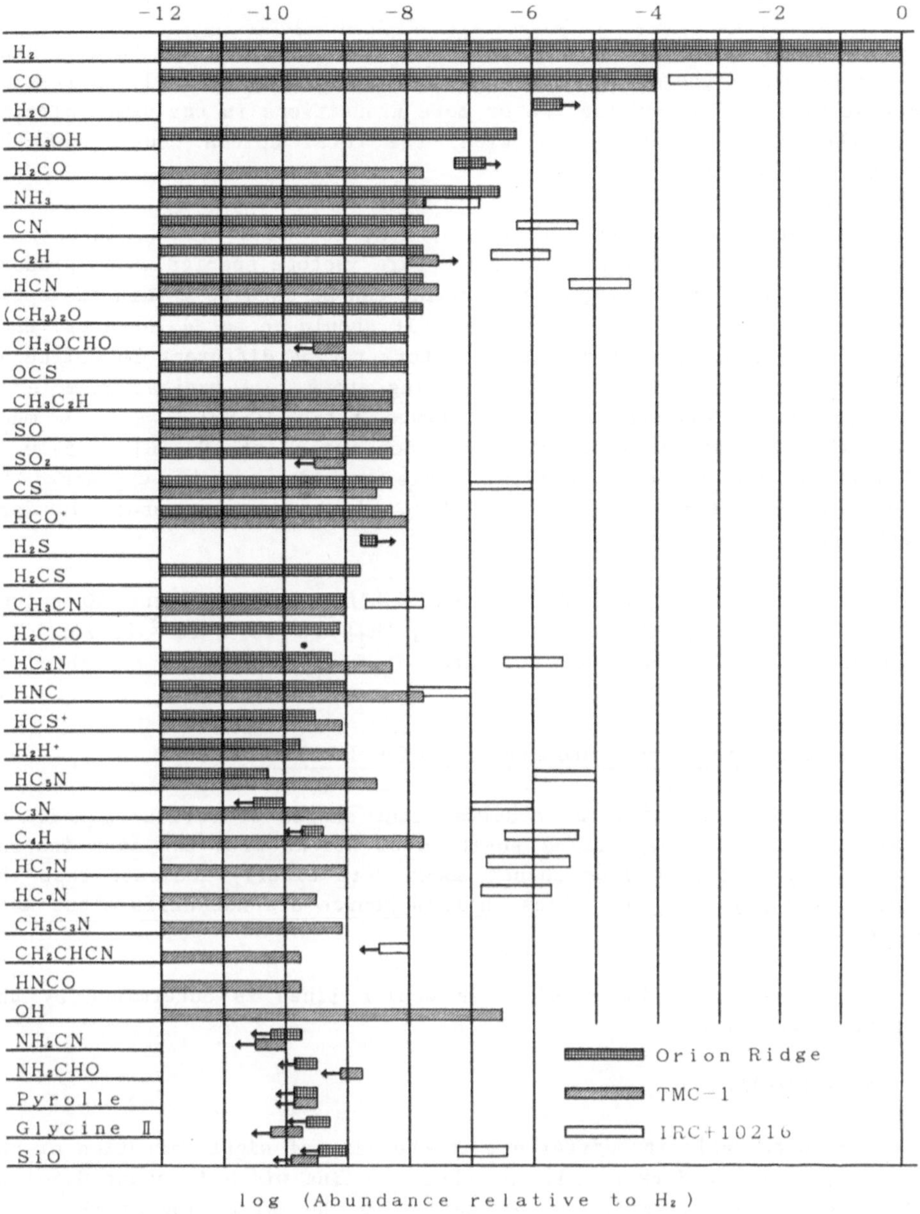

-12
-10
-8
-6
-4
-2
0
H2
CO
H2O
CH3OH
H2CO
NH3
CN
C2H
HCN
(CH3)2O
CH3OCHO
OCS
CH3C2H
SO
SO2
CS
HCO+
H2S
H2CS
CH3CN
H2CCO
HC3N
HNC
HCS+
H2H+
HC5N
C3N
C4H
HC7N
HC9N
CH3C3N
CH2CHCN
HNCO
OH
NH2CN
NH2CHO
Pyrolle
Glycine II
SiO
Orion Ridge
TMC-1
IRC+10216
log (Abundance relative to H2)

This gives a fraction of the molecular abundance. The excitation should be considered to obtain the total column density. The total column density of the molecule N_M in the optically thin limit is estimated by

$$N_M = N_u[\sum_i g_i \exp\{-E_i/kT\}]/[g_u \exp\{-E_u/kT\}]$$

Here i denotes an energy level of the molecule. The equation assumes that the excitation of the molecule can be characterized by a single excitation temperature T. By observing two or more transitions in the same molecule under the same excitation condition, the total column density of the molecule can be estimated.

For abundant molecules such as CO, HCN, CS, HCCCN, etc., line emissions are often optically thick. Those rare isotope species which produce optically thin lines are a good tool which can be used to obtain the optical depth of the main isotope species. It should be noted, however, that chemical selection effects may make isotope ratios different in molecular species from those among the corresponding atoms. A typical example is the case of $[C_3HD]/[C_3H_2]$. The observed ratio $[C_3HD]/[C_3H_2]$ is much larger than the standard "cosmic abundance ratio" of [D]/[H], 1.5×10^{-4}, and exceeds 0.1 in some sources. This anomaly can be explained by ion-molecule reactions if $[e^-] \doteqdot 3\times10^{-7}$, which is not unusual in dark clouds (Bell et al.[6]).

The column density of H_2 is obtained from the observed CO column density by assuming the value of $[CO]/[H_2]$ to be 8×10^{-5} [2,7]. The relative molecular abundances estimated in various sources are shown in Fig. 5.

2.3. Molecular Line Shapes and Dynamics of Molecular Gas

Figure 6 illustrates the observed line shapes of various molecules, obtained from the same central position of Ori-KL with the same instruments. Very different line shapes shown here clearly indicate that the chemical and physical conditions in this source are not uniform but much varied.

The width $\Delta\nu$ of interstellar molecular lines is determined by the Doppler width, i.e.,

$$\Delta\nu = (\nu/c)\Delta v$$

where Δv is the velocity difference in the line of sight direction of the cloud. In typical dark clouds the observed line width is about 0.5-1 km s^{-1}. The Gaussian velocity width due to the thermal motion of the cloud gas

$$\Delta v_t = 0.22\sqrt{(m_H/m_M)T_k} \text{ [km s}^{-1}]$$

gives a line width of about 0.2 km s^{-1} in the case of CO in a dark cloud with T_k = 10 K. Therefore the observed line width is much wider than the thermal width even in the quiet dark clouds, and a large scale velocity gradient (rotation or contraction of the cloud) or an internal micro turbulence should be responsible for a significant part of the observed

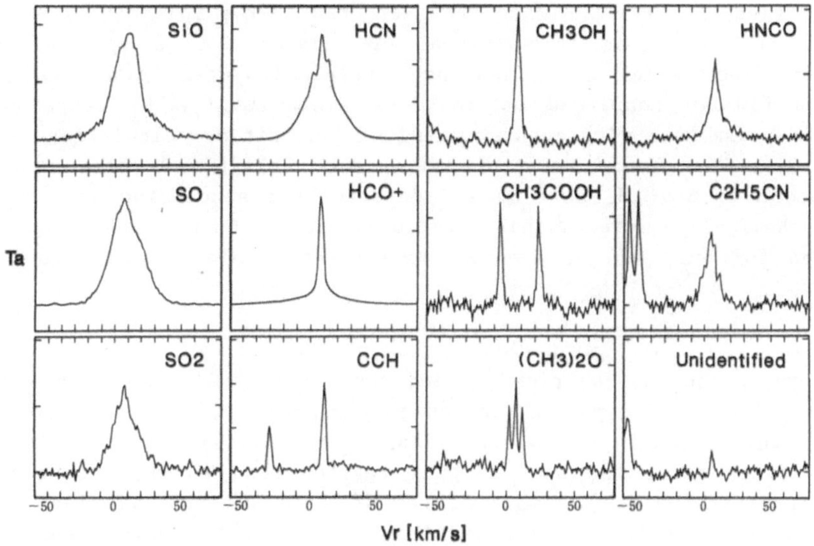

Fig. 6. Line shapes of various molecules observed towards the position IRC2 of Ori-KL. See also Fig. 11. The vertical axis is the antenna temperature in arbitrary units, and the horizontal axis is the radial velocity to the same scale. (Data from Nobeyama 45-m telescope.)

line width. In the case of star forming regions lines show a much wider width and a manifold shape as shown in Fig. 6. This is mainly due to various types of active motion of the gas during the star forming process. The bipolar molecular flow with a velocity of 10-100 km s^{-1} is the most spectacular. Also a rotating gas disk around proto stars and a rapidly expanding shock front produced by strong UV radiation from the newly born star contribute partly to the line width. These will be discussed in Section 4.

3. MOLECULES IN DARK CLOUDS

3.1. Chemistry in Dark Clouds

Observed chemical characteristics in TMC-1, a typical quiet dark cloud which has been studied well, are shown in Fig. 5. There exists a group of molecules with abundances of about 10^{-8} of H_2. They are CN, HCN, HCO^+, SO and CS. Their abundances in TMC-1 are almost the same as in Ori-ridge, a typical dense cloud core accompanied with massive star forming regions. CH_3OH, NH_3, H_2CO and OH are much more abundant in Orion than in TMC-1. These molecules are probably related to the strong heating and dust evapolation due to the star formation processes. On the other hand, the carbon chain molecules like HC_3N, HC_5N, C_3N, CCH and C_4H are more abundant in TMC-1 than in Orion. One of the extreme cases is a recently discovered carbon chain molecule CCS. It has an abundance of 10^{-8} relative to H_2 in TMC-1 and is widely found in dark clouds. It could not be detected, however, in Orion ridge and similar star forming cloud cores at all (Saito et al.[8]). Namely, the carbon molecules seem to characterize the chemistry in TMC-1 and dark clouds of similar type.

Figure 7 shows a spectrum of one recently detected long carbon chain, the C_6H radical. The C_6H was first detected in the molecular line survey in the 20 GHz region at Nobeyama. It appeared as a pair of double emission lines which could not be attributed to any known molecules, and was finally confirmed to be a new molecule C_6H by comparing the observed parameters with quantum chemical (ab initio) calculations (Suzuki et al.[9]). It should be noted that carbon chain molecules containing only one H or S (C_nH and C_nS) are detected for successive numbers of n, while chemically more stable molecules HC_nN, CH_3C_nN and CH_3C_nH are detected for only odd or even n numbers (see Table 1, and also Section 5.2).

TMC-1 and other quiet dark clouds of similar type show more or less chemistry in the "carbon rich" medium where [C] > [O]. The carbon chains are a typical indication of carbon rich chemistry, as they also can be seen in the carbon rich red supergiant star IRC+10216 (see Section 4.3). On the other hand, the star forming regions tend to show "oxygen rich" chemistry (Section 4.1). Some of the dark clouds such as L134N present an intermediate molecular composition. In L134N the carbon chain molecules are less abundant than in TMC-1, while NH_3, SO and SO_2 have a reverse trend[3]. TMC-1 and L134N have a similar physical condition with $n(H_2)$ of about 10^4 cm^{-3} and T_k of 10 K. They have no luminous heating source either. The difference in the relative abundance of gas phase elements in these clouds, which is possibly caused by the selective absorption of atoms and molecules onto dust surface, is likely to be responsible for the variation in chemistry. It is probable that O has a low density in gas phase because H_2O molecules take most of O and are adsorbed on the dust grains as ice. H_2O molecules in the low temperature clouds cannot be observed from the ground with mm-wave telescopes, but they are believed to be extremely abundant. The physical environment such as UV radiation field and the past disturbances (shock wave, etc.) are also likely to lead to the chemical difference.

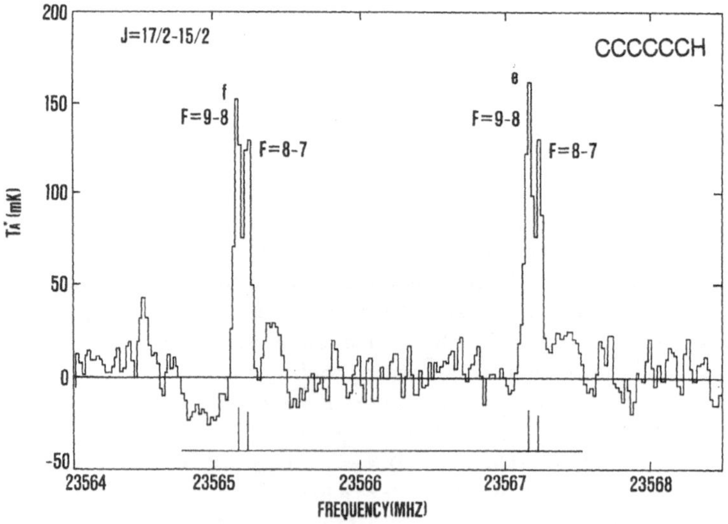

Fig. 7. Spectra of C_6H (taken from Suzuki et al.[9]).

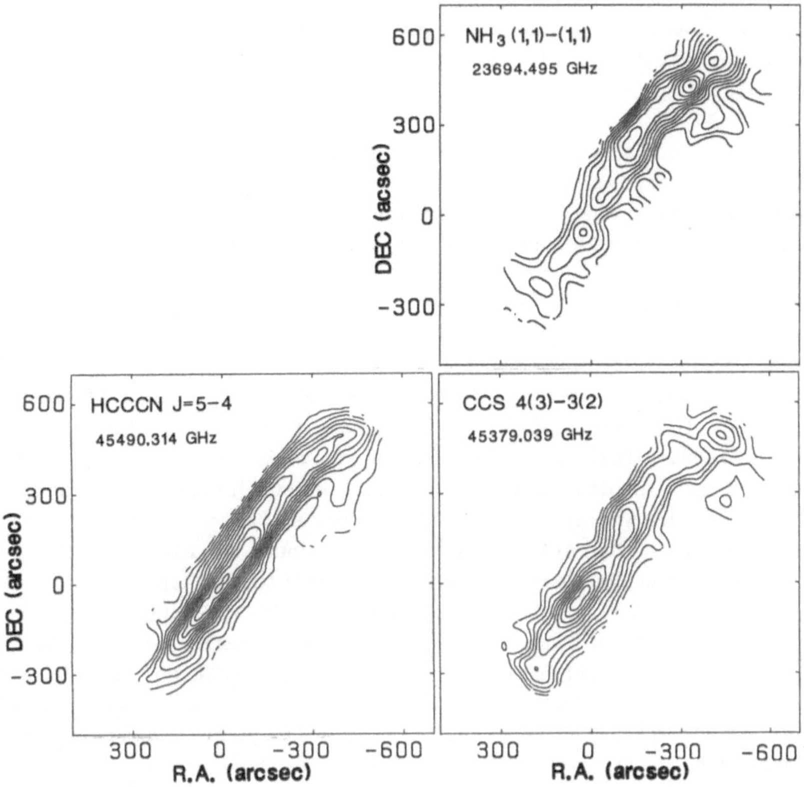

Fig. 8. A comparison of distributions of NH_3, HCCCN and CCS intensities in Taurus molecular cloud 1 (TMC1). The velocity range integrated for is 5.35-6.35 km/s (from Hirahara et al.[10]).

Recent detailed observations with large mm-wave telescopes have revealed that a clear difference in chemistry exists even in the same dark cloud. Figure 8 shows such a difference among the distributions of NH_3, HC_3N and CCS in TMC-1, obtained by a high resolution multi-line observation with the Nobeyama 45-m telescope (Hirahara et al.[10]). The distributions of NH_3 and CCS clearly show anticorrelation with each other. That of HC_3N is more or less flat due to the high optical depth of the HC_3N transition. Also note that the dark cloud very often looks like a filament as shown here. The mechanism required to form such elongated structure is not known exactly yet.

A considerable difference is also seen among the distributions of CS, HCO^+, and SO molecules in L134N [3]. It is not apparent why such a chemical inhomogeneity exists in the dark clouds. The UV radiation penetrating from outer space and therefore the density distribution in the cloud might play an important key role (see Section 5.2).

3.2. Molecular Line Survey for Dark Clouds

Dark clouds are a basic component in the interstellar space. To understand the chemical processes and history of dark clouds is an important aim of astrochemistry. A spectral line survey over a wide frequency range with one instrument is a powerful method used to study interstellar chemistry, because it provides rich and unbiased information. Such a survey often reveals unidentified lines which can be attributed to unknown molecules.

However such a survey for dark clouds is much more difficult than that for star forming regions or for red giant stars, because of very narrow line widths and low temperature in the dark clouds. Therefore it is necessary to have a spectrometer with high frequency resolution and large output channel numbers enough to cover as wide a frequency range as possible and a sensitive receiver frontend with wide instantaneous bandwidth.

A wide frequency survey for dark clouds has been performed at NRO since 1984. The first phase of the survey for TMC-1 in 36-50 GHz is almost finished. This is being extended to 10-36 GHz. As a result of the survey many unidentified lines have been detected, as well as new transitions of known molecules. Some unidentified lines were so strong that an intensive investigation has been done to identify them (Suzuki et al.[11]). Finally seven strong unidentified lines were revealed to be the transitions of two new molecules, CCS and CCCS (Fig. 9, [8]. [12], [53]). These molecules are a new series of carbon chains containing S. Some other new species, C_6H, CH_2CN and HC_3HO were also identified in conjunction with this survey (Fig. 10, Irvine et al.[13], Saito et al.[9,14]). As a result, in the frequency range of 36-50 GHz, 56 transitions of 19 molecules and 9 rare isotope species were detected, and some 30 other lines remain unidentified. Among the 19 molecules, seven are carbon chains, one is a ring molecule and one is an ion. About 40 molecules in total have been detected so far in TMC-1.

To understand the chemical history of dark clouds we still need more knowledge about the chemical composition. In particular we know very little about the complex organic molecules in dark clouds. Transitions in a large complex molecule tend to be considerablly weak, because of large partition functions even in the low kinetic temperature of the dark cloud. The weak unidentified lines detected in TMC-1 are likely to be transitions in such ring or complex organic molecules, which may not be very rare in the clouds compared with simple chain molecules.

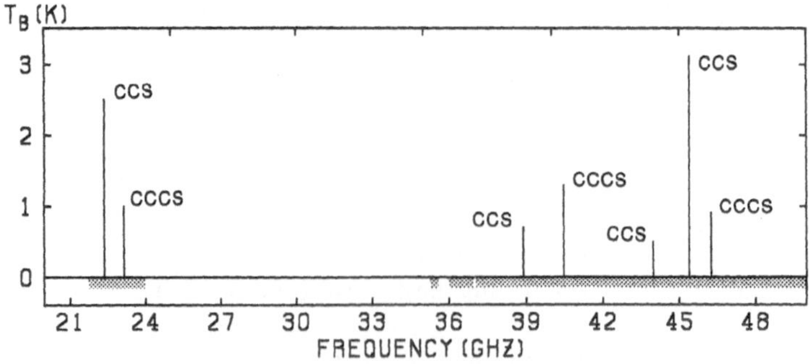

Fig. 9. Schematic spectra of CCS and CCCS molecules observed in the NRO molecular line survey for TMC1 (from Kaifu et al.[12]).

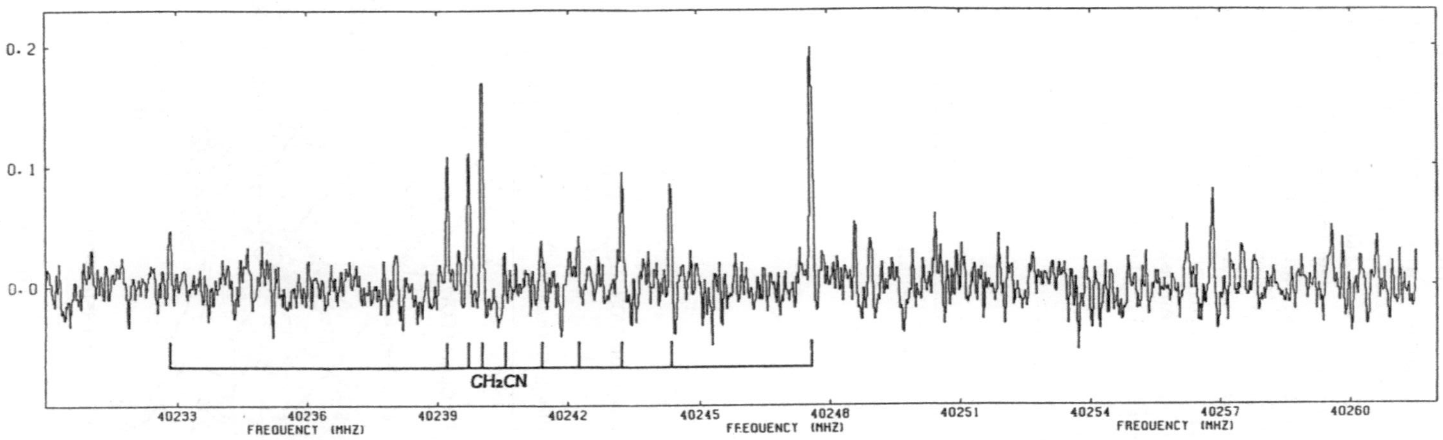

Fig. 10. Spectra of the CH_2CN radical observed with the NRO high-resolution AOS, including the hyperfine components derived from laboratory measurements.

4. MOLECULES IN STAR FORMING REGIONS AND IN AGED STARS

4.1. Ori-KL and Massive Star Forming Regions

Ori-KL is a rich source of radio emission of molecules in various physical and chemical conditions. A powerful infrared source IRC-2 is sitting at the center of the Ori-KL, where the strongest lines are observed from many molecules. The IRC-2 is interpreted as a massive (about 50 M⊙) protostar formed in the dense molecular cloud core. The core is embedded in a filamentaly shaped cloud which lies behind a huge plasma (HII) region known as the Orion Nebula. Four O and B type stars in the center of the Orion Nebula are heating sources of the nebula. The Ori-KL is the nearest region of massive star formation (distance being 500 pc), and hence has been extensively studied by many authors (see references in Hjalmarson[15], and [2,3]). Detailed studies of the chemical and physical conditions have been made based on the frequency survey towards IRC-2 by Johansson et al.[16], Olofsson[17], Ohishi et al.[18], and Blake et al.[19].

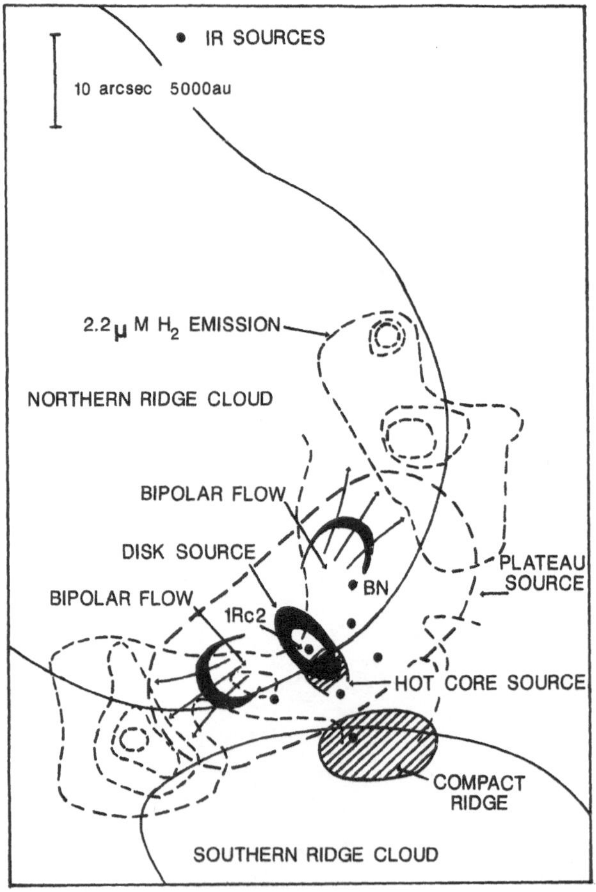

Fig. 11. A schematic picture of the Ori-KL region. The star-forming activity is centered on the source IRC2 (taken from Irvine et al.[2]).

Figure 11 illustrates the central part of the star forming region investigated with various mm-wave molecular lines and infrared observations.

The southern and northern ridges, which are a part of a molecular cloud filament elongated in the north-south direction, show the typical nature of chemistry in a dense cloud core (see Fig. 5). Though not seen in Fig. 11, a rapidly rotating massive ($M \fallingdotseq 10^3$ M⊙) disk is observed to surround the central source IRC-2 (Hasegawa et al.[20]. Such a cold, dense molecular disk is called protostellar disk.

The "disk source" shown in Fig. 11 is also called a "plateau source" because the molecular lines observed there are very broad, as presented in the profiles of the SiO, SO, SO_2 and HCN lines in Fig. 6. The width indicates the high velocity expansion ($V_e \fallingdotseq$ 10-15 km s^{-1}) probably due to a strong stellar wind from IRC-2 to the outer rotating disk mentioned above. The source is also called an "expanding doughnut." The chemistry in this region is characterized by O- and S-containing molecules; CS, SiO, SO, SO_2, OCS, H_2S, HDO, etc. Shock disruption of dust cores, followed by the neutral reactions in the cooling post shock region is thought to supply those refractory elements[19].

A very high velocity (about 100 km s^{-1}) bipoler flow is observed with CO and weakly with HCO^+. The hydrogen emission line at 2 mm shows that the shocked region is heated up to 2000 K by the high velocity bipolar flow.

The N-containing and fully hydrogenized species like HCN, NH_3, CH_3CN and C_2H_5CN are the characteristic molecules in the "hot core" source, which is adjacent to the expanding "plateau" source. The line width is typically 10 km s^{-1}. The rotational temperature and the density observed are quite high, i.e., 150-300 K and $\geq 10^7$ cm^{-3}, respectively. It is suggested that the massive release of grain mantle components explains the chemistry in this region[19].

Abundant CH_3OH, $(CH_3)_2O$, and $HCOOCH_3$ molecules are observed particularly in the "southern condensation" sources. These molecules also exhibit a high excitation temperature of 200 K, though the line widths are narrow (3-5 km s^{-1}, see Fig. 6). Such high excitation and peculiar chemistry implies that this region is also affected by the heating of the dust grain and H_2O, NH_3 and other species released would produce new chemical conditions.

The H_2O and SiO masers are also outstanding phenomena in the Ori-KL region. Both maser emissions are popular in the Mira-Type variables which are known to show the O-rich chemistry. The H_2O maser is also an indicator of massive star formation and observed as a group of many maser spots rapidly expanding from the site of star formation as a whole. The maser line of H_2O is the 6_{16}-5_{23} transition in high energy levels and can not be observed in normal molecular clouds. The maser phenomena suggest, however, that H_2O is very abundant in moleculer clouds, though most of them may be adsorbed on the dust surface in the form of ice.

The SiO maser, which also exhibits expanding motion in Ori-KL, is not as abundant as the H_2O maser. The former has been detected only very recently in the star forming regions other than Ori-KL, i.e., W-51 and Sgr B2, (Ukita et al.[21]).

There are huge regions of multiple massive star formation such as W-49 and W-51. They are a powerful source of H_2O maser and emit various molecular lines. They are quite distant (6-7 kpc). It is difficult to resolve the star forming phenomena in these sources. A mm-wave interferometer would improve this situation in great deal.

The Sgr-B2 is a huge complex of massive molecular clouds and HII regions in the Galactic center. This is the place where most of the interstellar molecules can be observed in spite of the long distance (8 kpc). In particular it is characterized by such saturated complex organic molecules as C_2H_5OH, C_2H_5CN, CH_3NH_2, NH_2CHO, etc. The molecular clouds in Sgr-B2 are dense (10^6 cm^{-3}) and warm (40-60 K). They are probably affected by the activity of the galactic central region through UV or X-ray photons, cosmic ray and active star formation.

4.2. Photostellar Disk in the Flow System

Unlike the site of massive star formation, the region of low mass star formation does not show spectacular molecular emissions. Instead they provide good samples of elementary processes of star formation, and even some evidence of the formation of planetary systems has been observed indirectly. This is partly because the low mass star (with a mass less than a few $M_\odot$) does not emit strong UV radiation which evaporates the surrounding matter cloud to form HII regions, and partly because they do not produce energetic effects which cause significant changes in the chemical process as described in Section 4.1.

The molecular line studies have revealed a number of unexpected phenomena by observing these regions in detail. First, it was found around most protostellar candidates that molecular gas flows from the central luminous infrared source (so-called protostars) towards two opposite directions (Snell et al.[22], a review paper by Lada[23]. The size, mass and flow velocity of these bipolar flows are of the order of 1 pc, 1 to 100 $M_\odot$, and 10 to 100 km s^{-1}, respectively. The flows are observed mainly in CO emission which represents relatively less dense($n(H_2) \fallingdotseq 10^2$-$10^3$ cm^{-3}) gas due to its small electric dipole moment of 0.1 Debye. Also HCO^+, HCN and CS flows are observed in the massive cloud cores, indicating that such flows contain flagments of dense (10^4-10^5 cm^{-3}) gas.

Small and flat disk features are also observed around many protostars with the bipolar flows by means of high resolution observations of CS, NH_3, HCN and HCO^+ molecular lines. The disk is often rotating, and has physical parameters as follows: size $\fallingdotseq$ 0.1 pc, density $\fallingdotseq 10^4$-10^6 cm^{-3}, mass $\fallingdotseq$ 1-10^3 $M_\odot$ (Kaifu et al.[24], review papers by Kaifu[25], Rodriguez [26]). The protostellar disk should be physically related to the molecular flow, and probably the significant fraction of flow material is supplied from the disk.

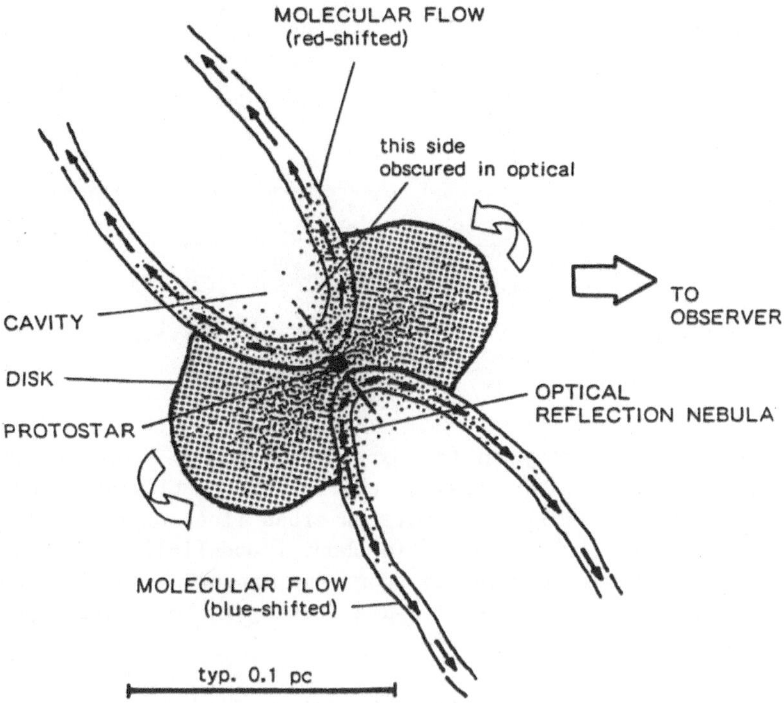

Fig. 12. A schematic model of the protostellar flow/disk system (from Kaifu[25]).

A schematic model of the protostellar disk and flow system is shown in Fig. 12. The disk is formed in the cloud core by self contraction. As the central condensation grows, it becomes hot and starts to emit strong infrared emission as a protostar.

Sometime during this stage the bipolar flow is accelerated and the extra mass, extra energy and angular momentum are ejected to the interstellar space. Thus the bipolar flows can be regarded as a mechanism which makes a successful formation of stars and planetary systems possible. Though the size of the disk observed is 100 times larger than that of the solar system, there is some observational evidence that thin, small dust disks are formed in the central part of the molecular disk, and a planet could be formed in this region (Bally et al.[27], Nagata et al.[28], Smith and Terrile[29]).

4.3. Molecules in the Expanding Envelope of Red Giant Stars

The expanding envelope of red giant/supergiant stars is also the source of various molecular lines. As mentioned in Section 1, this stellar envelope can not be called interstellar matter and the molecules observed there should be called "circumstellar molecules" in an exact sense.

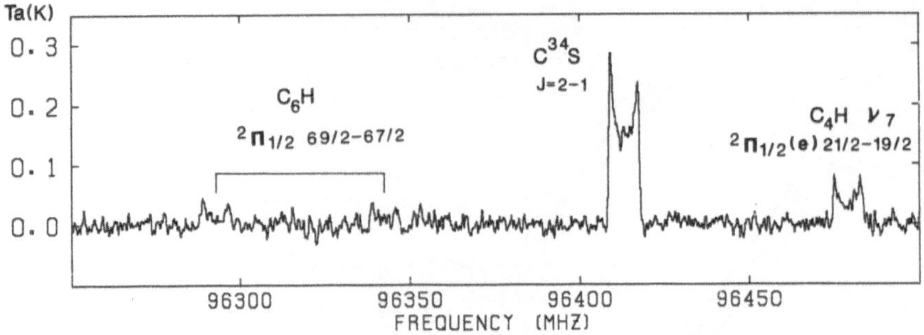

Fig. 13. Sample spectra of IRC+10216. (Data from Nobeyama 45-m telescope.)

However the chemistry in the expanding stellar envelope exhibits a very wide similarity to interstellar chemistry. Particularly the carbon-rich star (typically IRC+10216) emits numerous lines of carbon-containing molecules very similar to those in dark clouds[16]. The line shape observed in these stars apparently shows the expanding shell phenomena by its large width (about 20 km s^{-1}) and a rectangular or inverse paraboloid profile (Fig. 13).

On the other hand the oxygen-rich stars (typically Mira variables) emit very weak CO line but strong H_2O and SiO lines, which indicate O-rich chemistry. There is a group of red giant stars which are called OH/IR stars and include many Mira variables. These are infrared-bright giant/supergiant stars and characterized by OH maser emissions as well as H_2O and SiO masers. An extensively observed OH/IR star OH231.8+4.2 shows SO_2, SO, H_2S, CS, OCS, HCO^+ and other molecular emissions, which also strongly suggest O-rich chemistry (Morris et al.[30]).

Recently Cernicharo and Guelin[31] detected emission lines of metal halides NaCl, KCl, AlCl and AlF in IRC+10216. These molecules seem to be concentrated in the inner part of the expanding shell, suggesting that they are formed in the chemical equilibrium condition and then condense onto grains. The metal elements are significantly deficient in the interstellar space and no metal containing molecule has been detected in the interstellar clouds. Thus the above-mentioned detection of metal halides in IRC+10216 is a direct evidence that the most metal elements are condensed on dust grains as has been widely believed. The chemical reactions in these stellar expanding envelopes are analyzed by a thermodynamic equilibrium model in the C-rich or O-rich atmospheres (Tsuji[32]), and partly by photochemical processes and possibly by ion-molecule reactions in the outer regions, too (Bieging and Riew[33]). More details are given in the review papers by Tsuji[34], Olofsson[35], and Zuckerman[36].

5. FORMATION AND EVOLUTION OF INTERSTELLAR MOLECULES

5.1. Gas Phase Reaction in Dark Clouds

Very low gas density and low kinetic temperature in the interstellar cloud do not allow three-body collisions and chemical equilibrium such as in the terrestrial atmosphere. It is widely accepted that the main chemical processes is the formation of interstellar molecules are ion-molecule reactions (Herbst and Klemperer[37], Watson[38]).

First ion-molecule reactions do not require a third body that carries away the excess energy in the reaction. Even two-body collisions do not occur very often; roughly once a year with H or H_2, 10^{-4} times a year with an abundant heavy element such as C, N, and O. Within the life time of 10^7 to 10^8 years of dark clouds, however, those reactions can produce complex species. Second, most ion-molecule reactions do not require an activation energy and proceed even in the cold (10-30 K) cloud. However the reaction requires an ionization process first.

A rough scheme of the reaction process is as follows:

(1) Ionization

$$H_2 + C.R. \rightarrow H_2^+ + e + C.R.$$

$$(H^+ + H + e + C.R.)$$

Here the most effective cosmic ray (C.R.) is a proton with an energy of about 100 MeV. It penetrates into even the dense dark cloud with $n(H_2)L = 10^{24}$ cm^{-2} and ionizes H_2 molecules with an ionization rate of 10^{-17} s^{-1}. In diffuse dark clouds, stellar UV photons are more effective than the C.R.

(2) Activation of neutral species

$$H_2^+ + H_2 \rightarrow H_3^+ + H,$$

$$H_3^+ + A \rightarrow AH^+ + H_2$$

The first process is fast because H_2 is very abundant.

(3) Hydrogenation

$$AH^+ + H_2 \rightarrow AH_2^+ + H, \text{ etc.}$$

This process is also fast.

(4) Complication

$$AH_m^+ + B \rightarrow ABH_n^+ + pH_2 + qH, \text{ etc.}$$

(5) Neutralization $\quad AH_m^+ + M \rightarrow AH_m + M^+$

recombination $\quad AH_m^+ + e \rightarrow AH_m$

etc.

A computer simulation with ion-molecule reaction networks has been carried out by many authors (Suzuki [39], Leung et al.[40], Tarafder et al.[41], Miller et al.[42]). They are compared with the results of observations and predict interstellar chemistry. For reactions with no measured reaction rate available, the Langevin rate coefficient of 10^{-9} cm^3 s^{-1} is generally adopted.

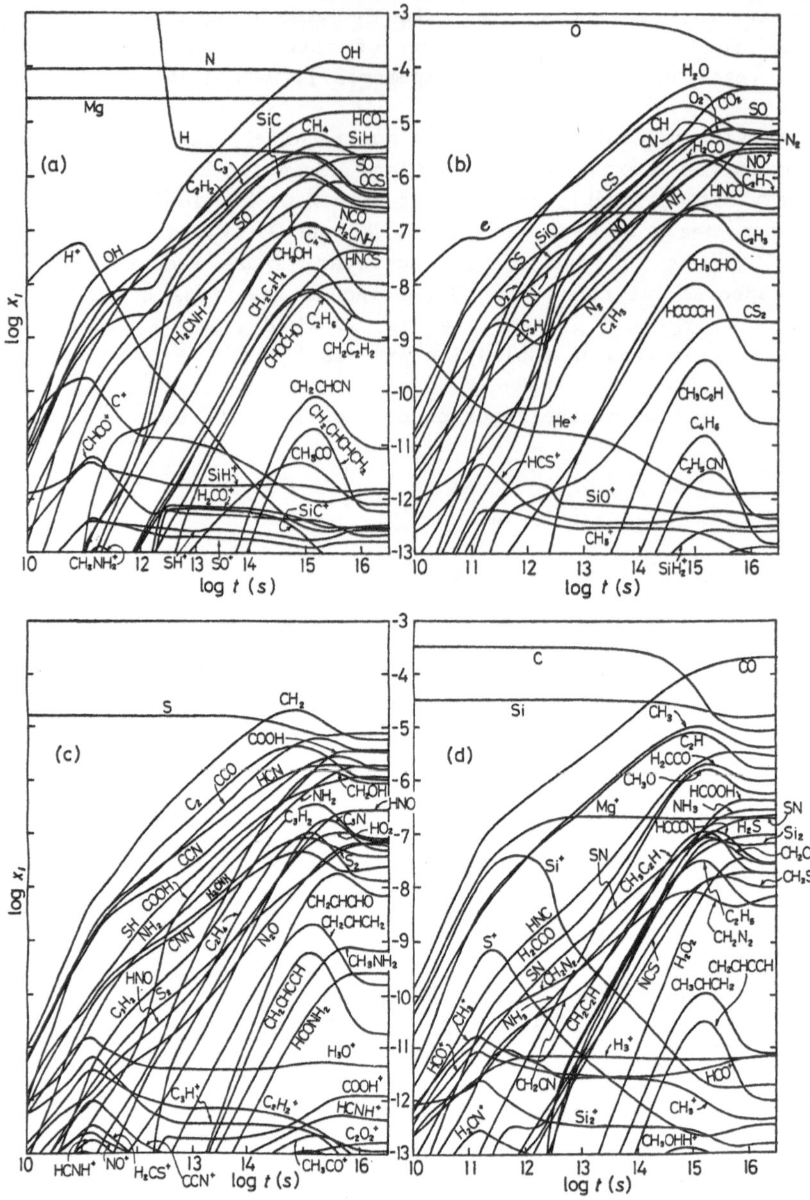

Fig. 14. (a)-(d) Relative abundance of molecule, $x_i=n_i/n_H$(n_H being the total nucleon number density taken to be 10^5 cm^{-3}), as a function of log t, where t is the time. Initially all elements are locked up in atomic form (taken from Suzuki[38]).

Time-dependent calculations by Suzuki[39,43] and Tarafdar et al.[41] show that in dense clouds the chemical processes reach the stable phase in 10^8-10^9 years. The time scale is longer than the age of molecular clouds (10^7-10^8 years) as shown by the diagrams in Fig. 14. Therefore the chemical evolution should be under process in the observed clouds and the cloud history could be studied through the chemical condition of the cloud.

Suzuki[39] also pointed out that carbon is a key element representing the chemical history of the cloud as shown in Fig. 14. The C atom becomes CO in 10^{16} s (3×10^{8} y) and most of the carbon-containing molecules other than CO tend to be in their maximum abundance at around the time of 10^{15} s. This calculation adopted a fixed $n(H_2)$ value of 10^5 cm^{-3}.

The recent advances in the study of gas-phase reaction schemes, such as the temperature dependence of rate coefficients (Smith and Adams[44]) and the role of radiative association (Barlow et al.[45]), will make it possible to establish a model of chemical evolution connected with the physical history of interstellar cloud.

5.2. Growth of Carbon Chains and Organic Molecules in Dark Clouds

A considerable fraction of interstellar carbons are embedded in a large variety of carbonaceous molecules. Highly unsaturated long carbon chain molecules are most remarkable. The longest chain discovered so far is $HC_{11}N$ (Bell and Matthews[46]).

The abundance of cyanopolyacetylene molecules HC_nN (n = 3,5,7,9,11) in TMC-1 decreases monotonically with increasing n. A similar tendency is observed in CH_3C_nH (n = 2,4) (Irvine et al.[3]; Snyder et al.[61]). Linear hydrocarbon radicals C_nH (n = 2,3,4,5,6) and linear polycarbon monosulphide and possibly monoxide molecules C_nS and C_nO (n = 2,3) are observed for both odd and even n numbers. They also show decreasing abundance with increasing n. However the species with even n is much more abundant than those with odd n, by about one order of magnitude([8,9], Yamamoto et al.[47], Ohishi et al., private communication, references in Irvine et al.[3]). On the other hand C_2N has not yet been detected and should be less abundant than C_3N. These facts, together with the estimate that the chemical evolution takes an equal or longer time than the cloud age, suggest that the carbon chains C_nX (X = H, S, O, N) are in a transient phase and ultimately form more stable observed molecules like HC_nN, CH_3C_nH, and also molecules stable but having no allowed rotational transitions such as HC_nH and $H_2C_nH_2$. In other words, the carbon chain molecules could be a good quantitative probe of the history of various dark clouds.

How do these long carbon grow? An interesting possibility is that the pure carbon chain grows first in the partially ionized diffuse phase of dark clouds, as proposed by Suzuki[48]. The Suzuki scheme (Fig. 15)

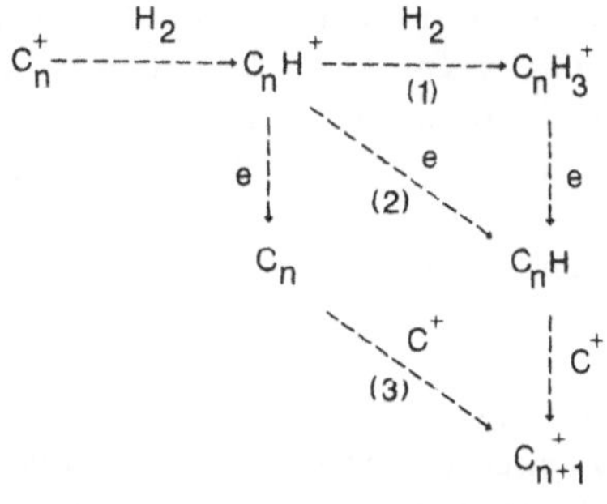

Fig. 15. A scheme of carbon chain growth (from Suzuki[48]).

involves three steps: (1) The hydrogenation of C_n^+, (2) the neutralization to produce C_n, and (3) the radiative association with C^+ to produce C_{n+1}^+ (path (3) in Fig. 15). As the branching ratio of path (2) is small compared with that of recombination (2), and as path (1) has a smaller reaction rate than the recombination in the partially ionized region where the electron density is high, the main path in Fig. 15 should be (3). Though the radiative association and the recombination are not yet known well, the scheme seems to be promising for the growth of pure carbon chains with n = 3-9 (Bohme[49]). In this model, it is also proposed that the higher density, lower temperature and faster contraction would provide favorable conditions for the chain growing[48].

A number of possible mechanisms for the formation of hydrocarbon chains, cyanopolyacetylene and other carbon chain molecules have been investigated by various authors (e.g., Bohme[49] Bohme et al.[50]).

The recent detection of cyclic C_3H_2 (Matthews et al.[51]; Thaddeus et al.[52]) and cyclic C_3H (Yamamoto et al.[53]) demonstrated the possibility that a variety of ring molecules might exist widely in the dark clouds, though the observed lines tend to be much weaker than the transitions of simple chain molecules.

The most plausible formation process of $c\text{-}C_3H_n$ is the dissociative recombination of the $C_3H_3^+$ ion. The $C_3H_3^+$ has two possible structures; the cyclic and linear forms. Both of them are efficiently formed by the radiative association

$$C_3H^+ + H_2 \rightarrow C_3H_3^+ + h\nu$$

with the branching ratio of production of $c\text{-}C_3H_3^+$ to production of $l\text{-}C_3H_3^+$ being 1:1 (Adams and Smith[54]). Actually the observed abundances of $c\text{-}C_3H_2$ and $c\text{-}C_3H$ in TMC-1 are nearly equal to those of $l\text{-}C_4H$ and $l\text{-}C_3H$, respectively. Since the C_3H^+ is a starting ion in the successive synthesis of carbon chain molecules and as the above-mentioned radiative association is the main reaction of C_3H^+, we may predict that about half of the chemical evolution of large hydrocarbons and related molecules originates from $c\text{-}C_3H_3^+$ (Vrtilek et al.[55], Suzuki[56]).

The complex organic molecules, such as saturated hydrocarbons, are also an important aim of more sensitive observations. Because of their internal motions and/or hyperfine structure, as well as their large moment of inertia, their transitions are diluted considerably. A good example is CH_2CN, which was identified only very recently in dark clouds[13,14]. The estimated abundance of CH_2CN in TMC-1 is the same as that of CCS, which is one of the main organic molecules in TMC-1, though the observed line intensity of CH_2CN were ten times weaker than that of CCS (Figs. 9 and 10).

From unidentified interstellar infrared emission features and visible interstellar diffuse absorption bands, it has been proposed that polycyclic aromatic hydrocarbon (PAH) might exist widely in the interstellar cloud (Leger et al.[57], Allamandola et al.[58], van der Zwet et al.[59], Leger et al.[60]). Recent infrared observations over a wide range of wavelengths also suggest that a number of emission or absorption features occur by the "ice" of H_2O, NH_3, CO, H_2S, CH_3OH, etc., on dust grains.

These rapid improvements in observational and experimental studies are beginning to bridge the gap between the chemistry in the gas phase and that in the solid phase in the interstellar clouds.

6. PERSPECTIVES

Apparently the study of interstellar molecules can be divided into two major fields. One is the field where we investigate various astrophysical phenomena by using the molecular lines as a very useful probe. In particular the formation process of stars has been revealed by this method. Further improvements of the observational instruments, particularly the mm-wave interferometers and sub-mm-wave instruments, will open this new window wider toward the distant external galaxies and the planetary systems in the Solar neighborhood. The second field is that of astrochemistry, where the chemical evolution in various interstellar matter would be revealed with further collaboration among radio and infrared astronomy, astrophysical theories, molecular physics both in laboratory and theory, solid state physics and other related fields. The molecular evolution in dark clouds from simple molecules to complex organic ones, dust grains and finally stars and planets will lead us to the understanding of physical evolution of interstellar matter and finally of the history of terrestrial lives, i.e., the understanding of matter circulation and evolution in the Galaxy, at least in the solar neighborhood.

The author is greatly indebted to Professor Kazuo Takayanagi for his continuing encouragement to the study of interstellar molecules in Japan. The mm-wave astronomy, especially the observational study of interstellar molecules in Japan owes a great deal to his deep and wide interests and supports to this new field since its very early stage.

The author also express deep thanks and offer his condolences to the late Dr. Hiroko Suzuki (1947-1987), who was a very distinguished astrophysicist, the most active staff member of Nobeyama Radio Observatory and was our best colleague and friend. She was one of the pioneers of astrochemistry, and was at the center of the Nobeyama molecular line survey, which detected many important molecules in dark clouds. She died on 22 November, 1987, the result of a car accident. Her great contributions to astrochemistry and radio astronomy, and her very active life will remain in our memory forever.

REFERENCES

1. S. Weinreb, A.H. Barret, M.L. Meeks, and J.C. Henry, Nature 200, 829 (1963).
2. W.M. Irvine, F.P. Schloerb, A. Hjalmarson, and E. Herbst, in: Protostars and Planets II (D.C. Black and M.S. Matthews, eds.), Univ. Arizona Press, p.579 (1985).
3. W.M. Irvine, P.F. Goldsmith, and A. Hjalmarson, in: Interstellar Processes (D.J. Hollenbach and H.A. Thronson, eds.), Reidel, p.561 (1987).
4. N. Kaifu, in: URSI Symposium on Millimeter and Submillimeter Astronomy (NRO Report 64) (1984).
5. N. Kaifu and Y. Chikada, in: URSI Symposium on Millimeter and Submillimeter Astronomy (NRO Report 63) (1984).

6. M.B. Bell, L.W. Avery, H.E. Matthews, P.A. Feldman, and J.K.G. Watson, Astrophys. J. 326, 924 (1988).
7. M. Guelin, in: Molecular Astrophysics - States of the Arts and Further Directions (G.H.F. Dierckson, W.F. Huebner, and P.W. Langhoff, eds.), Reidel, p.23 (1985).
8. S. Saito, K. Kawaguchi, S. Yamamoto, M. Ohishi, H. Suzuki, and N. Kaifu, Astrophys. J. 317, L115 (1987).
9. H. Suzuki, M. Ohishi, N. Kaifu, S. Ishikawa, T. Kasuga, K. Kawaguchi, and S.Saito, Publ. Astron. Soc. Japan 38, 911 (1986).
10. Y. Hirahara, Thesis, Faculty of Science, University of Tokyo (1989).
11. H. Suzuki, N. Kaifu, T. Miyaji, M. Morimoto, and M. Ohishi, Astrophys. J. 282, 197 (1984).
12. N, Kaifu, H. Suzuki, M. Ohishi, T. Miyaji, S. Ishikawa, T. Kasuga, M. Morimoto, and S. Saito, Astrophys. J. 317, L111 (1987).
13. W.M. Irvine, P. Friberg, A. Hjalmarson, S. Ishikawa, N. Kaifu, K. Kawaguchi, S.C. Madden, H.E. Matthews, M. Ohishi, S. Saito, H. Suzuki, P. Thaddeus, B.E. Turner, S. Yamamoto, and L.M. Ziurys, Astrophys. J. 334, L107 (1988).
14. S. Saito, S. Yamamoto, W.M. Irvine, L.M. Ziurys, H. Suzuki, M. Ohishi, and N. Kaifu, Astrophys. J. 334, L113 (1988).
15. A. Hjalmarson, in: Sub-Millimeter Astronomy (P.A. Shver and K. Kjar, eds.), ESO Conf. Workshop Proc. No.22, p.285 (1985).
16. L.E.B. Johansson, C. Andersson, J. Ellder, P. Friberg, A. Hjalmarson, B. Hoglund, W.M. Irvine, H. Olofsson, and G. Rydbeck, Astron. Astrophys. 130, 227 (1984).
17. H. Olofsson, Astron. Astrophys. 134, 36 (1984).
18. M. Ohishi, N. Kaifu, H. Suzuki, and M. Morimoto, Astrophys. Spacesci. 118, 405 (1986).
19. G.A. Blake, E.C. Sutton, C.R. Masson, and T.G. Phillips, Astrophys. J. 315, 621 (1987).
20. T. Hasegawa, N. Kaifu, J. Inatani, M. Morimoto, Y. Chikada, H. Hirabayashi, H. Iwashita, K. Morita, A. Tojo, K. Akabane, Astrophys. J. 283, 117 (1984).
21. N. Ukita, T. Hasegawa, N. Kaifu, K. Morita, S. Okumura, H. Suzuki, M. Ohishi, and M. Hayashi, in: Star Forming Regions (M. Peimbert and J. Jugaku, eds.), Reidel, p.178 (1987).
22. R.L. Snell, R.B. Loren, and R.L. Plambeck, Astrophys. J. 239, L17 (1980).
23. C.J. Lada, Ann. Rev. Astron. Astrophys. 23, 267 (1985).
24. N. Kaifu, S. Suzuki, T. Hasegawa, M. Morimoto, J. Inatani, K. Nagane, K. Miyazawa, Y. Chikada, T. Kanzawa, and K. Akabane, Astron. Astrophys. 134, 7 (1984).
25. N. Kaifu, in: Star Forming Regions (M. Peimbert and J. Jugaku, eds.), Reidel, p.275 (1987).
26. L.F. Rodriguez, in: Star Forming Regions (M. Peimbert, J. Jugaku, eds.), Reidel, p.239 (1987).
27. J. Bally, R.L. Snell, and R. Predmore, Astrophys. J. 272, 154 (1983).
28. T. Nagata, S. Sato, and Y. Kobayashi, Astron. Astrophys. 119, L1 (1983).
29. B.A. Smith and R.J. Terrile, Science 226, 1421 (1985).
30. M. Morris, S. Guilloteau, R. Lucas, and A. Omont, Astrophys. J. 321, 888 (1987).
31. J. Cernicharo and M. Guelin, Astron. Astrophys. 183, L10 (1987).
32. T. Tsuji, Astron. Astrophys. 23, 411 (1973).
33. J.H. Bieging and N.-Q. Rieu, Astrophys. J. 324, 516 (1988).
34. T. Tsuji, in: Astrochemistry (M.S. Vardya and S.P. Tarafdar, eds.), p.409 (1987).
35. H. Olofsson, in: Late Stages of Stellar Evolution (S. Kwok and S. Pottasch, eds.), Reidel, p.149 (1987).
36. B. Zuckerman, in: Astrochemistry (M.S. Vardya and S.P. Tarafdar, eds.), Reidel, p.345 (1987).

37. E. Herbst and W. Klemperer, Astrophys. J. 185, 505 (1973).
38. W.D. Watson, Astrophys. J. 188, 35 (1974).
39. H. Suzuki, Prog. Theor. Phys. 62, 936 (1976).
40. C.M. Leung, E. Herbst, and W.F. Huebner, Astrophys. J. (Suppl.) 56, 231 (1984).
41. S.P. Tarafdar, S.S. Prasad, W.T. Huntress, Jr., K.R. Villere, and D.C. Black, Astrophys. J. 289, 220 (1985).
42. T.J. Miller, and A. Freeman, Mon. Not. Roy. Astron. Soc. 207, 405 (1984).
43. H. Suzuki, S. Miki, K. Sato, M. Kiguchi, and Y. Nakagawa, Prog. Theor. Phys. 56, 1111 (1976).
44. D. Smith and N. Adams, in: Swarms of Ions and Electrons (T.D. Mard and Foworka, eds.), Springer, p.194 (1984).
45. S.E. Barlow, G.E. Dunn, and M. Schauer, Phys. Rev. Lett. 52, 902 (1984).
46. M.B. Bell and H.E. Matthews, Astrophys. J. 291, L63 (1985).
47. S. Yamamoto, S. Saito, K. Kawaguchi, N. Kaifu, H. Suzuki, and M. Ohishi, Astrophys. J. 317, L119 (1987).
48. H. Suzuki, Astrophys. J. 272, 579 (1983).
49. D.K. Bohme, in: Reaction Rate Coefficients in Astrophysics (T.J. Miller and D.A. Williams, eds.), Reidel, p.00 (1988).
50. D.K. Bohme, S. Wlodek, and A.B. Raksit, Can. J. Chem. 65, 2057 (1987).
51. H.E. Matthews and W.M. Irvine, Astrophys. J. 298, L61 (1985).
52. P. Thaddeus, J.M. Vrtilek, and C.A. Gottliev, Astrophys. J. 229, L63 (1985).
53. S. Yamamoto, S. Saito, M. Ohishi, H. Suzuki, S. Ishikawa, N. Kaifu, and A. Murakami, Astrophys. J. 322, L55 (1987).
54. N.G. Adams and D. Smith, Astrophys. J. 317, L25 (1987).
55. J.M. Vrtilek, C.A. Gottlieb, and P. Thaddeus, Astrophys. J. 314, 716 (1987).
56. H. Suzuki, S. Yamamoto, S. Saito, M. Ohishi, N. Kaifu, S. Ishikawa, and A. Murakami, in: IAU 3rd Asian Pacific Regional Meeting Peking (1987).
57. A. Leger and J.L. Puget, Astron. Astrophys. 137, L5 (1985).
58. L.J. Allamandola, A.G.G.M. Tielens, and J.R. Barker, Astrophys. J. 290, L28 (1985).
59. G.P. van der Zwet and L.J. Allamandola, Astron. Astrophys. 146, 76 (1985).
60. A. Leger and L. d'Hendecourt, Astron. Astrophys., 146, 81 (1985).
61. L.E. Snyder, T.L. Wilson, C. Henkel, P.R. Jewell, and C.M. Walmsley, Bull A.A.S. 16, 959 (1984).

ATOMIC AND MOLECULAR DATA NEEDED IN SPACE, FUSION, AND RELATED RESEARCHES

H. Tawara

Institute of Plasma Physics,* Nagoya University
Nagoya 464-01
Japan

1. INTRODUCTION

In this chapter we describe the present situation of atomic and molecular (AM) collision data which are relevant to the understanding of physical and chemical processes in many applications such as in interstellar clouds, in the earth's atmosphere, and in plasma-related fields, and how we can find these data. The most important physics and chemistry involving atomic and molecular collision processes are discussed in detail in the previous chapters.

As seen there, AM data necessary to understand the whole features involved in such processes as in astrochemistry or plasmas are still scarce. In particular, collision processes accompanying the change of the internal energy of collision partners have just begun to be investigated in detail in recent years with the advent of powerful and frequency-tunable lasers by which the internal energies are controlled and specified.

Here we would like to stress just how broad the energy range of atomic and molecular collisions is that is necessary for fully understanding collision processes involved in various applications. In applications to particle detectors, for example, we need data for collisions of very high energy up to 10^{10} eV/amu or more. On the other hand, the detailed mechanisms of these detector features are based upon the production and

* Presently National Institute for Fusion Science

behavior of very low energy electrons, with a few eV or less produced in such high energy collisions. For an understanding of collision processes involving ions, atoms and molecules, the collision energy required is as low as 10^{-6} to 10^{-7} eV/amu corresponding to the liquid helium temperature range.

Finally, some important directions in collision research involving ion, atom, molecules and radicals are discussed.

2. ASTROCHEMISTRY AND ATMOSPHERIC CHEMISTRY

Collision data involving atoms, molecules, radicals and their ions in gaseous phases are particularly important in their applications to astrochemistry. Collisions play a key role in the formation and destruction of various species in the interstellar clouds or atmospheres and consequently affect their composition. Furthermore, collision data are required in understanding the evolution of the universe. In the interstellar clouds or atmospheric regions, charge transfer and particle transfer collisions between atoms, molecules, radicals and their ions as well as collisions with photons and electrons play an important role. Another important collision partner is the cosmic rays. Also metal atoms and metallic ions, though their abundance is quite small compared with other atoms or molecules, play a role in formation and destruction of ions, atoms and molecular species in the interstellar clouds.

The cosmic rays, which are emitted from, for example, the sun or supernova into interstellar space, play an important role in the formation of various ions and molecules. In fact, ionization of H and H_2, the most abundant species in the interstellar clouds, by the cosmic rays, whose energy may range from a few MeV to 1 GeV or more, initiates all the chemical processes involving ion, atom, molecule and radical. Because of their high energy, cosmic rays can penetrate deep into the clouds or atmosphere where ion-molecule collisions occur predominantly. Ionization by high energy cosmic rays itself is well understood theoretically and experimentally.

It should be noted that secondary electrons, which are produced through collisions of molecules, atoms or ions with these energetic cosmic rays and whose energy and intensity distributions are strongly dependent upon target species, are still energetic enough to modify the abundance of atom or molecule species in the clouds through additional collisions. Therefore, they should be taken into account in ionization by the cosmic rays though they are quickly thermalized through a series of collisions, with only a small fraction having high energy. Correspondingly, the final energy distributions of electrons, which are dependent upon the species themselves and their density distribution, are also key parameters in further ionization or recombination processes.

Although UV radiation, whose wavelength of interest in interstellar problems ranges from a few hundred Å to a few thousand Å, tends to be significantly absorbed inside the dense clouds, it still plays an important role in ionization and dissociation of molecules in the outer (diffuse) region of the clouds. Through ionization and dissociation processes, molecules are destructed into different channels. Information on these channels and its accuracy is still limited. In dissociation of molecules, the effects of vibrationally and rotationally excited states induced by photon absorption are also expected to be important.

As mentioned before, in interstellar chemistry, an important role is played by cosmic rays. In dense interstellar clouds where no UV radiation can penetrate, ionization is sustained by the energetic cosmic rays and, in fact, the cosmic rays initiate the most important processes of formation and destruction of various molecules. For example, C-H and C-C chemistry in the interstellar clouds begins with the ionizing collision of He, one of the most abundant species, with the cosmic rays (CR)

$$CR + He \rightarrow He^+$$

followed by dissociative charge transfer with CO

$$He^+ + CO \rightarrow He + CO^+ \rightarrow He + C^+ + O$$

resulting in the production of C^+ ions.
In turn, these C^+ ions collide with molecular hydrogen resulting in synthesis of hydrocarbons and their ions, CH_n and CH_n^+, starting with the following radiative association process

$$C^+ + H_2 \rightarrow CH_2^+ + h\nu$$

or proton transfer process

$$H_3^+ + C \rightarrow H_2 + CH^+$$

Similar collisions between C^+ ion and hydrocarbon molecule

$$C^+ + CH_4 \rightarrow C_2H_4^+$$

result in the production of other types of hydrocarbons, for example, C_2H_n and $C_2H_n^+$. Indeed the yield from these reactions is known to be very small at low temperatures with the rate coefficient being of the order of 10^{-15}-10^{-16} cm^3/s. Once C-H bond formation is completed, higher order hydrocarbon molecules and their ions are produced quickly because most of the other collision processes involved in the synthesis of hydrocarbons are very rapid. Some experimental results have been reported on the synthesis of these hydrocarbons. However, investigations are very limited for hydrocarbons of higher order than C_2H_n molecules.

In order to simulate the evolution of the interstellar clouds involving C-H, C-C, C-O, C-N, O-H, N-H and N-O compounds, Prasad and Huntress[1] include more than 1400 collision processes for about 140 species resulting in production and destruction of atoms, molecules, radicals and their ions. In fact, most recently Anicichi and Huntress[2] have presented a list of the rate coefficients for photon, electron, ion, atom, molecule collisions of more than 2500 processes which seem to be relevant in forming interstellar clouds.

In some cases, heating of the interstellar clouds by shock waves is known to result in the enhancement of some endothermic chemical reactions and sometimes can explain extra-ordinarily intense ions or molecules. For example, the inclusion of the endothermic collision

$$C^+ + H_2 \rightarrow CH^+ + H$$

with the energy defect of 4640 K, which is induced by relative streaming of these two particles and results in formation of hydrocarbon molecular ions, should give an important effect and modify structures of shocks and radiation spectrum from them[3].

These molecular ions disappear, with much higher probabilities, through dissociative recombination (attachment) process such as

$$e + CH^+ \rightarrow C + H.$$

Combining the two processes above, that is, neutralization processes of C^+ ions results in significant modification in modelling. Indeed in collisions of molecules, ratios among products resulting from different dissociating channels, so-called branching ratios, are particularly important in the production of complex molecules and also in determining the abundance of ions, atoms or molecules in the interstellar space, in their clouds and other environments.

Dissociative recombination of molecular or radical ions such as H_3^+ ions by electrons is known to be strongly dependent upon their initial vibrational states and their distributions which may be very different in interstellar and laboratory plasmas[4]. In laboratory experiments, these cross sections are known to be dependent upon excitation modes[5]. In fact, the cross sections or rate coefficients at very low energies, determined by the merging beam technique, are much larger than those obtained with the ground state beam[6]. However, we must be careful in using data based upon this method because the beam used may contain some unknown fraction of molecules in vibrationally excited states.

This is particularly important in cross sections for dissociative recombination processes and has been shown in an experiment by Adams et al.[7] who suggest a significant dependence of the cross sections for H_3^+ ions on their excited states. In this particular case of $e + H_3^+$ recombination process, the cross sections or rate coefficients at 100 K are different by almost two orders of magnitude in the merging beam technique and the flowing afterglow (FALP) technique where the ions are totally relaxed into the ground state. This agrees with the theoretical prediction that the rate coefficients for the ground state become diminishingly small at low temperatures. This large difference is understood to be due to the effect of a significant fraction of beam components in vibrationally excited states in the original beam in that laboratory merging beam experiment[6].

Though data for dissociative recombination of cluster ions with low energy electrons are still less reliable because of experimental difficulties of establishing environments where only an electronically specified single species of cluster ions under consideration is present, the cross sections or rate coefficients seem to be very large, being of the order of 10^{-6} cm^3/s or more.

More reliable data for these electron-ion recombination processes are necessary, in particular for species in vibrationally and/or electronically excited states. Furthermore, the identification of the electronic states of the neutral products from such processes is important and has been largely neglected up to now, in analyzing the collision processes which follow.

The situation is worse in positive ion-negative ion recombination processes. The radiative recombination between them is of minor importance in most cases. On the other hand, mutual neutralization and three-body recombination processes are important sources for loss of ionization. Rate coefficients for the processes are sparse, in particular for recombination between cluster ions. Although a very limited amount of experimental data indicate that they seem to be insensitive to the species of the cluster ions involved and have a weak temperature dependence, the accumulation of reliable data is necessary.

Thus, it is most important to develop techniques of controlling and identifying the electronic states of particles involved before and after collisions. One technique is the use of lasers which can excite or deexcite a specified electronic state and is already in use in many applications.

In understanding the observed results in astrochemistry, not only ion-molecule or molecule-molecule collisions but also collisions between molecule and radical or radical and radical are important. In particular, a number of new radicals have been observed recently in interstellar space through observation of their microwaves. They are relatively short-lived in usual laboratory experimental conditions and, therefore, had not been easily observed there before. Some microwaves emitted from these radicals or molecules have recently been confirmed in laboratory experiments and their molecular structures are progressively understood. At present, collision data involving these radicals have not been determined in laboratories. At the moment, it seems that the production of some radicals such as C_nH, HC_nN, C_nS, C_nO, C_nN with sufficient intensities is not possible under well-controlled experimental conditions.

It should be pointed out that a role is played by metallic ions which are mainly originated from meteors and can become core ions in forming some heavy clusters or their ions in the atmosphere. Such metallic species or ions are abundant in some regions. However, investigations involving such metallic ions are still sparse.

The cluster ions are also important entities in determining the abundance of various species in some spaces. For example, $H_3O^+(H_2O)_n$ cluster ions with n up to 20 have been confirmed in recent space mass spectroscopy. In the formation of clusters, in particular of large mass clusters, three-body association processes are believed to play a role.

The negatively charged atomic, molecular or cluster ions are present in some region where ion-ion neutralization between the oppositely charged cluster species plays a key role in the loss of ionization. Such negative ions are produced most dominantly through three-body attachment processes including electrons, with an additional fraction from dissociative attachment of an electron. Meanwhile, these negative ions are mostly lost through associative detachment processes which result in the formation of neutral species.

At present, AM collision data involving such metallic ions and positive and negative cluster ions are limited. Furthermore, there should be other, minor but important, species which remain to be identified and whose production mechanisms have to be investigated. Indeed the importance of negatively charged ions has been largely neglected until recently.

Furthermore, an understanding of the collisions between hetero-molecular clusters, which are believed to play a role in the formation, evolution and destruction of large molecules, is very limited at the moment, compared with those of homo-molecular clusters.

Another important aspect of clusters is the fact that they can be the core or nucleus in the formation of grains, droplets or aerosols in the space and thus play a role in the production of very complex molecules or radicals. Their collisions with surfaces should play a role in adsorption or recombination for ions.

The energy or temperature dependence of rate coefficients is also one of the most important factors in understanding the production and destruction of interstellar clouds and atmospheric environments. In fact, in those cases where the temperature variation with altitude, for example, is significant, the knowledge of the temperature dependence of rate coefficients is a prerequisite in the modelling of chemistry in space. According to data obtained up to now, processes with large rate coefficients at room temperatures tend also to have large rate coefficients at low temperatures, showing either a weak temperature dependence or no temperature dependence at all. On the other hand, those processes with small rate coefficients at room temperatures have rate coefficients which increase with decreasing temperatures. Closely related to this problem, we note that only a small amount of data is available on the temperature dependence of branching ratios in dissociative processes.

In particular, information on collisions involving radicals is quite limited and their characteristics such as lifetimes and their formation and destruction through collisions with atoms or molecules should be investigated in laboratories under well-defined conditions. This means that AM data for collisions among the state-selected species are urgently needed. This is particularly important in collisions such as dissociative recombination processes whose cross sections are known to be strongly dependent on their, for example, vibrationally excited states, as mentioned before. In most observations made so far, no identification of the electronic states of ions, atoms, molecules or radicals before and after collisions has been made, except for a very few cases which have begun to be studied only recently.

Here we stress the fact that astrochemistry changes structures of the interstellar clouds and shocks and at the same time the structures themselves influence astrochemistry. We still do not know the detailed and precise density distributions (in space and time) for many species of atoms or molecules in a number of environments except for some of those near the earth's atmosphere. In fact, as shown in recent studies of linear chain carbon radicals such as C_nH, which were first observed through the astronomical telescope, they have immediately stimulated new experiments in laboratories and such radicals, which had been previously unnoticed, have been found relatively easily in laboratories. Laboratory observations of these molecules result in more detailed investigations and the understanding of physical and chemical properties of these molecular radicals such as their fine and hyperfine structures.

Furthermore, AM data in gas phases is not sufficient to understand the behavior of the interstellar or atmospheric clouds. Though out of the scope of this book, it should be pointed out that data for collisions of molecules or radicals with surfaces or grain dusts in space, which are composed mainly of carbonaceous or silicaceous species, are essential for a full description of the evolving universe. Even in laboratories, reliable experimental data involving collisions of molecules with surfaces are very limited and a full understanding of collision processes between molecules and surfaces is far from complete at present.

3. MOLECULAR PROCESSES IN PLASMA-RELATED FIELDS

Tokamak Plasma and Carbonization Processes

In tokamak and other types of magnetic confining fusion apparatus, the most significant energy losses from high temperature plasmas are believed to come from radiations (mostly in the X-ray region) emitted from impurity ions which capture an electron into their excited state and then decay radiatively into the ground state or a lower state. As radiation losses are known to be proportional to the square of the atomic number of ions, then, it has been realized that graphites and carbon-coated layers are good candidates for plasma-facing materials. Presently in most tokamaks graphites are used for many purposes such as limiters, armors or divertors. And for better treatments of carbon layers over a wide area of the inner surfaces of the first wall, in-situ carbon-coating techniques (sometimes called carbonization process) are also being developed. For example, under heavy irradiation of atomic hydrogens or other impurity ions in plasmas, graphites or coated carbon layers are sputtered away. In this sputtering under hydrogen bombardment, not only carbon atoms but also various hydrocarbon molecules such as CH_4 are produced through (synergistical or non-linear) chemical reactions which are known to be enhanced at the elevated temperatures of 500 - 900 C region (this enhancement is often called chemical sputtering)[8]. They first enter relatively cold plasmas and are ionized by them and finally arrive at the hottest region of the plasmas. On the other hand, in carbonization processes, CH_4 gases, pure or mixed with hydrogens or other gases, are often used under glow discharges or high frequency discharges in order to, in situ, produce and deposit proper carbon layers on the inner surfaces of the plasma vessel. There are a number of unknown processes to be understood. For example, the characteristics of carbon layers (such as ratios of carbon to hydrogen, which are a quite important parameter in hydrogen recycling processes through surface to plasma and plasma to surface, electric conductivities or hardness) coated in surfaces are not known. In fact, little understanding has been obtained up to now. There, not only collision processes of electrons, atomic hydrogens and molecular hydrogens but also a number of collisions of various hydrocarbon molecules and their radicals and their collisions with surfaces surely play a key role.

To understand the behavior of plasmas containing a significant amount of carbons and carbonization processes themselves and also the characteristics of coated carbon layers, we need AM data including hydrocarbon molecules and their radicals as well as hydrogen molecules. Probably some of the data already compiled for analysis of the interstellar clouds can be used. One of big differences is the presence of bulk surfaces of carbons in tokamak plasmas. Thus, in tokamak and related processes involving graphites and carbons, collisions of ions, atoms, molecules or radicals with surfaces play a very significant role, compared with those in the interstellar clouds.

Plasma Material Processing

We feel a strong need for AM data related to plasma material processing phenomena where a lot of information concerning, in particular, the metastable or radical species should play a role. Compared with ions, molecules or radicals in the interstellar clouds, those in laboratory or industrial plasmas contain more excited state species because the collision times are generally much shorter. Effects of these excited state species should be more significant in laboratories. Indeed, such data are limited in quantity as well as in quality at present, because of technical difficulties in preparing these species in well-defined electronic states with sufficient intensities. Presently, these applications seem to be strongly dependent upon their "brute" force at best. Though collisions of these species with surfaces of the apparatus should play a role, their effects in plasma processing are not known yet.

In the production of negative hydrogen ions, it is known both theoretically and experimentally that hydrogen molecules in vibrationally excited states play a role. However, the contribution of collisions of such molecules with surfaces in production of such vibrationally excited molecules is not well recognized because no experimental data for such collisions are available yet, though "dirty" surfaces, whose physical and chemical properties are not well characterized, are believed to play a key role there.

4. DATA CENTERS FOR ATOMIC AND MOLECULAR DATA

In some fields, in particular in interdisciplinary fields such as fusion plasma research, a tremendous amount of information or data is necessary over a wide range of various subfields ranging from basic AM data and surface collision data to bulk material data and further to superconductors, irradiation effects and tritium behaviors. The data must be very accurate and reliable. On the other hand, compilation and more importantly evaluation of such data are quite time-consuming and requires a lot of man-power. Therefore, the fusion plasma research community, for example, is trying to establish data centers in order to do such jobs. However, no single data center can perform all the necessary data activities even in a limited data field such as the relatively well-established one of atomic and chemical physics. The Atomic Data Unit at the International Atomic Energy Agency, created in 1976 as a part of its Nuclear Data Section, is trying to coordinate international collaboration and sharing of data activities related to fusion plasma research programs among data centers world-wide. The data centers form the so-called IAEA AM Datacenter Network and concentrate their effort in compiling and evaluating mainly AM data and surface data. This Network is now trying to establish some common formats for data transfer and exchange among the data centers and other users.

In the following, we list some of the most important data centers which publish reports on atomic and molecular data, either compiled or evaluated, and from which one may obtain copies :

Chinese Nuclear Data Center, Institute of Atomic Energy, Beijing, The People's Republic of China.

The Queen's University of Belfast, Belfast BT7 1NN, Northern Ireland, UK.

CRE "E. Clementel" ENEA, Via Mazzini 2, I-40138 Bologna, Italy.

Culham Laboratory, UK Atomic Energy Authority, Abingdon, Oxon, OX14 3DB, UK.

Atomic Data Unit, Nuclear Data Section, International Atomic Energy Agency, Wagramerstrasse 5, A-1400 Vienna, Austria.

The Institute of Space and Astronautical Science, Sagamihara, Kanagawa 229, Japan.

Japan Atomic Energy Research Institute, Tokai-mura, Ibaraki 319-11, Japan.

Data Center, Joint Institute for Laboratory Astrophysics, University of Colorado, Boulder, Colorado 80309, USA.

Kurchatov Institute of Atomic Energy, Moscow 123182, USSR.

Data and Planning Center, National Institute for Fusion Science, Nagoya 464-01, Japan.

Atomic and Plasma Radiation Division, Center for Radiation Research, US-NBS, Gaithersburg, Maryland 20899, USA.

Controlled Fusion Atomic Data Center, Oak Ridge National Laboratory, Oak Ridge, Tennessee 37831, USA.

Laboratoire de Physique des Gaz et des Plasmas, Batiment 212, Université de Paris-Sud, 91405 Orsay Cedex, France.

Princeton Plasma Physics Laboratory, Princeton University, Princeton, New Jersey 08544, USA.

Most of the above institutes publish compilations of important AM data relevant to their own research projects. Some are mostly devoted to data related to fusion programs and others to astrophysical aspects. The regions of interest in energy or temperature or the species of interest are sometimes different. Even in fusion-oriented programs we need not only AM data of highly stripped ions at relatively high energies which are concerned mainly with high velocity or high temperature for diagnostics of core plasmas on the one hand but also AM data involving molecular processes which are concerned mainly with the low temperature edge plasmas on the other hand. These data are often stored in the so-called databases in most of the above institutes or data centers and are easily accessible through computer terminals via direct linkage or satellite linkage.

We should add the fact that the most important contributions come not only from these data centers but also from individual groups.

5. ATOMIC AND MOLECULAR DATA COMPILATION AND EVALUATION ACTIVITIES —AN EXAMPLE AT RIC/IPP, NAGOYA UNIVERSITY

AM data activities at the Research Information Center (RIC), IPP/Nagoya (presently Data and Planning Center, National Institute for Fusion Science), began well before its formal establishment in 1978. Around 1972, eleven atomic physicists came together to do this work and formed a data study group on atomic processes in plasmas, with the emphasis on the compilation of AM data related to atomic hydrogen in collisions with photons, electrons, hydrogens and their ions. The results of their activities have been published as an internal report (IPPJ-DT-48) of IPP in 1975 (in Japanese). This report and a succeeding similar report (IPPJ-DT-50) became known worldwide when the First Advisory Group Meeting on Atomic and Molecular Data for Fusion under the organization of the International Atomic Energy Agency was held at Culham in 1976. Later one of them (IPPJ-DT-48) was translated into English to achieve greater circulation among the plasma as well as the atomic physics communities. The outcome of this meeting was the identification and recommendation of activities on AM data requirements for fusion research [see Physica Scripta 37, No.2 (1978)]. Since then, AM data activities at RIC follow this line, with close collaboration with the Atomic Data Unit, IAEA, and other AM data centers around the world.

In addition to two scientific members of staff specialized in AM and plasma physics, activities at RIC have been and are still supported by volunteer groups of atomic and molecular physics specialists, with the collaboration of plasma physics specialists who give their advice and suggestions on their needs for AM data. Task force groups are aften organized, depending on the data needed. Sometimes working groups are also organized under international collaboration programs.

Another important activity at RIC is to find new problems which might be, from the physics point of view, interesting topics and might contribute to understanding plasma behaviors. One of them, at present, is to find trends in atomic physics research under hot and dense plasma environments.

The financial funding of RIC for such AM data activities, marginal ~$3.000 per year, excluding the running cost of RIC, is used for the traveling expenses of the working groups. These activities often result in the production of some reports (as IPPJ-AM) which are categorized into the following :

1) AM data compilation and evaluation
2) Proceeding of international, domestic or institute workshops or meetings to identify data needs
3) Review reports to summarize the present understanding of AM processes and related topics and to recommend further activities in atomic physics community.

800 - 900 copies of each of these reports are printed and distributed free of charge to all those interested over the world. Up to now (January,

1989), RIC has published 62 IPPJ-AM reports which include our data compilation and related activities at RIC/IPP, Nagoya University, in the past ten years. A variety of AM data are requisite in understanding, modeling and diagnosing fusion plasmas. In the core plasmas where the plasma temperature is high (1-20 keV) and the density is also high (10^{13} - $10^{14}/cm^3$), almost all the elements are highly ionized. Although the main plasmas themselves consist of hydrogens or their isotopes, there are always some inherent impurity elements such as carbon or oxygen and also some other heavy metal elements of structural materials such as iron, nickel and molybdenum. There, the most important AM data are closely related to collision processes such as ionization, excitation, electron transfer, and atomic structure data such as the energy levels, transition energies, transition probabilities, of these highly ionized ions. At the same time, we need AM data involving molecular hydrogens. Though molecular hydrogens are usually present near the walls of plasma devices or the peripheral regions of plasmas and sometimes in the scraping-off plasmas and are far away from the main plasmas, it has been realized recently that they play a far more significant role through diffusion and other collision processes in determining qualities of high temperature core plasmas and, consequently, in realizing fusion reactions. Depending on the boundary plasmas themselves, their densities and temperatures range from 10^{10} - $10^{13}/cm^3$ and 1 - 1000 eV, respectively. Therefore, AM data are required for hydrogens, atomic and molecular, in collision with electrons as well as those in collisions between heavy atoms or molecules which include some chemical reactions as well as atomic collisions. AM data involving atomic and molecular hydrogens are found to be relatively well known and documented. One example of the compiled and evaluated data for various excitation

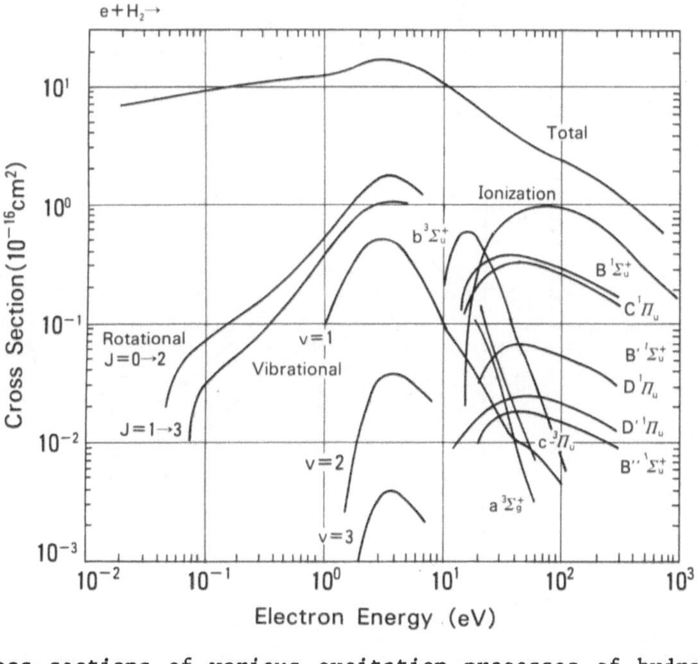

Fig. 1. Cross sections of various excitation processes of hydrogen molecules by electron impact as functions of the electron energy[5].

processes involving molecular hydrogens is shown in Fig. 1 over the electron energy of 10^{-2} to 10^3 eV[5]. Here total and ionization cross sections are also shown for comparison. Such a comparison of cross sections among possible collision processes is quite important in order to know what is the most dominant in collisions. For example, recent measurements of the cross sections for excitation to $b^3\Sigma_u^+$ state of H_2 molecules resulting in the production of two atomic hydrogens in the ground state have shown that previous values of Corrigan, which had been cited frequently, are too large by 30 %.

Not only the cross sections for individual collision processes but also those for some combined processes (so-called effective cross sections) such as photon emissions are sometimes needed for applications as well as basic research. In Fig. 2 the cross sections for photon emission resulting from various excitation processes of molecular hydrogens are shown. It should be noted that the cross sections for Lyman-alpha emission shown are normalized to the value by Shemansky et al. and may still be subject to change because some of the most recent measurements seem to be slightly smaller[9].

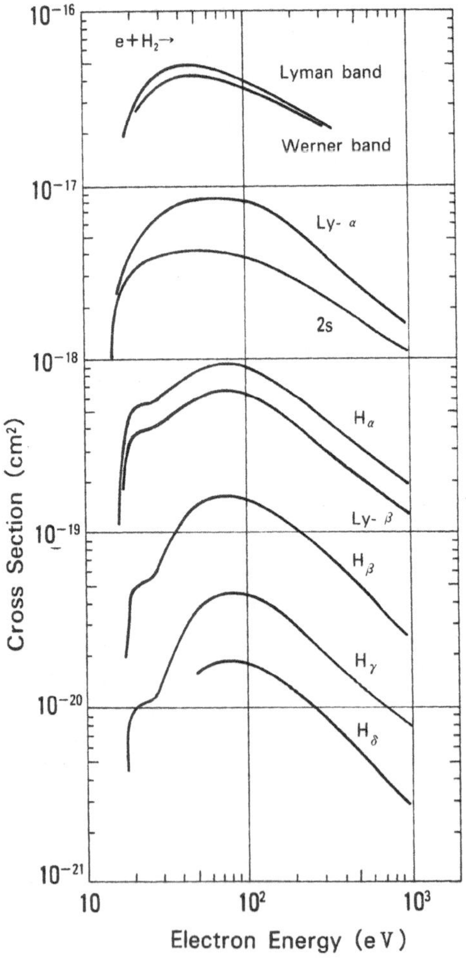

Fig. 2. Effective cross sections for photon emission from molecular hydrogens by electron impact[5].

However, data involving collisions with surfaces, which are not avoidable in laboratory plasma experiments including fusion plasmas, are really scarce. In particular, if these species are in their excited states or radicals, no reliable AM data exist yet.

Recently, as will be mentioned later, a number of databases have become available. Therefore, bibliographic databases in specified research fields can now be constructed using such databases, though again much man-power is necessary. At our Institute, all the bibliographic databases, INSPEC (see Section 7), from 1969 to the present are stored and their information amounts to more than 2.5 million records. To get quick access to this database, information related to atomic and molecular processes and to plasma and fusion processes is extracted from INSPEC as AM (3.9×10^5 records) and FUSION (3.5×10^5 records) subdatabases. Also a similar bibliographic database (5.9×10^4 records) of Oak Ridge National Laboratory is included in our database. Another database from IAEA (see Section 7) is planned to be installed soon in our database.

Based upon the bibliographic databases collected, compilation of numerical AM data can be, in principle, performed with much expertise and with laborious effort. On the other hand, evaluation of such numerical data is another but more serious issue. In fact, before going to evaluation, we have to first "evaluate" each data point if it is really reliable or not.

To do this, Dunn of JILA, University of Colorado, suggested that the following points should be checked for experimental data : (1) metastables in beam, (2) complete beam collection, (3) space charge modulation, (4) pressure modulation, (5) accuracy of beam-overlapping measurement, (6) accuracy of detector (sensitivity or efficiency) calibration, (7) pulse height discrimination, (8) dead time correction, (9) instrument (energy) calibration, (10) linearity of instruments, (11) anisotropy, (12) uniformity of detector sensitivity, (13) finite lifetime correction, (14) path-length correction, (15) effects of external fields and (16) others.

On the other hand, for theoretical data, the following points should be taken into account : (1) target wave function, (2) exchange effect, (3) coupling between states, (4) resonance effect, (5) asymptotic behavior, (6) consistency with isoelectronic sequences, (7) history of the methods and (8) others.

Probably no single paper, either theoretical or experimental, exists with the full description of the items mentioned above. Most papers give only a brief discussion on errors or uncertainties involved in their work. Indeed systematic errors can not be evaluated precisely if the same methods are used to determine data points even though the data seem to be in agreement with each other. Thus, it is found from our experience that the reliability of data is strongly dependent upon who has taken the data. Few scientists bear seriously in mind those points shown above and describe clearly the strong and also the weak points in their work. Generally their data are reliable within their uncertainty limit and some of them are "bench mark" data for relative calibration or normalization.

Evaluation of numerical data does indeed need much more expertise both in theory and experiment. Some averaging procedures over all data points, which in most cases are available only over a limited energy region or other parameter regions, by combining their corresponding weights and the resulting analytical fitting through data points are widely used. However, such formulas have to be used carefully to interpolate or extrapolate the necessary data. Great care should be exercised, in particular in extrapolation toward the outsides of the data available.

If some asymptotic behaviors are expected to be valid, we can rely upon those at higher and lower energies which can be combined to perform interpolation or extrapolation. There are empirical formulas to estimate cross sections for various collision processes. Until recently, total cross sections for electron transfer, for example, have been believed to be estimated for ions with the same ionic charge through scaling, though it is known that there is no such simple scaling for partial (nl) cross sections. Recent detailed calculations suggest that even for total cross sections no scaling can be expected to be valid, in particular at low energies because of the strong core effect of ions with the specific ionic charge. One such example of data evaluation for the $1s^2\ ^1s \rightarrow 1s2p\ ^3p$ excitation process of Li^+ ions by electron impact[10] is shown in Fig. 3 where the solid curve represents the evaluated collision strength fitted through available (theoretical) data as a function of the collision energy. Here the curve relies exclusively upon two close-coupling calculations, which are believed to be the most reliable, in the low (Christensen) and the high (Wyngaarden et al.) energy regions. However, no experimental confirmation for these data is available at the moment.

Here it should be stressed that one must be careful and indeed it is dangerous to use empirical formulas in order to simply extrapolate or interpolate data needed using data available without knowing their detailed

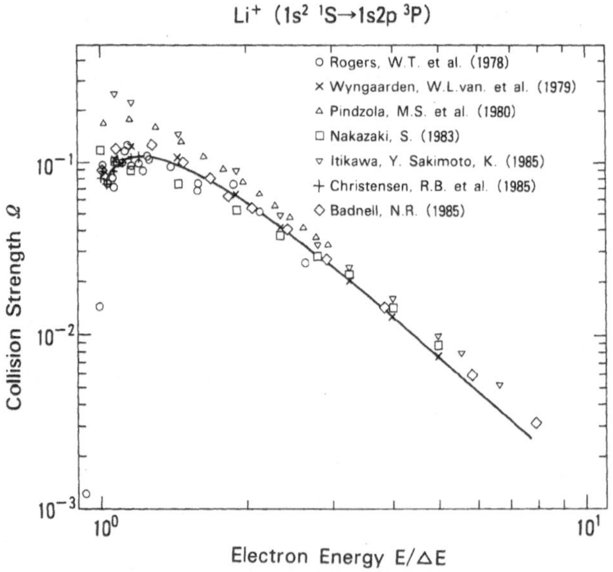

Fig. 3. Collision strength of the $1s^2\ ^1S \rightarrow 1s2p\ ^3p$ excitation of Li^+ ions by electron impact. The solid line represents the recommended values[10].

behavior. Unfortunately, such time-consuming evaluation work is not appreciated properly. Indeed many people still think that such data evaluation work could be performed part time. That is really not the case, as mentioned above. Another problem is how to maintain the scientific level and activities of the data evaluators, without which updating of data and their evaluation can not be performed properly.

Numerical data compiled and evaluated at RIC are normally stored in our computer system and are easily accessible. At the moment, RIC has developed the following numerical databases:

AMDIS (Atomic and molecular data interactive system): AM data for excitation and ionization by electron impact

CHART (Charge transfer): charge transfer cross sections for ion-atom and ion-ion collisions

BACKS (Backscattering of ions from solids): backscattering coefficients of light ions from solids

SPUTY (Sputtering yield): sputtering yields data for monatomic solids by ion impact

Some of them include the rate coefficients averaged over the Maxwellian velocity distributions as well as the cross sections themselves and also some convenient empirical formulas for estimating, or interpolating or extrapolating data. Indeed, in a number of applications, instead of cross sections, the rate coefficients which are averaged over the energy or temperature distribution (in most cases the Maxwellian distribution is assumed) are conveniently used.

However, we would like to stress the importance of the original cross section data as a function of the energy or temperature. Once a convenient analytic formula for the cross sections or rate coefficients is proposed, it is common that users tend not to care about the original data or their accuracy even if they transpire to be less reliable. In this case we have sometimes faced difficulties in checking the accuracy of the original data and in replacing them with a new version of more reliable data. Furthermore, if these distributions are non-Maxwellian, as is often the case, it is difficult to convert these known rate coefficients to those with different temperature distributions.

More importantly, it should be pointed out that the knowledge of a bunch of cross sections or rate coefficients for various processes alone can not tell you all the detailed mechanisms of production and destruction of the existing species, ions, atoms, molecules or their ions, and their abundance. One example is the fact that species in laboratories and space are usually not in the same internal electronic states and, consequently, the rate coefficients for particular processes may not be the same. Thus, straightforward applications of laboratory data to modeling of, for example, the evolution of the interstellar clouds often can not reproduce the actual situations there but, instead, may result in serious discrepancies between modeling and observations and this leads to misunderstanding. Great care should be exercised when using databases, by knowing the detailed experimental and observation conditions.

It should be remembered from our experience that such AM data activities sometimes result in finding some previously unrealized but most important and urgent topics which can be performed relatively easily with less funds and less technical and theoretical developments because experimental apparatus or theoretical techniques are already available. Such topics are often thought to be trivial but in truth can be found to contribute quite a lot to understanding plasmas and also atomic and molecular collision processes and structure problems themselves.

6. FUTURE TRENDS IN ATOMIC AND MOLECULAR RESEARCHES

Though a large number of experimental and theoretical investigations have been done so far, they are still scarce and do not cover a wide range of the species of atoms, molecules or radicals or their ions. The investigations do not provide the necessary data on collision processes for refining the modeling of physics and chemistry in planetary and interstellar environments and also in laboratories. In fact, abundances of some complex molecules in interstellar clouds, based upon models, for example, by Millar and Nejad[11] who used 560 reactions for 130 species and by Herbst and Leung[12] who used 2000 reactions for 200 species, are found to be significantly different (almost two orders of magnitude for HC_3N molecules) from each other. This large difference is partly due to the reactions used but also more strongly due to the branching ratios used of, for example, dissociative recombination processes of molecules or radicals by electrons or the negatively charged ions. Therefore, it may be concluded that, because of the low reliability of some of the AM data involved, the inclusion of a large number of collision processes does not always result in a true reproduction of the observed results.

As already mentioned above, we can summarize the future trends in research involving photons, electrons, atoms, molecules, radicals or ions which should contribute to understanding astrochemistry and other related phenomena as follows:

1) A large number of collision cross sections or rate coefficients should be determined for a number of species. In particular, these data should be investigated as a function of the temperature or energy of collision systems over a wide range, particularly at extremely low temperatures. In order to investigate cross sections at very low temperatures, for example, of a few 10 K, some quite promising techniques such as low temperature supersonic nozzle or cooled Penning trap techniques have been already developed. These have good sensitivities and rate coefficients of the order of 10^{-16} cm^3/s and below can be determined with these techniques. More importantly, cross sections should be measured for collision processes under the well-controlled specified internal energies of molecules or radicals involved. This is particularly important in collisions involving radicals or molecules in excited states. If their internal energies are not well defined, the observed rate coefficients might be different by orders of magnitude and then the modeling of the chemistry and physics might result in situations far different from the real world.

2) Another important parameter needed in the modeling of astrochemistry or other applications is branching ratios or fractions of various channels in ion, atom or molecule production resulting from photon, electron,

ion, atom, molecule or radical collisions. These branching ratios have a decisive importance in determining abundance of various species of atoms, molecules or radicals in the clouds and other environments. Also the energy or temperature dependence of the branching ratios should be investigated.

3) Because of their very small quantities in space, collisions involving metallic ions have not been well investigated up to now. However, some metallic ions such as transition metals are chemically very active. Thus, even a small amount of these metallic ions should sometimes play a key role in ion chemistry of the planetary or interstellar spaces.

4) Furthermore, collisions of neutral metal atoms should have similar importance in describing the evolution of these clouds. Until recently, the production of such metallic neutral atoms in laboratories was no easy task. With the recent development of reliable high power ion sources, we can now obtain neutral metal atoms with intensities sufficient for crossed-beam measurements. This is based upon neutralization of metallic ions in collisions with proper target gases or vapors. Here we should exercise care in choosing targets such that the ground state atoms are produced exclusively in such collisions, or their specified electronic states are selected. If they are not in the specified electronic states, the observed data would not be reliable anyway.

5) Formation and destruction mechanisms of large clusters and their ions should be investigated and furthermore their collision data are again one of the key parameters which should be determined. This is an interesting topic to pursue because many techniques related to cluster ions are being extensively developed presently in a number of laboratories.

6) In most of the modeling of chemistry in planetary or interstellar clouds, the effects of collisions of atoms, molecules, radicals or ions with surfaces or grains have not been taken into account properly. Although experimental work has been limited up to now, interactions with some surfaces should play a significant role. In particular, collisions of those in excited states with surfaces might significantly influence the fractions of various species of atoms, molecules or radicals in the clouds.

7) Though we did not discuss the significance of multiply (mostly doubly) charged ions in the interstellar, planetary or atmospheric environments, they should also play a key role in determining the abundance of atoms or molecules. A large number of the out-going channels, for example, in electron transfer processes are expected to exist and thus the presence of even a small quantity of the excited species results in a large difference of the observed cross sections. In fact some of them are known to be very fast, compared with those in the ground states.

8) Finally it should be stressed that we still need to find and identify some important ions which may exist, though they may not be in large quantities, but have not been observed yet, in particular in high

pressure regions. These may play a key role in understanding astrochemistry and atomospheric chemistry. Indeed such ions or molecules, even though their production rate coefficients might be small, result in, once they are produced, the formation of new species with much higher rate coefficients.

7. IMPORTANT ATOMIC AND MOLECULAR DATABASES

Bibliographical Databases

The most general and useful bibliographic information on AM data can be obtained through the following two databases:

INSPEC (International Information Services for the Physics and Engineering Communities published by the Institute of Electrical Engineers, UK)
This includes information on physics, astrophysics, electronics, plasma, etc., published since 1969. It is accessible through a computer.

CAS (Chemical Abstracts Service published by the American Chemical Society, USA)
This includes information on chemistry, physical chemistry, organic chemistry, applied chemistry, etc., published since 1979. It is also accessible through a computer.

Information included in INSPEC is also printed in book form as Physics Abstracts. Similar bibliographical information can be found in Current Physics Index (published by the American Institute of Physics) which contains references from relatively limited sources of Journals.

More specifically categorized bibliographical databases of collision processes can be found in GAPHYOR published by Laboratoire de Physique des Plasma, Universite Paris-sud. This database up to 1987 has been published recently as Gas-Phase Chemical Physics Data Base (J.L. Delcrois) Vols.1-3 (Elsevier, North-Holland, 1988).

A large amount of bibliographic information in specific fields is also summarized in different laboratories. For example:

Controlled Fusion Atomic Data Center, Oak Ridge National Laboratory (data for atomic and molecular collision processes, for surface-related topics and for plasma-related topics)
These databases are accessible through a computer and are also printed regularly in book form once a year as Bibliography on Atomic and Molecular Processes.

Atomic Data Unit, Nuclear Data Section, International Atomic Energy Agency (data for fusion plasma related topics)
Information is biannually published as International Bulletin on Atomic and Molecular Data for Fusion and finally printed as CIAMDA, the Computer Index to Atomic and Molecular Collision Data Relevant to Fusion Research. Up to now, two volumes, CIAMDA-80 (including the references from 1950 - 79) and CIAMDA-87(1979 - 87) have been printed.

Furthermore, bibliographic databases for very special topics are published at the data centers mentioned above.

Numerical Databases

Compared with bibliographical databases, the availability of numerical databases is quite limited because compilation and evaluation of numerical data are much more complicated and time-consuming and in addition a lot of specialized techniques and information are necessary. At the moment, no worldwide systematic survey has been reported on available numerical databases. In fact, even such surveys themselves require a lot of money and time in addition to man-power. We can see such an example in a long-standing effort, lasting more than 30 years, for evaluation and reevaluation of neutron reaction data for use in developing the most efficient nuclear fission reactors, which is organized by the International Atomic Energy Agency with the support from a host of countries.

Inherent problems to such work are the fact that some data tend to be out-of-date when their reports are completed. Thus an ever-lasting effort is essential for having good, reliable bibliographic and numerical databases. Such compiled and evaluated data, bibliographical or numerical, related with AM processes are sometimes published in the following regular journals:

Astronomy and Astrophysics (European Physical Society)
Astrophysical Journal (American Astronomical Society)
Atomic Data and Nuclear Data Tables (Academic Press)
International Journal of Mass Spectrometry and Ion Processes(Elsevier)
Journal of Physical and Chemical Reference Data (American Chemical Society and American Institute of Physics)
Physics Reports (North-Holland Physics)
Reviews of Modern Physics (American Institute of Physics)
Soviet Physics-USPEKHI (translated from USPEKHI Fizicheskikn Nauk, USSR, by American Institute of Physics).

There should be a lot of more databases available around the world. It seems to be important but yet difficult to trace all of the databases, though some organizations are trying to collect information on their availability. Indeed, there are a number of personal well-organized databases in very specialized fields. We really need some well-organized and continuous effort in compiling such databases worldwide.
Possibly we should have some journals or bulletins devoted to collecting information on the availability of reliable databases.

Acknowledgments The author would like to thank Dr. Y. Itikawa, the Institute of Space and Astronautical Science, for his useful suggestions and comments.

It is the author's great pleasure to dedicate this article to Prof. K. Takayanagi on the occasion of his retirement. Prof. Takayanagi was one of the most important promoters in establishing the Research Information Center, Institute of Plasma Physics, Nagoya University, which is now actively engaged in compilation, evaluation and dissemination of AM data to the plasma community and also to the atomic physics community. He noted the importance of atomic and molecular physics in plasma physics and engineering which plays a role in understanding the behavior of plasmas and made a systematic survey of requirements of AM data in plasma research by sending a questionnaire to the plasma community as well as to the atomic and molecular physics community in Japan around 1970. Following this survey, he, together with Prof. H. Suzuki, Sophia University, Tokyo, organized a data study group on atomic processes in plasmas to compile the necessary AM data. This resulted in the first volume of AM data compilation (Institute Report IPPJ-DT-48) published from the Institute of Plasma Physics, Nagoya University in 1975 and was followed by the second volume (IPPJ-DT-50) in 1976, both of which are highly appreciated by plasma and other communities. This activity is presently succeeded by the Research Information Center, Institute of Plasma Physics, Nagoya University, which was recently reorganized as Data and Planning Center, National Institute for Fusion Science.

REFERENCES

1. S.S. Prasad and W.T. Huntress, Astrophys. J. Suppl. Ser. 43, 1-35 (1980); Astrophys. J. 239, 151-165 (1980).
2. G. Anicichi and W.T. Huntress, Astrophys. J. Suppl. Ser. 62, 553-672 (1986).
3. D.R. Flower, G. Pineau des Forets, and T.W. Hartquist, Mon. Not. Roy. Astron. Soc. 216, 775-794 (1985).
4. D.R. Pineau des Forets, D.R. Flower, W.T. Huntress and, A. Dalgarno, Mon. Not. Roy. Astron. Soc. 230, 801-824 (1986).
5. H. Tawara, Y. Itikawa, Y. Itoh, T. Kato, H. Nishimura, S. Ohtani, H. Takagi, K. Takayanagi, and M. Yoshino, IPPJ-AM-46 (Inst. Plasma Phys., Nagoya Univ.), "Atomic data involving hydrogens relevant to edge plasmas" (1986).
6. H. Hus, F. Youssif, A. Sen, and J.B.A. Mitchell, Phys. Rev. A38, 658-663 (1988).
7. N.G. Adams, D. Smith, and E. Alge, J. Chem. Phys. 81, 1778-1784 (1984).
8. N. Itoh and K. Kamada, eds, Radiat. Effects 89, 1-148 (1985).
9. H. Tawara, Y. Itikawa, H. Nishimura and M. Yoshino, IPPJ-AM-55 (Inst. Plasma Phys., Nagoya Univ.), "Atomic data for hydrogens in collisions with electrons— addenda to IPPJ-AM-46" (1987).
10. T. Kato and S. Nakazaki, IPPJ-AM-58 (Inst. Plasma Phys, Nagoya Univ.), "Recommended data for excitation rate coefficients of helium atoms and helium-like ions by electron impact" (1988).
11. T.J. Millar and L.A.M. Nejad, Mon. Not. Roy. Astron. Soc. 217, 507-522 (1985).
12. E. Herbst and C.M. Leung, Astrophys. J. 310, 378-382 (1986).

CONTRIBUTORS

P.G. Burke
Department of Applied Mathematics and Theoretical Physics,
Queen's University of Belfast, Northern Ireland

A. Dalgarno
Harvard-Smithsonian Center for Astrophysics, Cambridge, Massachusetts,
U.S.A.

H. Ehrhardt
Fachbereich Physik, Universität Kaiserslautern, West Germany

M.R. Flannery
School of Physics, Georgia Institute of Technology, Atlanta, Georgia,
U.S.A.

F.A. Gianturco
Department of Chemistry, University of Rome, Italy

M. Inokuti
Argonne National Laboratory, Illinois, U.S.A.

N. Kaifu
Nobeyama Radio Observatory, National Astronomical Observatory, Japan

Y. Kaneko
Department of Physics, Tokyo Metropolitan University, Japan

I. Shimamura
RIKEN (Institute of Physical and Chemical Research), Wako, Japan

T. Shimazaki
Ames Research Center, NASA, Moffett Field, California, U.S.A.

H. Tawara
National Institute for Fusion Science, Nagoya, Japan

INDEX